STUDENT SOLUTIONS MANUAL TO ACCOMPANY THE SYSTEMATIC IDENTIFICATION OF ORGANIC COMPOUNDS

STUDENT SOLUTIONS MANUAL TO ACCOMPANY THE

SYSTEMATIC IDENTIFICATION

OF ORGANIC COMPOUNDS

Ninth Edition

CHRISTINE K. F. HERMANN

TERENCE C. MORRILL

RALPH L. SHRINER

REYNOLD C. FUSON

WILEY

Published by John Wiley & Sons, Inc., Hoboken, New Jersey.
Published simultaneously in Canada.

For general information on our other products and services or for technical support, please contact our Customer Care Department within the United States at (800) 762-2974, outside the United States at (317) 572-3993 or fax (317) 572-4002.

Wiley also publishes its books in a variety of electronic formats. Some content that appears in print may not be available in electronic formats. For more information about Wiley products, visit our web site at www.wiley.com.

Library of Congress Cataloging-in-Publication Data applied for:

Paperback: 9781119799856

Cover design by Wiley

Set in 9/13pt Ubuntu by Straive, Pondicherry, India

Table of Contents

Preface

For the first six editions of this textbook, no solutions manual was available. As some of the problems presented throughout the textbook are quite difficult, I wrote the solutions manual as an aid to users of the textbook. The textbook contains no problems in Chapters 1, 2, 3, and 12.

I would like to thank Terence Morrill for the answers to Problem Sets 6–20 in Chapter 13, which is available on a companion website. These answers had been in handwritten form for many years. I would like to thank Danielle Davis for her assistance in writing the solutions manual for the seventh edition. I would like to thank Stephen Pond for his patience and support during the preparation of this edition.

Christine K. F. Hermann
Radford University

Chapter 4

Preliminary Examination, Physical Properties, and Elemental Analysis

1. Use the values for a nonassociated liquid.

$$\text{corr bp} = 167 + \frac{760 - 650\text{mm}}{10\text{mm}}\left\{\left(\frac{0.56 - 0.50}{200 - 150} \times (167 - 150)\right) + 0.50\right\}$$

$$= 172.72$$

1,3-dichlorobenzene

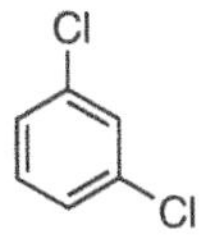

2. Use the values for an associated liquid.

$$\text{corr bp} = 180 + \frac{760 - 725\text{mm}}{10\text{mm}}\left\{\left(\frac{0.46 - 0.42}{200 - 150} \times (180 - 150)\right) + 0.42\right\}$$

$$= 181.55$$

2,3-dichloro-1-propanol

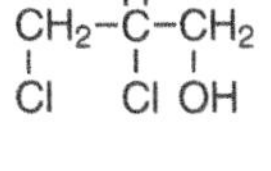

Student Solutions Manual to Accompany The Systematic Identification of Organic Compounds, Ninth Edition.
Christine K. F. Hermann, Terence C. Morrill, Ralph L. Shriner, and Reynold C. Fuson.
© 2023 John Wiley & Sons, Inc. Published 2023 by John Wiley & Sons, Inc.

3. 285 °C

4. 410 °C

5. Methyl acetate > methyl thioacetate

6. 2-Methyl-2-pentanol < 3-methyl-2-pentanol < 2-hexanol < 1-hexanol

7. $\text{sp gr}\,^{20}_{\ 4} = \dfrac{0.989}{0.834} = 1.186$

$\text{sp gr}\,^{20}_{\ 4} = \dfrac{0.989}{0.834} \times 0.99823 = 1.184$

methyl salicylate

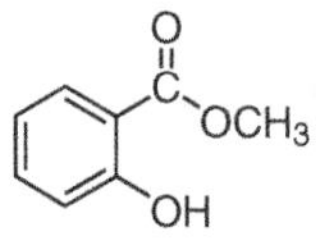

8. Bromomethane < bromoethane < 1-bromopropane < 1-bromobutane < 1-bromopentane < 1-bromohexane

9. 4-Octene > 3-octene > 2-octene > 1-octene

10. Ethyl propyl ether < 1-pentanamine < 1-pentanol < pentanal < methyl butanoate < pentanoic acid

11. $n_D^{20} = 1.430 + \left[(35 - 20)(0.00045)\right] = 1.437$

12. $[\alpha]_D^{25°} = \dfrac{26.6}{(2.5\,cm)\left(\dfrac{10\,cm}{1\,dm}\right)\left(\dfrac{0.8g}{50\,mL}\right)} = 66.5$, sucrose

13. Optical purity $= \dfrac{6.00 - 4.00}{6.00 + 4.00} \times 100 = 20.00\%$

Observed rotation $= +30.00 \times 0.200 = +6.00°$

$\%\ ee = \dfrac{+6.00}{+30.00} \times 100 = 20.00\%$

$\%\ \text{of}\,(+)\text{isomer} = \dfrac{4.00}{6.00 + 4.00} \times 100 = 40.00\%$

$\%\ \text{of}\,(-)\text{isomer} = \dfrac{6.00}{6.00 + 4.00} \times 100 = 60.00\%$

14. Methylene chloride with ether, toluene, or hydrocarbons as co-solvents

15. Methanol; ethanol

16. $\text{mg of C} = 10.71 \text{ mg of CO}_2 \times \dfrac{12.011 \text{ C}}{44.009 \text{ CO}_2} = 2.92 \text{ mg of C in sample}$

$\%C = \dfrac{2.92 \text{ mg of C}}{13.66 \text{ mg of sample}} \times 100 = 21.38\% \text{ of C}$

$\text{mg of H} = 3.28 \text{ mg of H}_2\text{O} \times \dfrac{2.016 \text{ H}}{18.015 \text{ H}_2\text{O}} = 0.367 \text{ mg of H in sample}$

$\%H = \dfrac{0.367 \text{ mg of H}}{13.66 \text{ mg of sample}} \times 100 = 2.69\% \text{ of H}$

$\%Br = \dfrac{3.46 \text{ mg of H}}{4.86 \text{ mg of sample}} \times 100 = 71.19\% \text{ of Br}$

$\%O = 100 - (21.40 + 2.69 + 71.19) = 4.72\% \text{ of O}$

$C = \dfrac{21.40}{12.011} = 1.782 \qquad \dfrac{1.782}{0.295} = 6.04$

$H = \dfrac{2.669}{1.0080} = 2.669 \qquad \dfrac{2.669}{0.295} = 9.05$

$Br = \dfrac{71.19}{79.904} = 0.891 \qquad \dfrac{0.891}{0.295} = 3.02$

$O = \dfrac{4.72}{15.999} = 0.295 \qquad \dfrac{0.295}{0.295} = 1.00$

Empirical formula = $C_6H_9Br_3O$
Empirical weight = 336.86
Molecular formula = $C_{12}H_{18}Br_6O_2$

17. a. $C_6H_{12}O_2$
$U = 6 + 1 - \frac{1}{2}(12) + \frac{1}{2}(0) = 1$
One double bond; or one ring

b. $C_5H_{10}Cl_2$
$U = 5 + 1 - \frac{1}{2}(12) + \frac{1}{2}(0) = 0$
No double bonds, triple bonds, or rings

c. $C_7H_{13}N$
$U = 7 + 1 - \frac{1}{2}(13) + \frac{1}{2}(1) = 2$
Two rings; or two double bonds; or one ring and one double bond; or one triple bond

d. $C_{12}H_{10}$
$U = 12 + 1 - \frac{1}{2}(10) + \frac{1}{2}(0) = 8$
Benzene plus three rings; or benzene plus three double bonds; or benzene plus two rings and one double bond; or benzene plus two double bonds and one ring; or benzene plus one triple bond and one double bond; or benzene plus one triple bond and one ring

Chapter 5

Classification of Organic Compounds by Solubility

1. **a.** N – alcohols, aldehydes, ketones, esters with one functional group and more than five but fewer than nine carbons, ethers, epoxides, alkenes, alkynes, some aromatic compounds (especially those with activating groups).

    b. S_1 – monofunctional alcohols, aldehydes, ketones, esters, nitriles, and amides with five carbons or fewer.

    c. A_2 – weak organic acids; phenols, enols, oximes, imides, sulfonamides, thiophenols, all with more than five carbons; β-diketones; nitro compounds with α-hydrogens.

2. **a.** Insoluble in water, soluble in 5% sodium hydroxide solution, soluble in 5% sodium bicarbonate solution.

    4-methylbenzoic acid

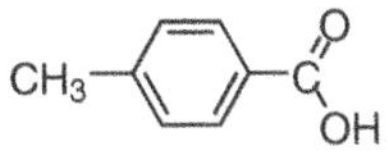

    b. Soluble in water and ether, litmus paper turned blue.

    butylamine $CH_3CH_2CH_2CH_2NH_2$

    c. Insoluble in water, 5% sodium hydroxide solution, and 5% hydrochloric acid solution, but soluble in 96% sulfuric acid solution.

    3-heptene $CH_3CH_2CH=CHCH_2CH_2CH_3$

Student Solutions Manual to Accompany The Systematic Identification of Organic Compounds, Ninth Edition. Christine K. F. Hermann, Terence C. Morrill, Ralph L. Shriner, and Reynold C. Fuson. © 2023 John Wiley & Sons, Inc. Published 2023 by John Wiley & Sons, Inc.

d. Soluble in water and ether; litmus paper is unchanged in color.

ethyl propanoate $\quad$ $CH_3CH_2\underset{\underset{O}{\|}}{C}-O-CH_2CH_3$

e. Soluble in water and ether; litmus paper is unchanged in color.

propanenitrile $\quad$ $CH_3CH_2C\equiv N$

f. Insoluble in water and 5% sodium bicarbonate solution, but soluble in 5% sodium hydroxide solution.

4-methylphenol $\quad$ $CH_3-\!\!\!\!\bigcirc\!\!\!\!-OH$

3. a. A_1

$\quad$ **b.** S_B

$\quad$ **c.** N

$\quad$ **d.** S_1

$\quad$ **e.** S_1

$\quad$ **f.** A_2

4. Both acyl halides and anhydrides react with water, so any results would be of the carboxylic acid that is formed from the hydrolysis of these compounds.

5. a. 1-Propanol is more soluble in water than diethyl ether. 1-Propanol can hydrogen bond with water, whereas diethyl ether does not hydrogen bond with water.

$CH_3CH_2CH_2OH \qquad CH_3CH_2OCH_2CH_3$

$\quad$ 1-propanol $\qquad$ diethyl ether

$\quad$ **b.** Pentanedioic acid is more soluble in water than butanedioic acid. Odd carbon number of dioic acids have less intracrystalline forces than even carbon number of dioic acids.

$HO\underset{\underset{O}{\|}}{\overset{CH_2}{C}}\overset{CH_2}{\underset{CH_2}{}}\underset{\underset{O}{\|}}{C}OH \qquad HO\underset{\underset{O}{\|}}{\overset{CH_2}{C}}\overset{\overset{O}{\|}}{\underset{CH_2}{C}}OH$

$\qquad$ pentanedioic acid $\qquad\qquad$ butanedioic acid

$\quad$ **c.** Methyl methacrylate is more soluble in water than poly(methyl methacrylate). Polymers are insoluble in water.

$CH_2=\overset{CH_3}{\underset{\underset{O}{\|}}{C}-OCH_3} \qquad \left[-CH_2-\overset{CH_3}{\underset{\underset{O}{\|}}{C}-OCH_3}-\right]_n$

$\qquad$ methyl methacrylate $\quad$ poly(methyl methacrylate)

d. Butanoic acid is more soluble in water than 2-Bromobutanoic acid. The addition of halogens increases the molecular weight and decreases the solubility.

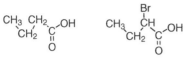

       butanoic acid   2-bromobutanoic acid

e. 2-Methylpropanoic acid is more soluble in water than butanoic acid. Branching lowers the boiling point and lowers intermolecular forces. Thus, a branched compound is more soluble in water than a straight chain molecule.

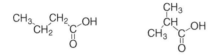

      butanoic acid   2-methylpropanoic acid

6. a. 1-Chlorobutane is insoluble in water since it is an haloalkane. It is insoluble in 5% sodium hydroxide solution, 5% hydrochloric acid solution, and 96% sulfuric acid solution. Thus 1-chlorobutane is in solubility class I.

    1-chlorobutane $CH_3CH_2CH_2CH_2Cl$

b. 4-Methylaniline is insoluble in water since it is an aniline. It is basic, due to the presence of the amino group, and therefore is insoluble in 5% sodium hydroxide solution, but soluble in 5% hydrochloric acid solution, and thus 4-methylaniline is in solubility class B.

    4-methylaniline H_2N—⟨benzene⟩—CH_3

c. 1-Nitroethane is insoluble in water since it is a weak organic acid. It is soluble in 5% sodium hydroxide solution, but insoluble in 5% sodium bicarbonate solution, and thus 1-nitroethane is in solubility class A_2.

    1-nitroethane $CH_3CH_2NO_2$

d. Alanine is an amino acid, thus is amphoteric, and soluble in water. Alanine is insoluble in ether and thus in solubility class S_2.

    alanine CH_3—$\underset{\underset{{}^+NH_3}{|}}{CH}$—$\overset{\overset{O}{\|}}{C}$—$O^-$

e. Benzophenone is insoluble in water because it has more than five carbon atoms. Since it is a neutral compound, containing only carbon, hydrogen, and oxygen, it is insoluble in water, 5% sodium hydroxide solution, and 5% hydrochloric acid solution. It is soluble in 96% sulfuric acid solution. Therefore, benzophenone is in solubility class N.

    benzophenone

f. As a strong organic acid, benzoic acid is insoluble in water, soluble in 5% sodium hydroxide solution, and soluble in 5% sodium bicarbonate solution. Therefore, benzoic acid is in solubility class A_1.

benzoic acid

g. As a saturated hydrocarbon, hexane is insoluble in water, 5% sodium hydroxide solution, 5% hydrochloric acid solution, and 95% sulfuric acid. Hexane is in solubility class I.

hexane $\quad\quad CH_3CH_2CH_2CH_2CH_2CH_3$

h. 4-Methylbenzyl alcohol is insoluble in water since it has more than five carbon atoms. It is a neutral compound with only carbon, hydrogen, and oxygen. It is insoluble in water, 5% sodium hydroxide solution and 5% hydrochloric acid solution. However, 4-methylbenzyl alcohol is soluble in 96% sulfuric acid and thus in solubility class N.

4-methylbenzyl alcohol $\quad CH_3 -\!\!\!\bigcirc\!\!\!- CH_2OH$

i. As an amine with less than six carbons, ethylmethylamine is soluble in water. It is soluble in ether and its aqueous solutions turn litmus paper blue, therefore it is basic. Ethylmethylamine is in solubility class S_B.

ethylmethylamine $\quad\quad CH_3NHCH_2CH_3$

j. Propoxybenzene is insoluble in water because it has more than five carbon atoms. It is a neutral compound with only carbon, hydrogen, and oxygen and is insoluble in 5% sodium hydroxide solution, and 5% hydrochloric acid solution. However, propoxybenzene is soluble in 96% sulfuric acid solution and in solubility class N.

propoxybenzene $\quad\quad \bigcirc\!\!- OCH_2CH_2CH_3$

k. Propanal is soluble in water because it has less than five carbons. As a neutral compound, it is soluble in ether and its aqueous solutions do not change the color of litmus paper. Propanal is in solubility class S_1.

propanal $\quad\quad CH_3CH_2 -\underset{\underset{O}{\|}}{C} - H$

l. As an aryl halide, 1,3-Dibromobenzene is insoluble in water, 5% sodium hydroxide solution, 5% hydrochloric acid solution, and 96% sulfuric acid. 1,3-Dibromobenzene is in solubility class I.

1,3-dibromobenzene

m. Propanoic acid is soluble in water because it contains less than five carbon atoms. It is soluble in ether, but its aqueous solutions turn litmus paper red. Propanoic acid is in solubility class S_A.

propanoic acid
$$CH_3CH_2-\underset{\underset{O}{\|}}{C}-OH$$

n. As a weak organic acid, benzenesulfonamide is insoluble in water. It is soluble in 5% sodium hydroxide solution but insoluble in 5% sodium bicarbonate solution. Benzenesulfonamide is in solubility class A_2.

benzenesulfonamide

o. 1-Butanol is soluble in water because it contains less than five carbon atoms. It is soluble in ether and its aqueous solutions do not change the color of litmus. 1-Butanol is in solubility class S_1.

1-butanol $\quad CH_3CH_2CH_2CH_2OH$

p. Methyl propanoate is soluble in water since it contains less than five carbon atoms. It is soluble in ether and its aqueous solutions do not change the color of litmus. Methyl propanoate is in solubility class S_1.

methyl propanoate
$$CH_3CH_2-\underset{\underset{O}{\|}}{C}-OCH_3$$

q. 4-Methylcyclohexanone is insoluble in water since it contains more than five carbon atoms. It is also insoluble in 5% sodium hydroxide solution and 5% hydrochloric acid solution, but soluble in 96% sulfuric acid. Therefore, 4-methylcyclohexanone is in solubility class N.

4-methylcyclohexanone $\quad CH_3-$

r. Since 4-Aminobiphenyl is an aniline, it is insoluble in water. It is insoluble in 5% sodium hydroxide solution, but soluble in 5% hydrochloric acid solution. 4-Aminobiphenyl is in solubility class B.

4-aminobiphenyl

s. Since 4-Methylacetophenone contains more than five carbon atoms, it is insoluble in water. It is also insoluble in 5% sodium hydroxide solution and 5% hydrochloric acid solution, but soluble in 96% sulfuric acid. Therefore, 4-methylacetophenone is in solubility class N.

4-methylacetophenone

t. Naphthalene is insoluble in water since it contains more than five carbon atoms. It is insoluble in 5% sodium hydroxide solution and 5% hydrochloric acid solution. However, naphthalene is soluble in 96% sulfuric acid, and therefore in solubility class N.

naphthalene

u. Phenylalanine is an amino acid, thus is amphoteric, and soluble in water. Phenylalanine is insoluble in ether and thus in solubility class S_2.

phenylalanine

v. Benzoin is insoluble in water since it has more than five carbon atoms. It is insoluble in water, 5% sodium hydroxide solution, and 5% hydrochloric acid solution. However, benzoin is soluble in 96% sulfuric acid, and therefore in solubility class N.

benzoin

w. As a strong organic acid, 4-hydroxybenzenesulfonic acid is insoluble in water. However, it is soluble in 5% sodium hydroxide solution and 5% sodium bicarbonate solution. 4-Hydroxybenzenesulfonic acid is in solubility class A_1.

4-hydroxybenzenesulfonic acid

7. Listed in order from least basic to most basic (see next question).

$$N, S_1 < B < S_B$$

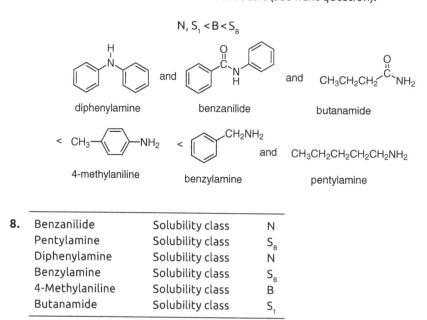

diphenylamine   and   benzanilide   and   butanamide

< 4-methylaniline   <   benzylamine   and   pentylamine

8.		
Benzanilide	Solubility class	N
Pentylamine	Solubility class	S_B
Diphenylamine	Solubility class	N
Benzylamine	Solubility class	S_B
4-Methylaniline	Solubility class	B
Butanamide	Solubility class	S_1

9. Listed in order from least soluble to most soluble in water.

    a. Smaller alcohols are more soluble.

$$CH_3CH_2CH_2CH_2OH \; < \; (CH_3)_2CHOH \; < \; CH_3CH_2OH \; < \; CH_3OH$$

        1-butanol   2-propanol   ethanol   methanol

    b. More hydroxyl groups increase solubility.

$$CH_3CH_2CH_2CH_3 \; < \; CH_3CH_2CH_2CH_2OH \; < \; HOCH_2CH_2CH_2CH_2OH$$

         butane   1-butanol   1,2-butanediol

    c. Branching increases solubility.

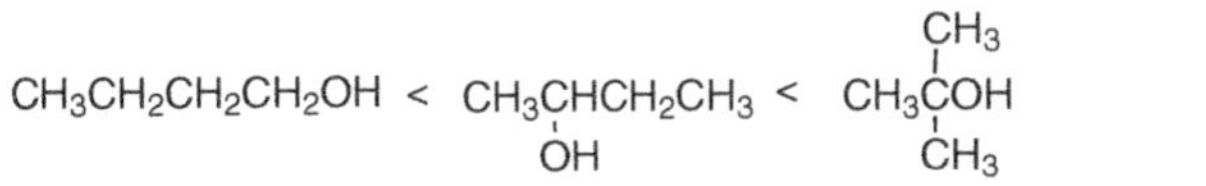

        1-butanol   2-butanol   2-methyl-2-propanol

    d. $N < S_1 - N < S_B < S_2$

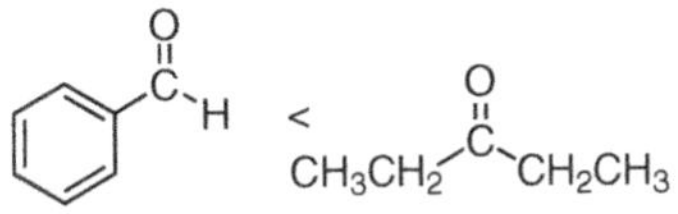

        benzaldehyde   3-pentanone

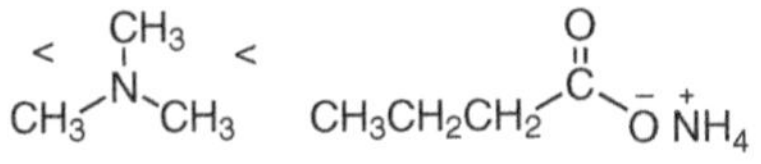

          triethylamine   ammonium butanoate

10. a.

Methanol	Solubility class	S_1
Isopropyl alcohol	Solubility class	S_1
Ethanol	Solubility class	S_1
1-Butanol	Solubility class	S_1

    b.

Butane	Solubility class	I
1,4-Butanediol	Solubility class	S_2
1-Butanol	Solubility class	S_1

    c.

1-Butanol	Solubility class	S_1
2-Methyl-2-propanol	Solubility class	S_1
2-Butanol	Solubility class	S_1

    d.

Ammonium butanoate	Solubility class	S_2
3-Pentanone	Solubility class	$S_1 - N$
Benzaldehyde	Solubility class	N
Trimethylamine	Solubility class	S_B

11. The following compounds are listed in order of decreasing activity.

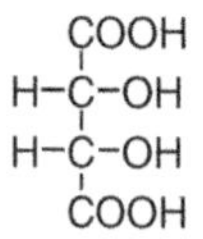

meso-tartaric
acid (S$_2$)

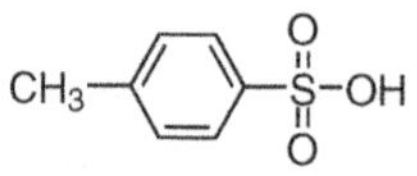

4-toluenesulfonic
acid (S$_A$)

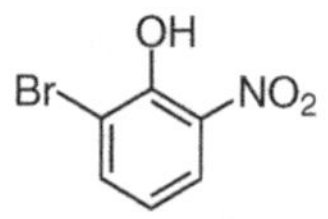

2-bromo-6-
nitrophenol (A$_1$)

and

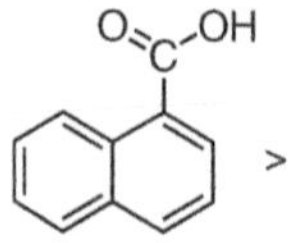

1-naphthoic
acid (A$_1$)

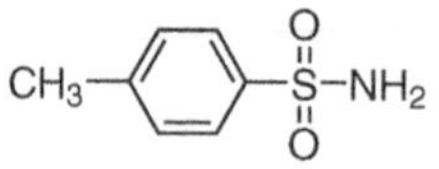

4-toluene-
sulfonamide (A$_2$)

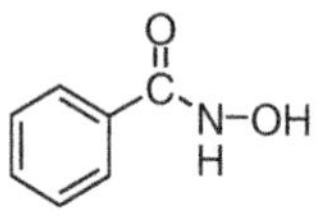

> benzohydroxamic
acid (MN)

and saccharin (MN)

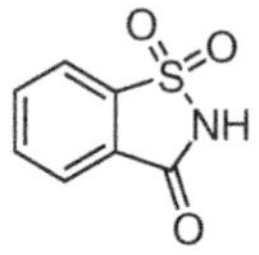

and

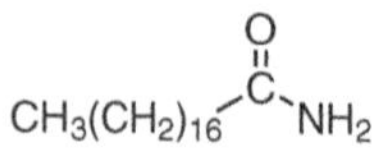

octadecanamide (MN)

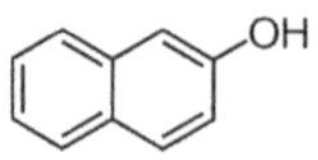

2-naphthol (N)

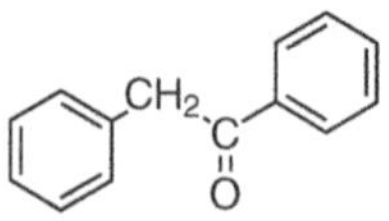

benzyl phenyl ketone (N)

Chapter 6

Separation of Mixtures

1. **a.** Pentane – volatile

   b. D-glucose – low volatility, with certain exceptions these compounds cannot be distilled at atmospheric pressure

   c. Diethylamine – readily distill, many compounds boil below 100°C

   d. 2-Butanone – readily distill, many compounds boil below 100°C

2. **a.** Pentane – volatile with steam

   b. D-Glucose – not volatile with steam

   c. Diethylamine – volatile with steam

   d. 2-Butanone – volatile with steam

Student Solutions Manual to Accompany The Systematic Identification of Organic Compounds, Ninth Edition.
Christine K. F. Hermann, Terence C. Morrill, Ralph L. Shriner, and Reynold C. Fuson.
© 2023 John Wiley & Sons, Inc. Published 2023 by John Wiley & Sons, Inc.

3.

Chlorobenzene, *N,N*-dimethylaniline, 2-naphthol,
2-methoxybenzaldehyde

Ether, | 5% HCl

Organic ↓   ↓ Aqueous

Chlorobenzene, 2-naphthol,   *N,N*-Dimethylanilinium
2-methoxybenzaldehyde   chloride

                                            ↓ 5% NaOH

5% | NaOH   *N,N*-Dimethylaniline

Organic ↓   ↓ Aqueous

Chlorobenzene,   Sodium 2-naphthol
2-methoxybenzaldehyde
                            ↓ 5% HCl

NaHSO₃   2-naphthol

Organic ↓   ↓ Aqueous

Chlorobenzene   Sodium bisulfite derivative
                 of 2-methoxybenzaldehyde

                        ↓ 5% HCl

                 2-Methoxybenzaldehyde

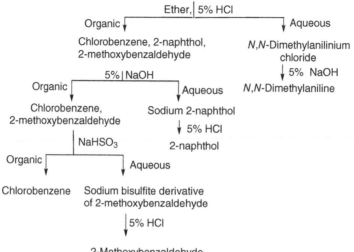

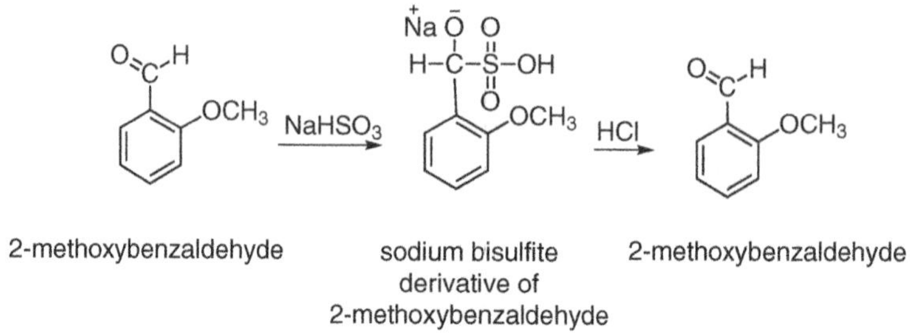

2-methoxybenzaldehyde	NaHSO₃	sodium bisulfite derivative of 2-methoxybenzaldehyde	HCl	2-methoxybenzaldehyde

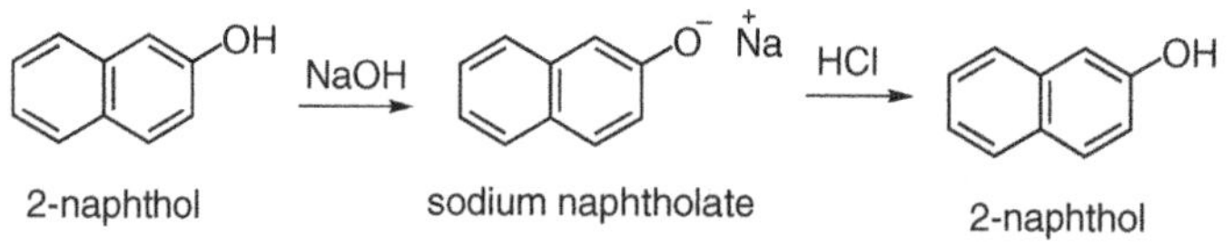

2-naphthol → NaOH → sodium naptholate → HCl → 2-naphthol

N,N-dimethylaniline → HCl → *N,N*-dimethylanilinium chloride → NaOH → *N,N*-dimethylaniline

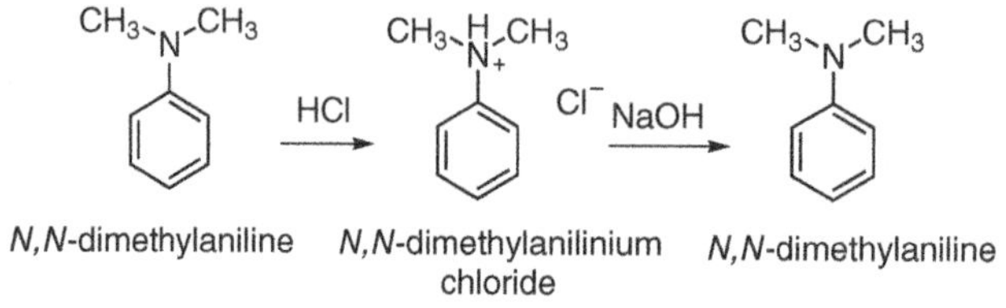

4.

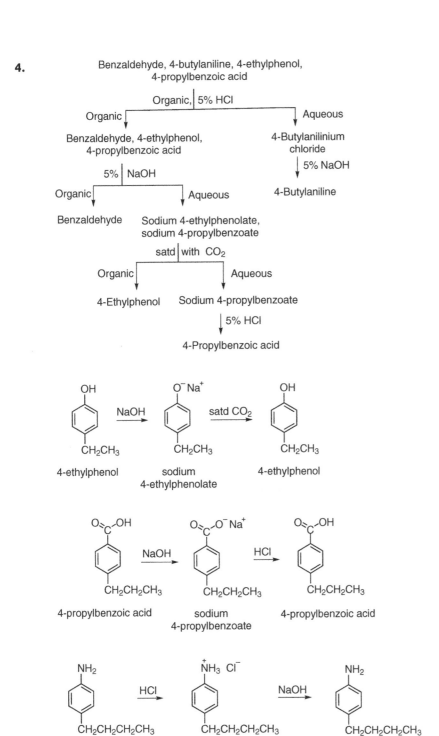

Benzaldehyde, 4-butylaniline, 4-ethylphenol,
4-propylbenzoic acid

Organic, 5% HCl

Organic → Benzaldehyde, 4-ethylphenol, 4-propylbenzoic acid

Aqueous → 4-Butylanilinium chloride

5% NaOH

4-Butylaniline

5% NaOH

Organic → Benzaldehyde

Aqueous → Sodium 4-ethylphenolate, sodium 4-propylbenzoate

satd with CO₂

Organic → 4-Ethylphenol

Aqueous → Sodium 4-propylbenzoate

5% HCl

4-Propylbenzoic acid

4-ethylphenol → NaOH → sodium 4-ethylphenolate → satd CO₂ → 4-ethylphenol

4-propylbenzoic acid → NaOH → sodium 4-propylbenzoate → HCl → 4-propylbenzoic acid

4-butylaniline → HCl → 4-butylanilium chloride → NaOH → 4-butylaniline

5. Since glucose is also insoluble in ether, it would be filtered out as Residue 1.

6. Lactic acid, piperidine, acetic acid, 2-propanol

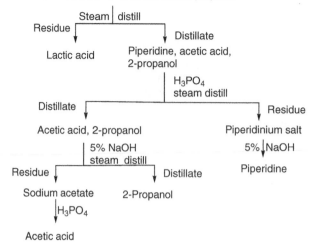

Steam | distill

Residue ↓

Lactic acid

Distillate

Piperidine, acetic acid, 2-propanol

H₃PO₄ steam distill

Distillate

Acetic acid, 2-propanol

Residue

Piperidinium salt

5% NaOH steam distill

Residue ↓

Sodium acetate

Distillate

2-Propanol

5% NaOH

Piperidine

H₃PO₄

Acetic acid

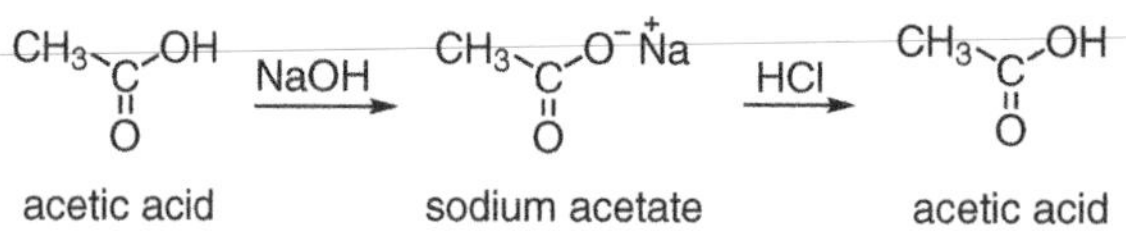

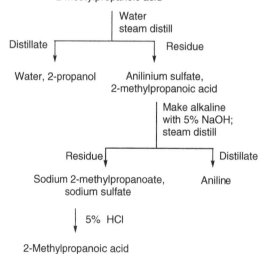

$CH_3-\underset{O}{\overset{}{C}}-OH$ acetic acid →NaOH→ $CH_3-\underset{O}{\overset{}{C}}-O^- \overset{+}{Na}$ sodium acetate →HCl→ $CH_3-\underset{O}{\overset{}{C}}-OH$ acetic acid

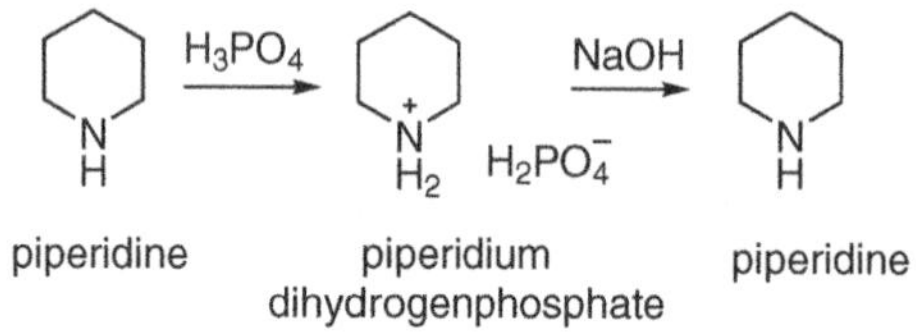

piperidine →H₃PO₄→ piperidium dihydrogenphosphate H₂PO₄⁻ →NaOH→ piperidine

7. a.   Water, 2-propanol, anilinium sulfate,
2-methylpropanoic acid

Water
steam distill

Distillate

Water, 2-propanol

Residue

Anilinium sulfate,
2-methylpropanoic acid

Make alkaline
with 5% NaOH;
steam distill

Residue ↓

Sodium 2-methylpropanoate,
sodium sulfate

Distillate ↓

Aniline

5% HCl

2-Methylpropanoic acid

b.

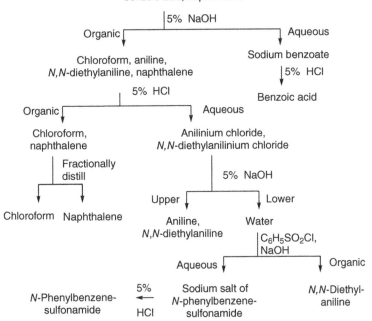

c.

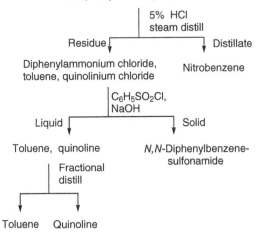

d.

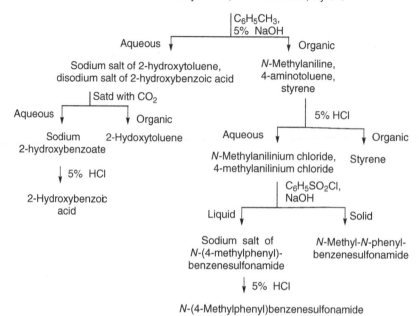

2-Hydroxytoluene, 2-hydroxybenzoic acid,
N-methylaniline, 4-aminotoluene, styrene

$C_6H_5CH_3$,
5% NaOH

Aqueous → | Organic

Sodium salt of 2-hydroxytoluene,
disodium salt of 2-hydroxybenzoic acid

N-Methylaniline,
4-aminotoluene,
styrene

Satd with CO_2

Aqueous → | Organic

Sodium
2-hydroxybenzoate

2-Hydroxytoluene

5% HCl

Aqueous → | Organic

N-Methylanilinium chloride,
4-methylanilinium chloride

Styrene

5% HCl

2-Hydroxybenzoic
acid

$C_6H_5SO_2Cl$,
NaOH

Liquid → | Solid

Sodium salt of
N-(4-methylphenyl)-
benzenesulfonamide

N-Methyl-N-phenyl-
benzenesulfonamide

5% HCl

N-(4-Methylphenyl)benzenesulfonamide

e.

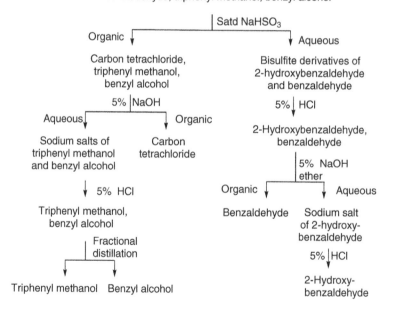

Carbon tetrachloride, 2-hydroxybenzaldehyde,
benzaldehyde, triphenyl methanol, benzyl alcohol

Satd $NaHSO_3$

Organic → | Aqueous

Carbon tetrachloride,
triphenyl methanol,
benzyl alcohol

Bisulfite derivatives of
2-hydroxybenzaldehyde
and benzaldehyde

5% NaOH

5% HCl

Aqueous → | Organic

Sodium salts of
triphenyl methanol
and benzyl alcohol

Carbon
tetrachloride

2-Hydroxybenzaldehyde,
benzaldehyde

5% NaOH
ether

5% HCl

Organic → | Aqueous

Triphenyl methanol,
benzyl alcohol

Benzaldehyde

Sodium salt
of 2-hydroxy-
benzaldehyde

Fractional
distillation

5% HCl

Triphenyl methanol   Benzyl alcohol

2-Hydroxy-
benzaldehyde

f.

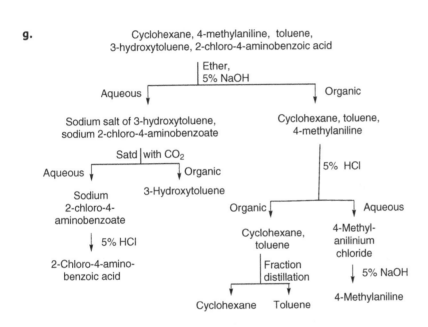

Diethyl ether, 3-pentanone, diethylamine, acetic acid

5% HCl

Organic → 3-Pentanone, diethyl ether, acetic acid

Aqueous → Diethylammonium chloride

5% NaOH → Diethylamine

3-Pentanone, diethyl ether, acetic acid → 5% NaOH

Aqueous → Sodium acetate

Organic → Diethyl ether, 3-pentanone

Sodium acetate → 5% HCl → Acetic acid

Diethyl ether, 3-pentanone → NaHSO₃

Aqueous → Sodium bisulfite derivative of 3-pentanone

Organic → Diethyl ether

Sodium bisulfite derivative of 3-pentanone → 5% HCl → 3-Pentanone

g.

Cyclohexane, 4-methylaniline, toluene, 3-hydroxytoluene, 2-chloro-4-aminobenzoic acid

Ether, 5% NaOH

Aqueous → Sodium salt of 3-hydroxytoluene, sodium 2-chloro-4-aminobenzoate

Organic → Cyclohexane, toluene, 4-methylaniline

Satd with CO₂

Aqueous → Sodium 2-chloro-4-aminobenzoate

Organic → 3-Hydroxytoluene

Sodium 2-chloro-4-aminobenzoate → 5% HCl → 2-Chloro-4-amino-benzoic acid

Cyclohexane, toluene, 4-methylaniline → 5% HCl

Organic → Cyclohexane, toluene

Aqueous → 4-Methyl-anilinium chloride

Cyclohexane, toluene → Fraction distillation → Cyclohexane, Toluene

4-Methyl-anilinium chloride → 5% NaOH → 4-Methylaniline

8. To solve this problem, the number of moles of each substance must be determined.

Compound	grams	MW	moles
2-Methyl-2-propanol	2.47	74.63	0.033
Benzyl alcohol	3.60	108.81	0.033
Benzaldehyde	3.53	106.79	0.033
Acetyl chloride	5.20	78.69	0.066
Acetophenone	4.00	120.92	0.033
N,N-Dimethylaniline	28.2	121.95	0.231

The acetyl chloride will react with the 2-methyl-2-propanol and the benzyl alcohol to form the 1,1-dimethylethyl acetate and the benzyl acetate, respectively.

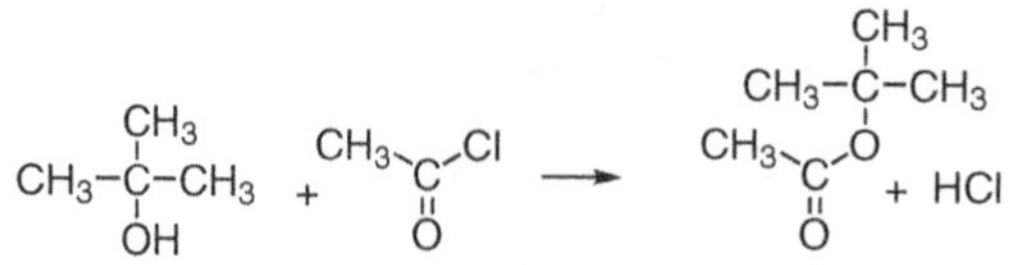

$$CH_3-\underset{\underset{OH}{|}}{\overset{\overset{CH_3}{|}}{C}}-CH_3 \; + \; CH_3-\underset{\underset{O}{\|}}{C}-Cl \longrightarrow CH_3-\underset{\underset{O}{\|}}{C}-O-\underset{\underset{CH_3}{|}}{\overset{\overset{CH_3}{|}}{C}}-CH_3 \; + \; HCl$$

2-methyl-2-propanol   acetyl chloride   1,1-dimethylethyl acetate (*t*-butyl acetate)

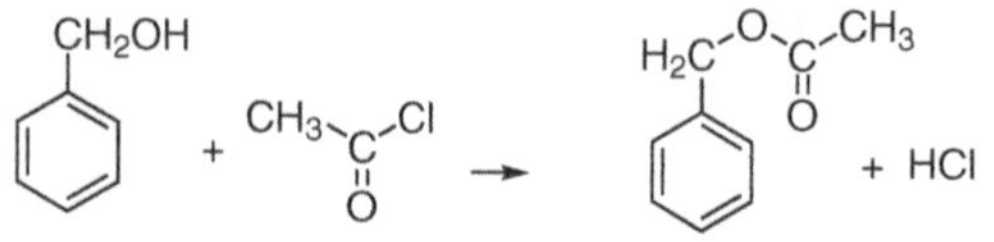

$$C_6H_5-CH_2OH \; + \; CH_3-\underset{\underset{O}{\|}}{C}-Cl \longrightarrow C_6H_5-CH_2-O-\underset{\underset{O}{\|}}{C}-CH_3 \; + \; HCl$$

benzyl alcohol   acetyl chloride   benzyl acetate

The hydrogen chloride, produced from the above reactions, reacts with *N,N*-dimethylaniline to form *N,N*-dimethylanilinium chloride.

$$C_6H_5-N(CH_3)_2 \; + \; HCl \longrightarrow C_6H_5-\overset{+}{N}H(CH_3)_2 \quad Cl^-$$

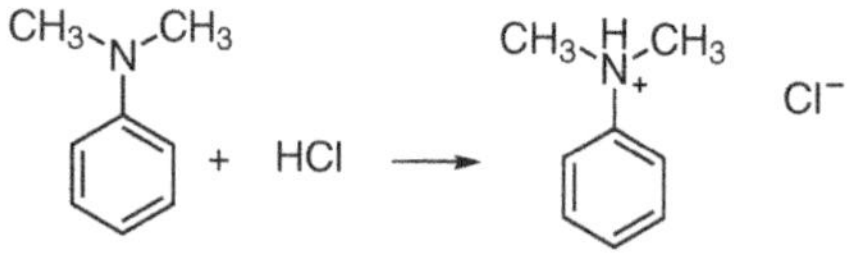

N,N-dimethylaniline   *N,N*-dimethylanilinium chloride

If complete reactions are assumed, then the following compounds would be present at the end of a week.

Compound	moles
1,1-Dimethylethyl acetate	0.033
Benzyl acetate	0.033
Benzaldehyde	0.033
Acetophenone	0.033
N,N-Dimethylanilinium chloride	0.066
N,N-Dimethylaniline	0.165

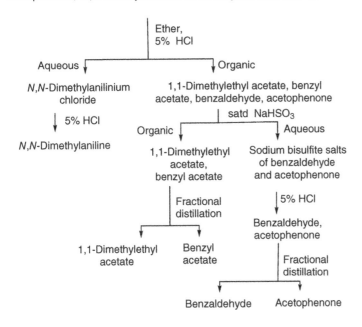

1,1-Dimethylethyl acetate, benzyl acetate, benzaldehyde, acetophenone, *N,N*-dimethylaniline, *N,N*-dimethylanilinium chloride

Ether, 5% HCl

Aqueous

N,N-Dimethylanilinium chloride

5% HCl

N,N-Dimethylaniline

Organic

1,1-Dimethylethyl acetate, benzyl acetate, benzaldehyde, acetophenone

satd NaHSO₃

Organic

1,1-Dimethylethyl acetate, benzyl acetate

Fractional distillation

1,1-Dimethylethyl acetate

Benzyl acetate

Aqueous

Sodium bisulfite salts of benzaldehyde and acetophenone

5% HCl

Benzaldehyde, acetophenone

Fractional distillation

Benzaldehyde   Acetophenone

9. **a.** Pentane – very nonpolar

   b. Butanoic acid – polar

   c. 2-Butanone – nonpolar

   d. D-Fructose – very polar

10. Pentane < 2-Butanone < butanonic acid < D-fructose

11. **a.** Since these compounds are alcohols, gas chromatography could be utilized. These compounds would be separated by bp.

    b. Extraction methods would be the preferred method of separating these compounds.

    c. Gas chromatography could be utilized to separate these compounds. These compounds would be separated by bp.

12. The compounds must be soluble in a solvent and/or be able to be volatized in order to be separated by GC or LC. Esters are more volatile and are more easily dissolved than carboxylic acids.

Chapter 7

Nuclear Magnetic Resonance Spectrometry

For some of the hydrogen and carbon calculations, the closest structure in the tables were used.

1. Compound with a formula of C_3H_7Br.

   Unsaturation number:

   $U = 3 + 1 - (\frac{1}{2} \times 8) + (\frac{1}{2} \times 0) = 0$
   No double bonds, triple bonds, or rings

   ^1H NMR spectrum:

	Chemical shift (Actual)	(Calculated)	Splitting	Integration	Interpretation
A	1.01	1.1	t	3H	CH_3 adjacent to CH_2
B	1.87	1.9	sext	2H	CH_2 adjacent to 5H
C	3.38	3.3	t	2H	CH_2 adjacent to CH_2

   Bromine must be at the end of a propane chain.

Student Solutions Manual to Accompany The Systematic Identification of Organic Compounds, Ninth Edition.
Christine K. F. Hermann, Terence C. Morrill, Ralph L. Shriner, and Reynold C. Fuson.
© 2023 John Wiley & Sons, Inc. Published 2023 by John Wiley & Sons, Inc.

Calculations:

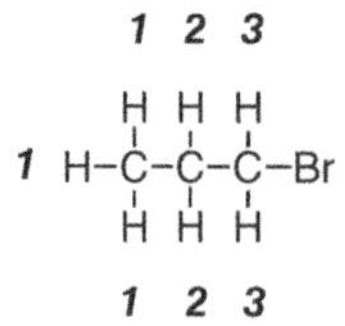

```
         1  2  3
         H  H  H
         |  |  |
  1   H--C--C--C--Br      1-bromopropane
         |  |  |
         H  H  H
         1  2  3
```

	1	2	3
	0.9	1.2	1.2
	0.2	0.7	2.1
	1.1	1.9	3.3

Labeled structure:

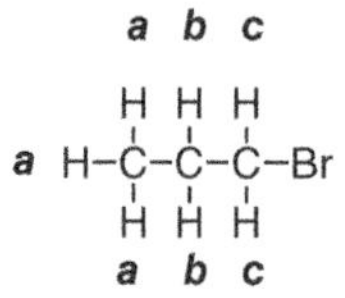

```
         a  b  c
         H  H  H
         |  |  |
  a   H--C--C--C--Br      1-bromopropane
         |  |  |
         H  H  H
         a  b  c
```

^{13}C NMR spectrum:

	Chemical shift		Interpretation
	(Actual)	(Calculated)	
A	12.96	12.6	Alkyl
B	26.22	27.1	Alkyl
C	35.94	35.6	Alkyl bromide

Calculations:

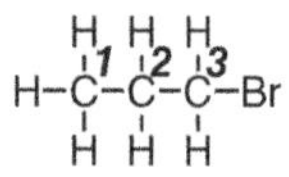

```
        H  H  H
        |1 |2 |3
   H----C--C--C--Br      1-bromopropane
        |  |  |
        H  H  H
```

1	2	3
15.6	16.1	15.6
−3	11	20
12.6	27.1	35.6

Labeled structure:

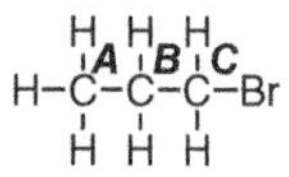

```
        H  H  H
        |A |B |C
   H----C--C--C--Br      1-bromopropane
        |  |  |
        H  H  H
```

2. Compound with a formula of $C_5H_{12}O$.

Unsaturation number:

$U = 5 + 1 - \frac{1}{2}(12) + \frac{1}{2}(0) = 0$

No double bonds, triple bonds, or rings

The compound must be an ether or alcohol.

¹³C NMR and DEPT spectra:

	Chemical shift		DEPT	Interpretation
	(Actual)	(Calculated)		
A	14.03	13.7	CH₃	Alkyl
B	18.91	17.6	CH₂	Alkyl
C	23.35	23.7	CH₃	Alkyl
D	41.44	44.5	CH₂	Alkyl
E	67.86	70.6	CH	Alcohol

One hydrogen is missing from DEPT. The compound must be an alcohol. Since all five carbons show up in ¹³C NMR, then the structure must be 2-pentanol.

Calculations:

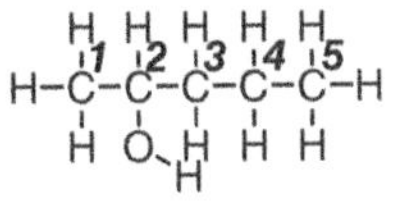

2-pentanol

1	2	3	4	5
13.7	22.6	34.5	22.6	13.7
10	48	10	−5	0
23.7	70.6	44.5	17.6	13.7

Labeled structure:

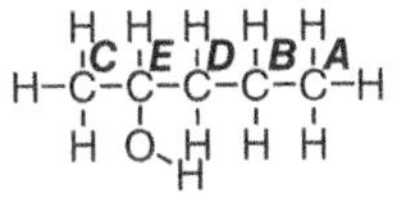

2-pentanol

3. Compound with a formula of $C_5H_{10}O$.

   Unsaturation number:

   $U = 5 + 1 - ½(10) + ½(0) = 1$
   One double bond or one ring
   The compound must be an aldehyde, ketone, ring or alkene with alcohol or ether.

   ¹H NMR spectrum:

	Chemical shift		Splitting	Integration	Interpretation
	(Actual)	(Calculated)			
a	0.90	0.9	t	3H	CH₃ adjacent to CH₂
b	1.57	1.5	sext	2H	CH₂ adjacent to 5H
c	2.10	2.1	s	3H	CH₃ isolated
d	2.38	2.4	t	2H	CH₂ adjacent to CH₂

   The chemical shifts are low indicating no alkene. A ring is also ruled out since methyls are present. Since there is no signal for an aldehyde, the structure must be a ketone – 2-pentanone.

Calculations:

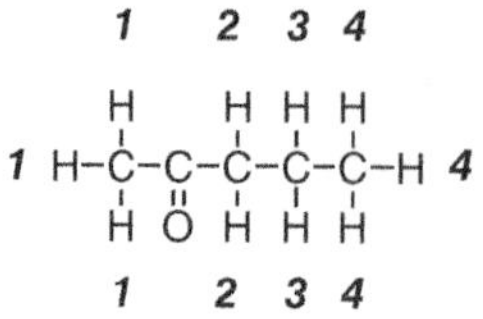

```
        1   2  3  4
        H     H  H  H
        |     |  |  |
1   H–C–C–C–C–C–H  4    2-pentanone
        |  ||  |  |  |
        H  O  H  H  H
        1     2  3  4
```

1	2	3	4
0.9	1.2	1.2	0.9
1.2	1.2	0.3	0
2.1	2.4	1.5	0.9

Labeled structure:

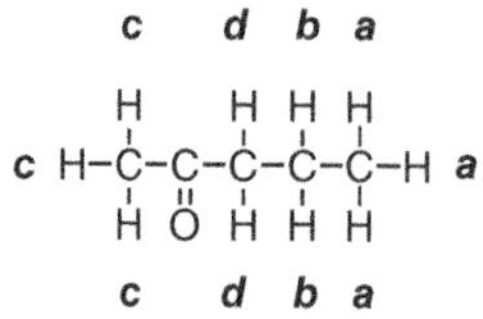

```
        c     d  b  a
        H     H  H  H
        |     |  |  |
c   H–C–C–C–C–C–H  a    2-pentanone
        |  ||  |  |  |
        H  O  H  H  H
        c     d  b  a
```

^{13}C NMR and DEPT spectra:

	Chemical shift		DEPT	Interpretation
	(Actual)	(Calculated)		
A	13.63	13.6	CH_3	Alkyl
B	17.25	17.1	CH_2	Alkyl
C	29.80	27.9	CH_3	Alkyl
D	45.63	45.6	CH_2	Alkyl
E	209.14	180–220	C	Ketone

Calculations:

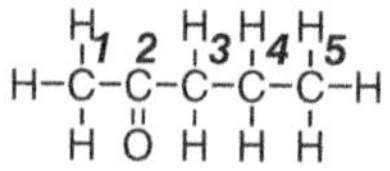

```
        H        H  H  H
        |1  2    |3 |4 |5
    H–C–C–C–C–C–H      2-pentanone
        |  ||  |  |  |
        H  O  H  H  H
```

1	2	3	4	5
−2.1		15.6	16.1	15.6
30		30	1	−2
27.9	180–220	45.6	17.1	13.6

Labeled structure:

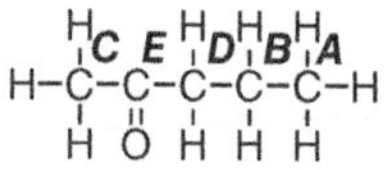

```
        H        H  H  H
        |C  E    |D |B |A
    H–C–C–C–C–C–H      2-pentanone
        |  ||  |  |  |
        H  O  H  H  H
```

4. Compound with a formula of $C_8H_8O_3$.

Unsaturation number:

$U = 8 + 1 - \frac{1}{2}(8) + \frac{1}{2}(0) = 5$
Benzene plus one ring or double bond

¹H NMR spectrum:

	Chemical shift (Actual)	(Calculated)	Splitting	Integration	Interpretation
A	3.94	4.0	s	3H	CH_3 isolated
B	6.85–6.89	6.74	m	1H	Aromatic
C	6.97–6.99	6.94	m	1H	Aromatic
D	7.42–7.47	7.33	m	1H	Aromatic
E	7.81–7.84	7.87	m	1H	Aromatic
F	10.76	4.5–7.7	s	1H	OH isolated

The structure is a disubstituted benzene. All aromatic peaks are multiplets, indicating that the compound is *o*-disubstituted. The first signal is at 3.94, indicating that a methyl might be attached to an oxygen. The last peak indicates a phenol more than an alcohol. These signals indicate methyl salicylate.

Calculations:

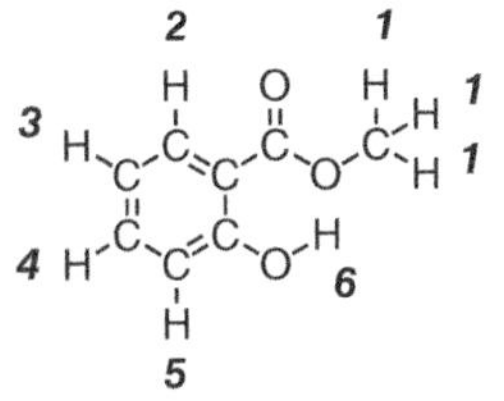

methyl salicylate

1	2	3	4	5	6
0.9	7.27	7.27	7.27	7.27	
3.1	0.74	0.07	0.20	0.07	
	−0.14	−0.40	−0.14	−0.50	
4.0	7.87	6.94	7.33	6.74	4.5–7.7

Labeled structure:

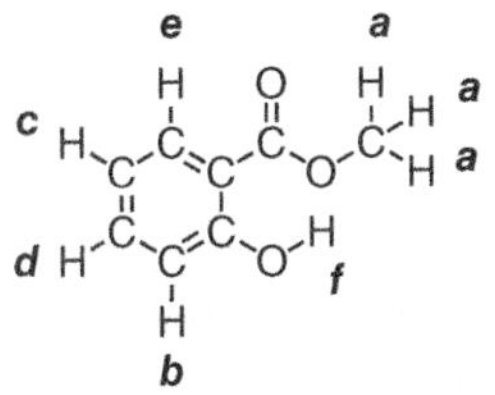

methyl salicylate

¹³C NMR and DEPT spectra:

	Chemical shift (Actual)	(Calculated)	DEPT	Interpretation
A	52.18	48.9	CH_3	Alkyl
B	112.31	115.9	CH	Aromatic
C	117.49	118.0	C	Aromatic
D	119.08	121.3	CH	Aromatic
E	129.83	131.3	CH	Aromatic
F	135.62	134.9	CH	Aromatic
G	161.53	156.8	C	Aromatic
H	170.51	150—185	C	Ester

The ¹³C NMR spectrum confirms a disubstituted benzene and an ester.

Calculations:

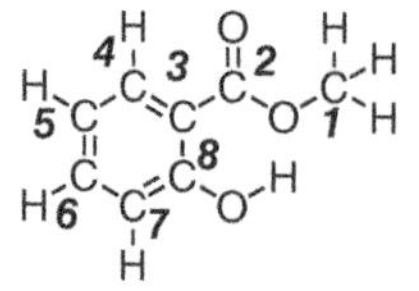

methyl salicylate

1	2	3	4	5	6	7	8
−2.1		128.7	128.7	128.7	128.7	128.7	128.7
51		2.0	1.2	−0.1	4.8	−0.1	1.2
		−12.7	1.4	−7.3	1.4	−12.7	26.9
48.9	150−185	118.0	131.3	121.3	134.9	115.9	156.8

Labeled structure:

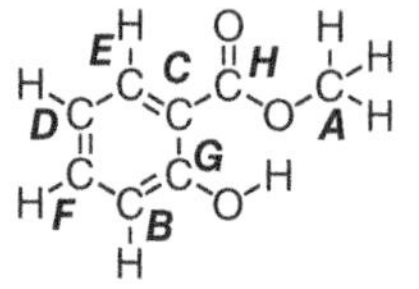

methyl salicylate

5. Compound with a formula of C_7H_9N.

Unsaturation number:

$U = 7 + 1 − \frac{1}{2}(9) + \frac{1}{2}(1) = 4$
Benzene

¹H NMR spectrum:

	Chemical shift		Splitting	Integration	Interpretation
	(Actual)	(Calculated)			
a	2.89	2.4	s	3H	CH_3 isolated
b	3.78	1.0−5.0	bs	1H	NH isolated
c	6.68−6.70	6.52	m	2H	aromatic
d	6.77−6.81	6.64	m	1H	aromatic
e	7.24−7.28	7.03	m	2H	aromatic

The compound contains a monosubstituted benzene, since five hydrogens are in the aromatic region. The only substituent is a methyl amino. The only possibility is *N*-methylaniline.

Calculations:

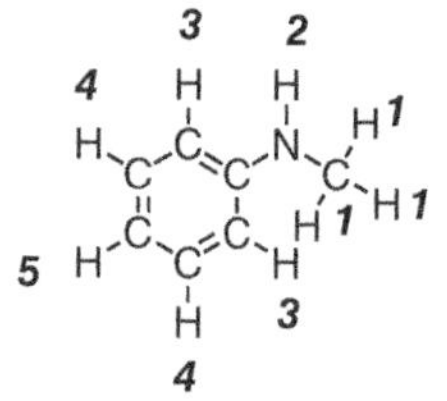

N-methylaniline

1	2	3	4	5
0.9		7.27	7.27	7.27
1.5		−0.75	−0.24	−0.63
2.4	1.0−5.0	6.52	7.03	6.64

Labeled structure:

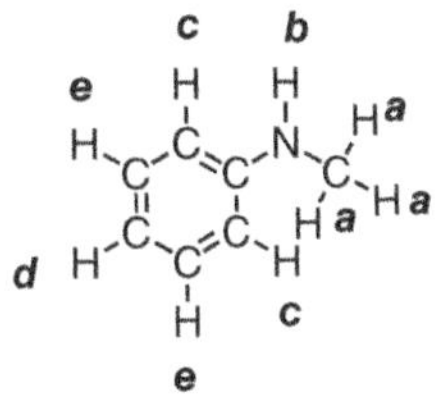

N-methylaniline

¹³C NMR and DEPT spectra:

	Chemical shift		DEPT	Interpretation
	(Actual)	(Calculated)		
A	30.87	34.9	CH₃	Amino
B	112.61	113.1	2 CH	Aromatic
C	117.46	116.9	CH	Aromatic
D	129.26	129.7	2 CH	Aromatic
E	149.22	151.3	C	Aromatic

Calculations:

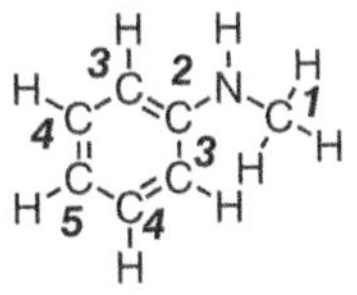

N-methylaniline

1	**2**	**3**	**4**	**5**
−2.1	128.7	128.7	128.7	128.4
37	22.6	−15.6	1.0	−11.5
34.9	151.3	113.1	129.7	116.9

Labeled structure:

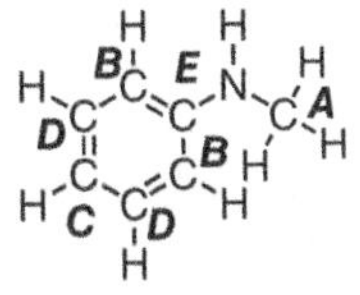

N-methylaniline

COSY spectrum:

	e 7.24–7.28	*d* 6.77–6.81	*c* 6.68–6.70	*b* 3.78	*a* 2.89
a 2.89					
b 3.78					
c 6.68–6.70	x				
d 6.77–6.81	x				
e 7.24–7.28		x	x		

H_a is adjacent to nothing.

H_b is adjacent to nothing.

H_c is adjacent to H_e.

H_d is adjacent to H_e.

H_e is adjacent to H_c and H_d.

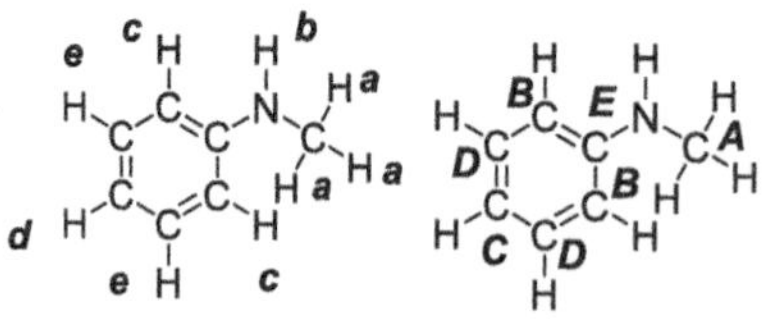

N-methylaniline

HSQC spectrum:

		e 7.24–7.28	d 6.77–6.81	c 6.68–6.70	b 3.78	a 2.89
A	30.87					x
B	112.61			x		
C	117.46		x			
D	129.26	x				
E	149.22					

H_a is attached to C_A.

H_b is attached to nothing.

H_c is attached to C_B.

H_d is attached to C_C.

H_e is attached to C_D.

6. Compound with a formula of $C_4H_8O_2$.

Unsaturation number:

$U = 4 + 1 - \frac{1}{2}(8) + \frac{1}{2}(0) = 1$

One double bond or one ring

^{1}H NMR spectrum:

	Chemical shift (Actual)	Chemical shift (Calculated)	Splitting	Integration	Interpretation
a	1.16	1.2	d	6H	2 CH$_3$ adjacent to CH
b	2.55	2.6	sept	1H	CH adjacent to 2 CH$_3$
c	12.09	10–13	s	1H	OH isolated

The compound contains an isopropyl group. The OH is in the carboxylic acid range. The only possible structure is isobutyric acid.

Calculations:

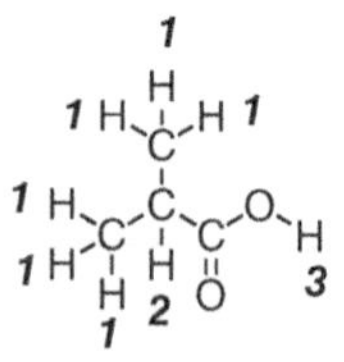

isobutyric acid

1	2	3
0.9	1.5	
0.3	1.1	
1.2	2.6	10–13

Labeled structure:

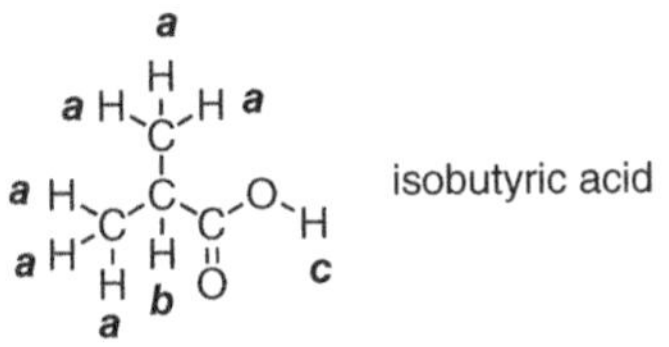

isobutyric acid

^{13}C NMR and DEPT spectra:

	Chemical shift		DEPT	Interpretation
	(Actual)	**(Calculated)**		
A	18.61	18.6	2 CH_3	Alkyl
B	33.85	37.1	CH	Alkyl
C	184.06	160–185	C	Carboxylic acid

Calculations:

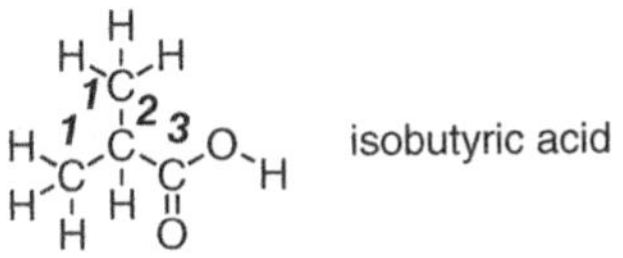

isobutyric acid

1	**2**	**3**
15.6	16.1	
3	21	
18.6	37.1	160–185

Labeled structure:

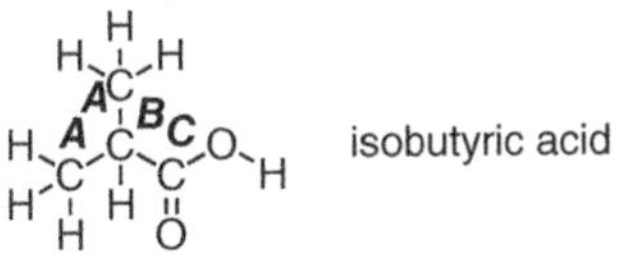

isobutyric acid

COSY

	c 12.09	**b** 2.55	**a** 1.16
a 1.16		x	
b 2.55			x
c 12.09			

H_a is adjacent to H_b.

H_b is adjacent to H_a.

H_c is adjacent to nothing.

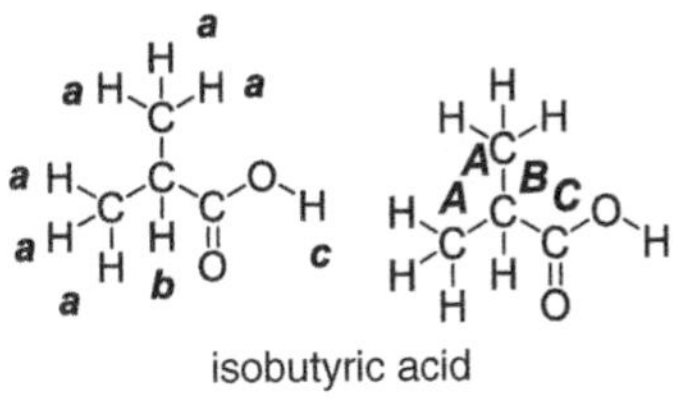

isobutyric acid

	c 12.09	b 2.55	a 1.16
A 18.61			x
B 33.85		x	
C 184.05			

H_a is attached to C_A.

H_b is attached to C_B.

H_c is attached to **nothing.**

7. Compound with a formula of $C_{10}H_{13}NO_2$.

Unsaturation Number:

$U = 10 + 1 - \frac{1}{2}(13) + \frac{1}{2}(1) = 5$

Benzene plus one double bond or one ring

^{1}H NMR spectrum:

	Chemical shift (Actual)	(Calculated)	Splitting	Integration	Interpretation
a	1.37	1.4	t	3H	CH_3 adjacent to CH_2
b	2.11	1.9	s	3H	CH_3 isolated
c	3.97	4.0	q	2H	CH_3 adjacent to CH_2
d	6.79	7.03	d	2H	Aromatic
e	7.36	7.52	d	2H	Aromatic
f	7.98	5.0–9.0	bs	1H	NH isolated

Two doublets in the aromatic is indicative of a *p*-disubstituted benzene. The CH_2 at δ 3.97 is probably next to an oxygen, so one of the groups attached to the benzene is an ethoxy group.

Calculations:

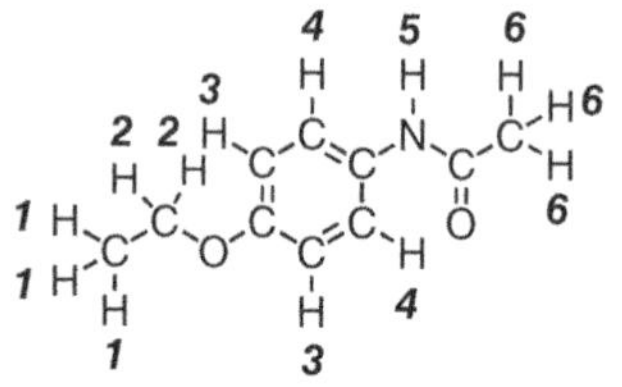

4-ethoxyacetanilide

1	2	3	4	5	6
0.9	1.2	7.27	7.27		0.9
0.5	2.8	−0.27	−0.08		1.0
		0.03	0.33		
1.4	4.0	7.03	7.52	5.0−9.0	1.9

Labeled structure:

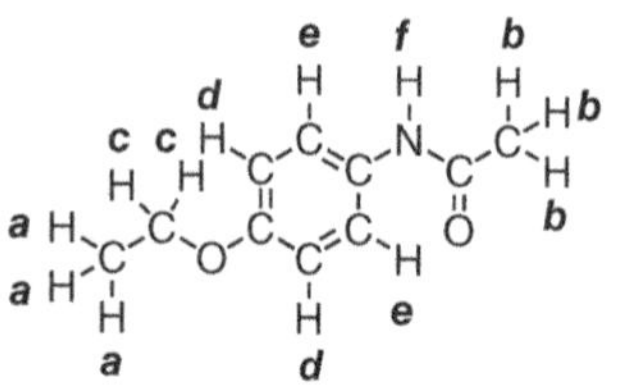

4-ethoxyacetanilide

¹³C NMR and DEPT spectra:

	Chemical shift (Actual)	(Calculated)	DEPT	Interpretation
A	14.83	13.9	CH_3	Alkyl
B	24.09	27.9	CH_3	Alkyl
C	63.68	63.9	CH_2	Alkyl
D	114.68	114.5	CH	Aromatic
E	122.11	119.8	CH	Aromatic
F	130.92	132.1	C	Aromatic
G	155.82	154.5	C	Aromatic
H	168.82	150–180	C	Amide

Calculations:

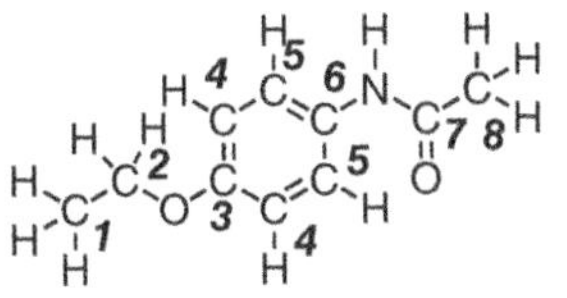

4-ethoxyacetanilide

1	2	3	4	5	6	7	8
5.9	5.9	128.7	128.7	128.7	128.7		−2.1
8	58	−5.6	0.2	−9.9	11.1		30
		31.4	−14.4	1.0	−7.7		
13.9	63.9	154.5	114.5	119.8	132.1	150–180	27.9

Labeled structure:

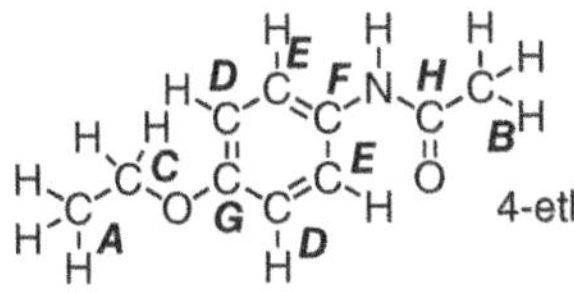

4-ethoxyacetanilide

COSY:

	f 7.98	*e* 7.36	*d* 6.79	*c* 3.97	*b* 2.11	*a* 1.37
a 1.37				x		
b 2.11						
c 3.97						x
d 6.79		x				
e 7.36			x			
f 7.98						

H_a is adjacent to H_c.

H_b is adjacent to nothing.

H_c is adjacent to H_a

H_d is adjacent to H_e.

H_e is adjacent to H_d

H_f is adjacent to nothing.

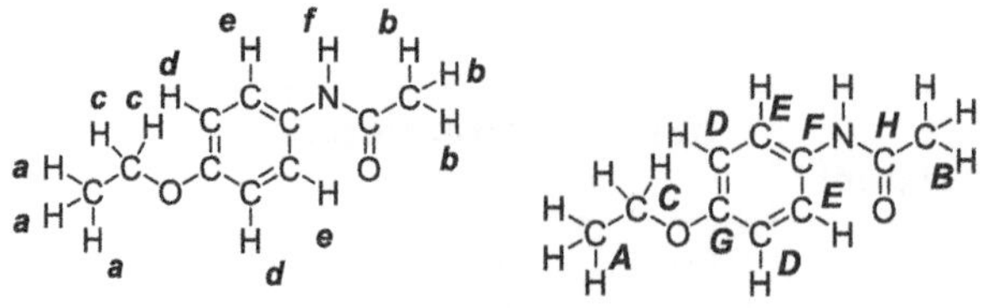

4-ethoxyacetanilide

HSQC:

	f 7.98	e 7.36	d 6.79	c 3.97	b 2.11	a 1.37
A 14.83						x
B 24.09					x	
C 63.68				x		
D 114.68			x			
E 122.11		x				
F 130.92						
G 155.82						
H 168.82						

H_a is attached to C_A.

H_b is attached to C_B.

H_c is attached to C_C.

H_d is attached to C_D.

H_e is attached to C_E.

8. Compound with a formula of $C_4H_8O_2$.

Unsaturation number:

$U = 4 + 1 - \frac{1}{2}(8) + \frac{1}{2}(0) = 1$
One double bond or one ring

^1H NMR spectrum:

	Chemical shift (Actual)	(Calculated)	Splitting	Integration	Interpretation
a	1.19	1.4	t	3H	CH_3 adjacent to CH_2
b	1.98	2.0	s	3H	CH_3 isolated
c	4.05	4.0	q	2H	CH_2 adjacent to CH_3

Given the high value of **c**, it would have to be next to an oxygen. The only possibility is ethyl acetate.

Calculations:

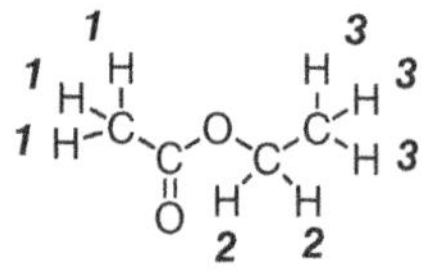

ethyl acetate

1	2	3
0.9	1.2	0.9
1.1	2.8	0.5
2.0	4.0	1.4

Labeled structure:

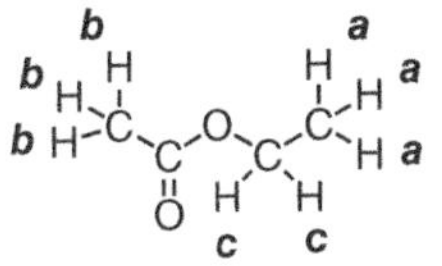

ethyl acetate

^{13}C NMR and DEPT spectra:

| | Chemical shift | | DEPT | Interpretation |
	(Actual)	(Calculated)		
A	14.10	11.9	CH_3	Alkyl
B	20.91	17.9	CH_3	Alkyl
C	60.28	56.9	CH_2	Ester
D	170.99	150–185	C	Ester

Calculations:

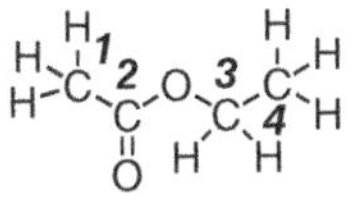

ethyl acetate

1	2	3	4
−2.1		5.9	5.9
20		51	6
17.9	150–185	56.9	11.9

Labeled structure:

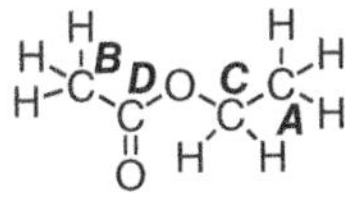

ethyl acetate

COSY:

	c 4.05	*b* 1.98	*a* 1.19
a 1.19	x		
b 1.98			
c 4.05			x

H_a is adjacent to H_c.

H_b is adjacent to nothing.

H_c is adjacent to H_a

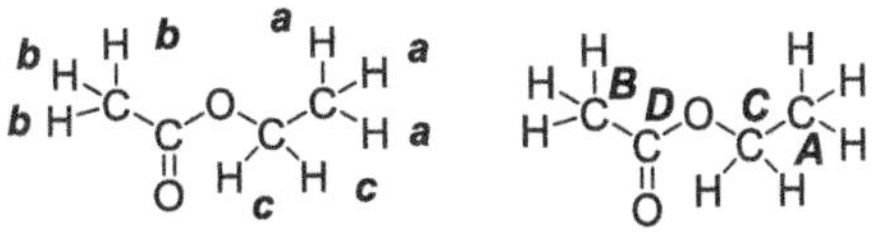

ethyl acetate

HSQC:

	c 4.05	b 1.98	a 1.19
A 14.10			x
B 20.91		x	
C 60.28	x		
D 170.99			

H_a is attached to C_A.

H_b is attached to C_B.

H_c is attached to C_C.

9. Compound with a formula of C_7H_5ClO.

Unsaturation number:

$U = 7 + 1 - (\frac{1}{2} \times 6) + (\frac{1}{2} \times 0) = 5$
Benzene plus one ring or double bond

^1H NMR spectrum:

	Chemical shift (Actual)	(Calculated)	Splitting	Integration	Interpretation
a	7.52	7.50	d	2H	Aromatic
b	7.83	7.79	d	2H	Aromatic
c	9.99	9.0–10.0	s	1H	CH isolated

Since there are two doublets in the aromatic, the compound is *p*-disubstituted. The peak at δ is indicative of an aldehyde. The structure is 4-chlorobenzaldehyde.

Calculations:

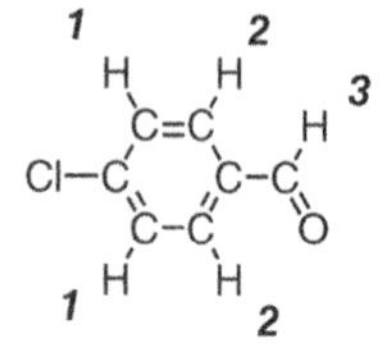

4-chlorobenzaldehyde

1	2	3
7.27	7.27	
0.02	−0.06	
0.21	0.58	
7.50	7.79	9.0 – 10.0

Labeled structure:

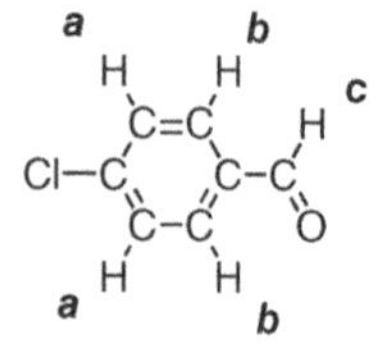

4-chlorobenzaldehyde

¹³C NMR and DEPT spectra:

	Chemical shift		DEPT	Interpretation
	(Actual)	**(Calculated)**		
A	129.48	129.7	2 CH	Aromatic
B	130.92	131.3	2 CH	Aromatic
C	134.73	135.4	C	Aromatic
D	140.97	140.4	C	Aromatic
E	190.86	175–220	CH	Aldehyde

Calculations:

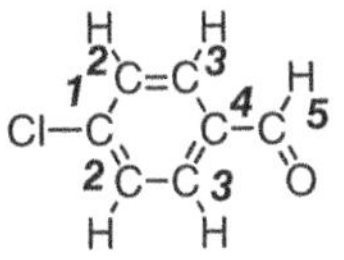

4-chlorobenzaldehyde

1	**2**	**3**	**4**	**5**
128.7	128.7	128.7	128.7	
6.2	0.4	1.3	−1.9	
5.5	0.6	1.3	8.6	
140.4	129.7	131.3	135.4	175–220

Labeled structure:

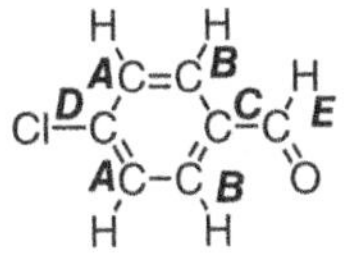

4-chlorobenzaldehyde

COSY:

	c 9.99	**b** 7.83	**a** 7.52
a 7.52		x	
b 7.83			x
c 9.99			

H_a is adjacent to H_b.

H_b is adjacent to H_a.

H_c is adjacent to nothing.

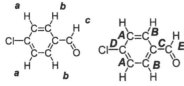

4-chlorobenzaldehyde

HSQC:

	c 9.99	*b* 7.83	*a* 7.52
A 129.48			x
B 130.92		x	
C 134.73			
D 140.97			
E 190.86			

H_a is attached to C_A.

H_b is attached to C_B.

H_c is attached to nothing.

10. Compound with a formula of $C_4H_{10}O$.

Unsaturation number:

$U = 4 + 1 - (\frac{1}{2} \times 10) + (\frac{1}{2} \times 0) = 0$

No double bonds, no triple bonds, or no rings

The compound must be an alcohol or ether.

¹H NMR spectrum:

	Chemical shift (Actual)	(Calculated)	Splitting	Integration	Interpretation
a	1.19	1.2	t	6H	2(CH_3 adjacent to CH_2)
b	3.46	3.3	q	4H	2(CH_2 adjacent to CH_3)

The only possibility is diethyl ether.

Calculations:

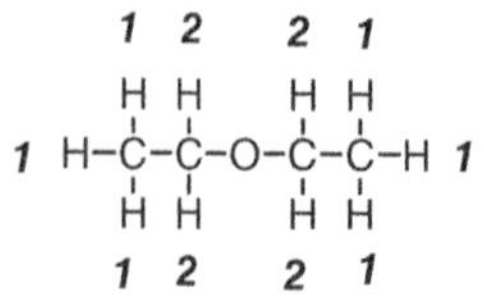

```
      1  2   2  1
      H H   H H
      | |   | |
1  H-C-C-O-C-C-H  1    diethyl ether
      | |   | |
      H H   H H
      1  2   2  1
```

1	2
0.9	1.2
0.3	2.1
1.2	3.3

Labeled structure:

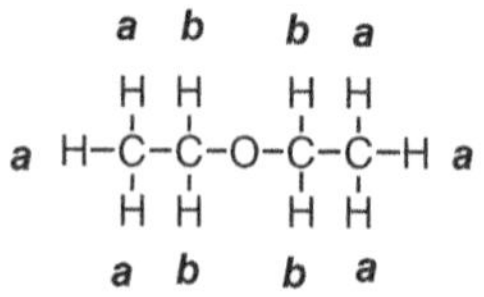

```
      a  b   b  a
      H H   H H
      | |   | |
a  H-C-C-O-C-C-H  a    diethyl ether
      | |   | |
      H H   H H
      a  b   b  a
```

^{13}C NMR and DEPT spectra:

	Chemical shift (Actual)	Chemical shift (Calculated)	DEPT	Interpretation
A	15.21	13.9	2 CH_3	Alkyl
B	65.81	63.9	2 CH_2	Ether

Calculations:

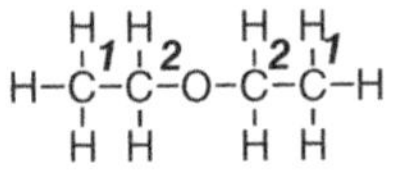

diethyl ether

1	***2***
5.9	5.9
8	58
13.9	63.9

Labeled structure:

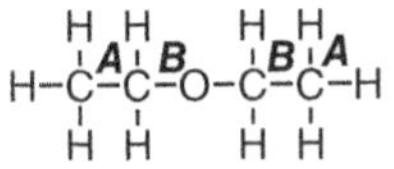

diethyl ether

COSY:

	b 3.46	***a*** 1.19
a 1.19	x	
b 3.46		x

H_a is adjacent to H_b.

H_b is adjacent to H_a.

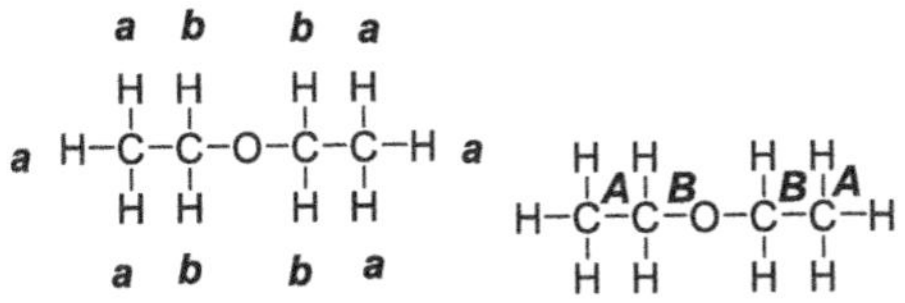

diethyl ether

HSQC:

	b 3.46	***a*** 1.19
A 15.21		x
B 65.81	x	

H_a is attached to C_A.

H_b is attached to C_B.

11. Compound with a formula of $C_7H_6O_2$.

Unsaturation number:

$U = 7 + 1 - (\frac{1}{2} \times 6) + (\frac{1}{2} \times 0) = 5$
Benzene plus one double bond or one ring

^1H NMR spectrum:

	Chemical shift		Splitting	Integration	Interpretation
	(Actual)	(Calculated)			
a	6.31	4.5–7.7	bs	1H	OH isolated
b	7.00	6.98	d	2H	Aromatic
c	7.84	7.71	d	2H	Aromatic
d	9.89	9.0–10.0	s	1H	CH isolated

The signal at δ 9.89 is indicative of an aldehyde. The signal at δ 6.31 is an OH. Two doublets in the aromatic indicates a *p*-disubstituted benzene. The only possibility is 4-Hydroxybenzaldehyde.

Calculations:

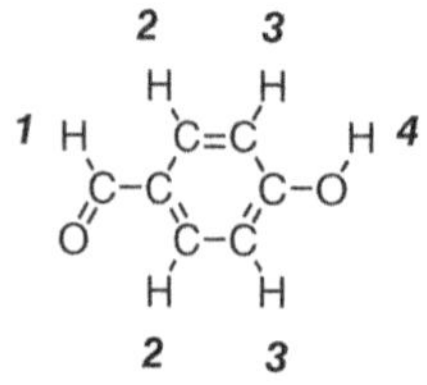

4-hydroxybenzaldehyde

	1	2	3	4
		7.27	7.27	
		−0.14	−0.50	
		0.58	0.21	
	9.0–10.0	7.71	6.98	4.5–7.7

Labeled structure:

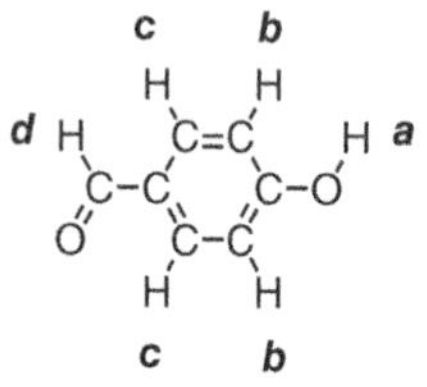

4-hydroxybenzaldehyde

^{13}C NMR and DEPT spectra:

	Chemical shift		DEPT	Interpretation
	(Actual)	(Calculated)		
A	116.03	116.6	2 CH	Aromatic
B	129.86	130.0	C	Aromatic
C	132.56	131.4	2 CH	Aromatic
D	161.61	161.1	C	Aromatic
E	191.30	175—220	CH	Aldehyde

Calculations:

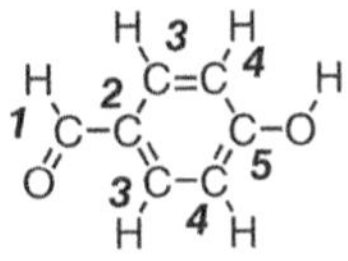

4-hydroxybenzaldehyde

1	*2*	*3*	*4*	*5*
	128.7	128.7	128.7	128.7
	8.6	1.3	0.6	5.5
	−7.3	1.4	−12.7	26.9
175−220	130.0	131.4	116.6	161.1

Labeled structure:

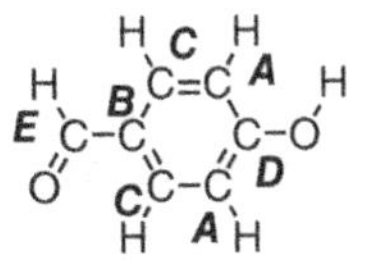

4-hydroxybenzaldehyde

COSY:

	d 9.89	*c* 7.84	*b* 7.00	*a* 6.31
a 6.31				
b 7.00		X		
c 7.84			X	
d 9.89				

H_a is adjacent to nothing.

H_b is adjacent to **H_c**.

H_c is adjacent to **H_b**.

H_d is adjacent to nothing.

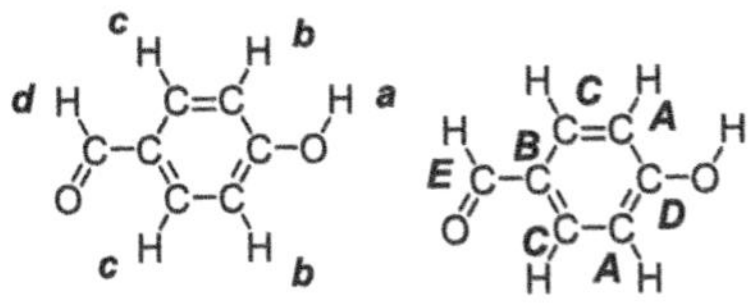

4-hydroxybenzaldehyde

HSQC:

	d 9.89	*c* 7.84	*b* 7.00	*a* 6.31
A 116.03			X	
B 129.86				
C 132.56		X		
D 161.61				
E 191.30				

H_a is attached to nothing.

H_b is attached to C_A.

H_c is attached to C_C.

H_d is attached to nothing.

H_e is attached to nothing.

12. Compound with a formula of $C_8H_9NO_2$.

Unsaturation number:

$U = 8 + 1 - (\frac{1}{2} \times 9) + (\frac{1}{2} \times 1) = 5$

Benzene plus one double bond or one ring

^1H NMR spectrum:

	Chemical shift (Actual)	(Calculated)	Splitting	Integration	Interpretation
a	1.97	1.9	s	1H	CH_3 isolated
b	6.67	6.80	d	2H	aromatic
c	7.33	7.44	d	2H	aromatic
d	9.12	4.5–7.7	s	1H	OH isolated
e	9.63	5.0–9.0	s	1H	NH isolated

The compound is a p-disubstituted aromatic. The OH isolated is a phenol. The NH isolated would be an amide. A good possibility is 4-acetaminophenol.

Calculations:

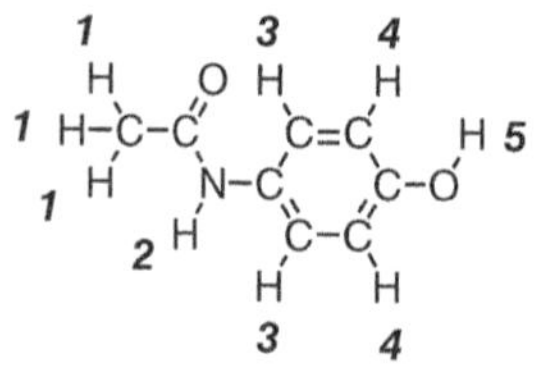

4-acetamidophenol

1	2	3	4	5
0.9		7.27	7.27	
1.0		0.33	0.03	
		−0.14	−0.50	
1.9	5.0−9.0	7.44	6.80	4.5−7.7

Labeled structure:

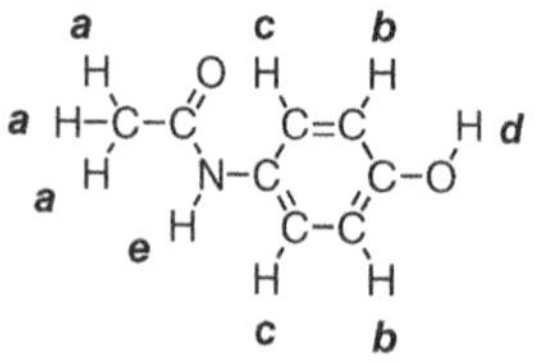

4-acetamidophenol

^{13}C NMR and DEPT spectra:

	Chemical shift (Actual)	Chemical shift (Calculated)	DEPT	Interpretation
A	24.18	27.9	CH_3	Alkyl
B	115.47	116.2	2 CH	Aromatic
C	121.33	120.2	2 CH	Aromatic
D	131.48	132.5	C	Aromatic
E	135.60	150.0	C	Aromatic
F	168.02	150–180	C	Amide

Calculations:

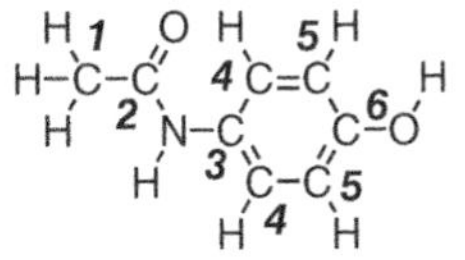

4-acetamidophenol

1	***2***	***3***	***4***	***5***	***6***
–2.1		128.7	128.7	128.7	128.7
30		11.1	–9.9	0.2	–5.6
		–7.3	1.4	–12.7	26.9
27.9	150–180	132.5	120.2	116.2	150.0

Labeled structure:

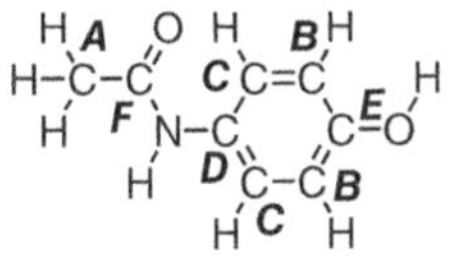

4-acetamidophenol

COSY:

	e 9.63	***d*** 9.12	***c*** 7.33	***b*** 6.67	***a*** 1.97
a 1.97					
b 6.67			x		
c 7.33				x	
d 9.12					
e 9.63					

H_a is adjacent to nothing.

H_b is adjacent to H_c.

H_c is adjacent to H_b.

H_d is adjacent to nothing.

H_e is adjacent to nothing.

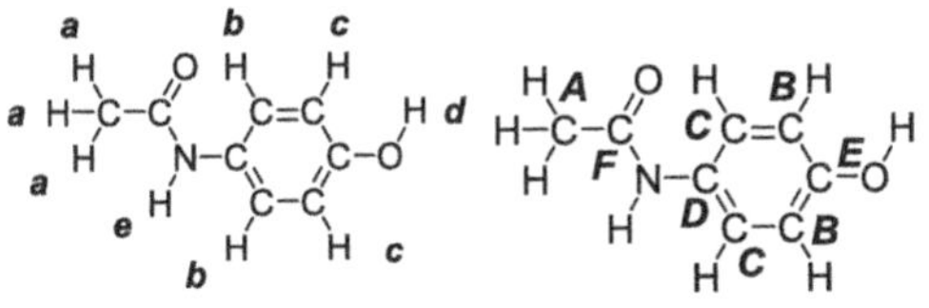

4-acetamidophenol

HSQC:

	e 9.63	*d* 9.12	*c* 7.33	*b* 6.67	*a* 1.97
A 24.18					x
B 115.47				x	
C 121.33			x		
D 131.48					
E 135.60					
F 168.02					

H_a is attached to C_A.

H_b is attached to C_C.

H_c is attached to C_B.

H_d is attached to nothing.

H_e is attached to nothing.

Chapter 8

Infrared Spectrometry

1. Compound with a formula of $C_4H_{10}O$.

   Unsaturation number:

   $U = 4 + 1 - (\frac{1}{2} \times 10) + (\frac{1}{2} \times 0) = 0$

   No double bonds, triple bonds, or rings

   With an unsaturation number of zero, an alcohol or ether is present.

Frequency	Bond	Compound type
775	O—H bend	Alcohol
920	C—H bend	$-CH(CH_3)_2$
1045	C—O stretch	Saturated primary alcohol
1372	C—H bend	$-CH(CH_3)_2$
1388	C—H bend	$-CH(CH_3)_2$
1469	C—H bend	$-CH_2$
2961	C—H stretch	Alkyl
3332	O—H stretch	Alcohol

   The broad peak at 3332 is indicative of an alcohol. Peaks are present to indicate an isopropyl group. The structure must be a primary alcohol with an isopropyl group. The only possibility is 2-methyl-1-propanol.

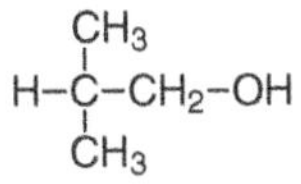

   $$
   \begin{array}{c}
   CH_3 \\
   | \\
   H-C-CH_2-OH \\
   | \\
   CH_3
   \end{array}
   $$

Student Solutions Manual to Accompany The Systematic Identification of Organic Compounds, Ninth Edition. Christine K. F. Hermann, Terence C. Morrill, Ralph L. Shriner, and Reynold C. Fuson.
© 2023 John Wiley & Sons, Inc. Published 2023 by John Wiley & Sons, Inc.

2. Compound with a formula of $C_4H_8O_2$.

Unsaturation number:

$U = 4 + 1 - (½ \times 8) + (½ \times 0) = 1$
One double bond or one ring

Frequency	Bond	Compound type
912	C—H bend	—$CH(CH_3)_2$
1235	C—O stretch	Carboxylic acid
1372	C—H bend	—$CH(CH_3)_2$
1388	C—H bend	—$CH(CH_3)_2$
1416	O—H bend	Carboxylic acid
1703	C=O stretch	Aliphatic carboxylic acid
2937	O—H stretch	Carboxylic acid
2973	C—H stretch	Alkyl

The broad signal at 2937 indicates a carboxylic acid. An isopropyl group is present. With the formula, only 2-methylpropanoic acid can be the structure.

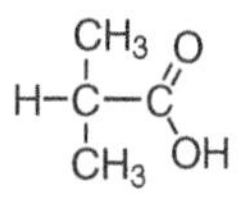

3. Compound with a formula of $C_7H_5NO_3$.

Unsaturation number:

$U = 7 + 1 - (½ \times 5) + (½ \times 1) = 6$
Benzene plus two double bonds, two rings, one double bond and one ring, or a triple bond

Frequency	Bond	Compound type
670	C—H out-of-plane bend	*m*-Disubstituted aromatic
811	C—H out-of-plane bend	*m*-Disubstituted aromatic
1352	N--O stretch	Aromatic nitro compounds
1396	C—H bend	Aldehyde
1448	C=C stretch	Aromatic
1533	N--O stretch	Aromatic nitro compounds
1582	C=C stretch	Aromatic
1690	C=O stretch	Aromatic aldehyde
2768	C—H stretch	Aldehyde
2832	C—H stretch	Aldehyde
3070	C—H stretch	Aromatic

The structure is a *m*-disubstituted benzene with an aldehyde group and a nitro group. The structure is 3-nitrobenzaldehyde.

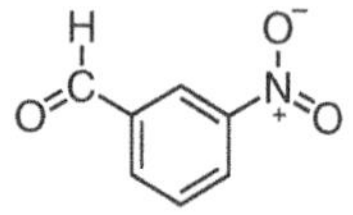

4. Compound with a formula of $C_4H_{10}O$.

Unsaturation number:

$U = 4 + 1 - (½ \times 10) + (½ \times 0) = 0$
No double bonds, triple bonds, or rings
  The compound is either an alcohol or an ether.

Frequency	Bond	Compound type
666	O—H bend	Alcohol
1073	O—H stretch	Saturated primary alcohol
1331	O—H bend	Primary alcohol
1380	C—H stretch	—CH$_3$
1424	O—H bend	Primary alcohol
1460	C—H bend	—CH$_2$
2933	C—H stretch	Alkyl
3328	O—H stretch	Alcohol

  The frequencies indicate an alcohol. Peaks for an isopropyl group are not present. A primary alcohol is present. The compound is 1-butanol.

$$CH_3\text{–}CH_2\text{–}CH_2\text{–}CH_2\text{–}O\text{–}H$$

5. Compound with a formula of C_4H_8O.

Unsaturation number:

$U = 4 + 1 - (½ \times 8) + (½ \times 0) = 1$
One double bond or one ring
  The compound is either a ketone, an aldehyde, or an alkene with an alcohol or ether.

Frequency	Bond	Compound type
1170	C—H bend	—CH$_2$
1170	C=O stretch and bend	Aliphatic ketone
1364	C—H bend	—CH$_3$
1460	C—H bend	—CH$_2$
1711	C=O stretch	Aliphatic ketone
2981	C—H stretch	Alkyl

  A strong carbonyl signal is present at 1711 cm^{-1}. The spectrum does not show any C–H stretch for an aldehyde, so the structure is 2-butanone.

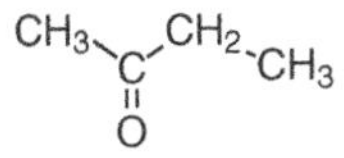

6. Compound with a formula of $C_8H_8O_3$.

Unsaturation number:

$U = 8 + 1 - (½ \times 8) + (½ \times 0) = 1$
Benzene plus one ring or double bond.

Frequency	Bond	Compound type
698	O—H bend	Phenol
755	C—H out-of-plane bend	o-Disubstituted aromatic
1214	C—O stretch	Phenol
1251	C—O stretch	Ester of aromatic acid
1331	C—O stretch	Phenol
1440	C—H bend	—CH$_3$
1440	C=C stretch	Aromatic
1586	C=C stretch	Aromatic
1674	C=O stretch	Aromatic ester
2957	C—H stretch	Alkyl
3183	C—H stretch	Aromatic
3183	O—H stretch	Phenol

The compound contains an o-disubstituted benzene. The peaks do not show a carboxylic acid or aldehyde. A phenol and an aromatic ester are present. The compound is methyl salicylate.

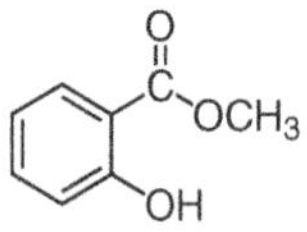

7. Compound with a formula of C$_8$H$_9$NO$_2$.

Unsaturation number:

$U = 8 + 1 - (½ × 9) + (½ × 1) = 5$
Benzene plus one ring or double bond.

Frequency	Bond	Compound type
683	O—H bend	Phenol
809	C—H out-of-plane bend	p-Disubstituted aromatic
1226	C—O stretch	Phenol
1368	C—O stretch	Phenol
1437	C=C stretch	Aromatic
1562	N—H bend	Secondary amide
1611	C=C stretch	Aromatic
1647	C=O stretch	Anilide
2879	C—H stretch	Alkyl
3110	C—H stretch	Aromatic
3110	O—H stretch	Phenol
3321	N—H stretch	Secondary amide

The compound is a p-disubstituted aromatic with a hydroxy and an amide attached. The structure is p-acetamidophenol.

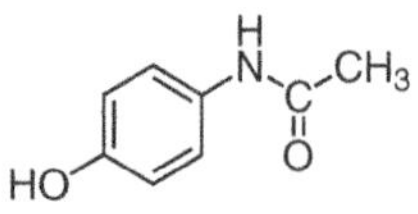

8. Compound with a formula of C_7H_9N.

Unsaturation number:

$U = 7 + 1 - (\tfrac{1}{2} \times 9) + (\tfrac{1}{2} \times 1) = 4$
Benzene

Frequency	Bond	Compound type
690	C—H out-of-plane bend	Monosubstituted aromatic
746	C—H out-of-plane bend	Monosubstituted aromatic
1319	C—N stretch	Secondary amide
1448	C—H bend	—CH_3
1448	C=C stretch	Aromatic
1505	N—H bend	Secondary amide
1602	C=C stretch	Aromatic
2876	C—H stretch	Alkyl
3050	C—H stretch	Aromatic
3417	N—H stretch	Secondary amide

The structure contains a secondary amide and must be monosubstituted. The only possible structure is *N*-methylaniline.

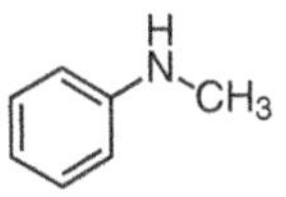

Chapter 9

Mass Spectrometry

1. C_3H_4O

$$\%[M+1]=[1.09\times3]+[0.0115\times4]+[0.0381\times1]=3.35$$

$$\%[M+2]=\left[\frac{(1.09\times3)^2}{200}\right]+[0.205\times1]=0.258$$

C_4H_8

$$\%[M+1]=[1.09\times4]+[0.0115\times8]=4.45$$

$$\%[M+2]=\left[\frac{(1.09\times4)^2}{200}\right]=0.0950$$

$C_2H_4N_2$

$$\%[M+1]=[1.09\times2]+[0.0115\times4]+[0.365\times2]=2.96$$

$$\%[M+2]=\left[\frac{(1.09\times2)^2}{200}\right]=0.0238$$

Student Solutions Manual to Accompany The Systematic Identification of Organic Compounds, Ninth Edition.
Christine K. F. Hermann, Terence C. Morrill, Ralph L. Shriner, and Reynold C. Fuson.
© 2023 John Wiley & Sons, Inc. Published 2023 by John Wiley & Sons, Inc.

2. $C_6H_{13}Br$

$$\%[M+1]=[1.09\times6]+[0.0115\times13]=6.69$$

$$\%[M+2]=\left[\frac{(1.09\times6)^2}{200}\right]+[97.3\times1]=97.5$$

$C_6H_{13}Cl$

$$\%[M+1]=[1.09\times6]+[0.0115\times13]=6.69$$

$$\%[M+2]=\left[\frac{(1.09\times6)^2}{200}\right]+[32.0\times1]=32.2$$

3. 1-Butanol

m/z	Fragment

31   H$_2$C=OH$^+$

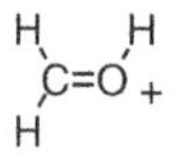

41    $^+CH_2$—CH=CH$_2$

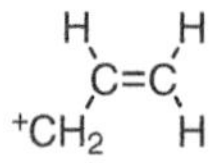

43    $CH_3CH_2CH_2^+$

45    $^+CH_2CH_2OH$

56   (base peak)

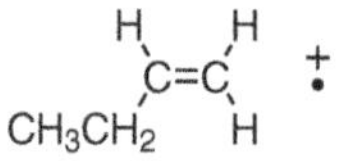

74    $CH_3CH_2CH_2CH_2OH\overset{+}{\cdot}$ (molecular ion)

Several fragments are due to loss of water.

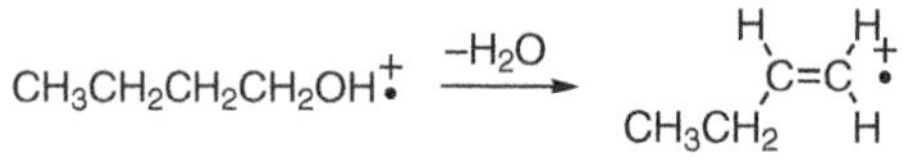

$CH_3CH_2CH_2CH_2OH\overset{+}{\cdot}$ $\xrightarrow{-H_2O}$

m/z 74   56

$+\,CH_2CH_2CH_2OH$ $\xrightarrow{-H_2O}$

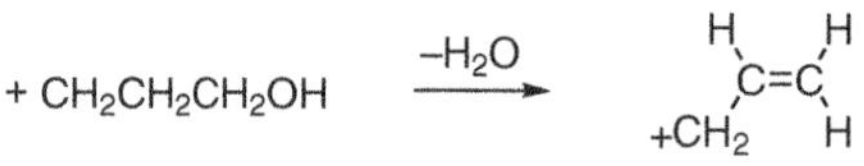

m/z 59   41

4. 2-Pentanone

m/z	Fragment

28

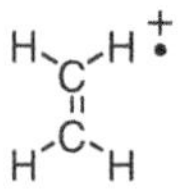

43    $CH_3-C\equiv O^+$ $^+CH_2CH_2CH_3$ (base peak)

m/z	Fragment

58

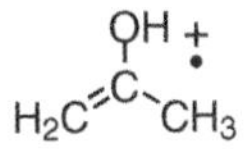

$$\underset{H_2C}{\overset{\displaystyle OH}{\diagdown}}\overset{+}{\underset{\displaystyle \bullet}{\underset{C}{\diagup}}}\overset{}{\underset{CH_3}{}}$$

71

$$\overset{+}{O}\equiv C\text{-}CH_2CH_2CH_3 \qquad CH_3\underset{\displaystyle \parallel}{\underset{O}{\diagdown}}\overset{}{\underset{C}{\diagup}}CH_2CH_2^{+}$$

86

$$CH_3\underset{\displaystyle \parallel}{\underset{O}{\diagdown}}\overset{}{\underset{C}{\diagup}}CH_2CH_2CH_3 \overset{+}{\underset{\bullet}{}} \qquad \text{(molecular ion)}$$

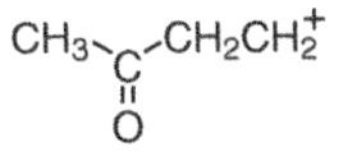

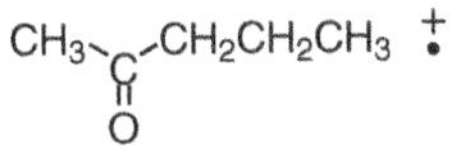

Some species are due to McLafferty rearrangement.

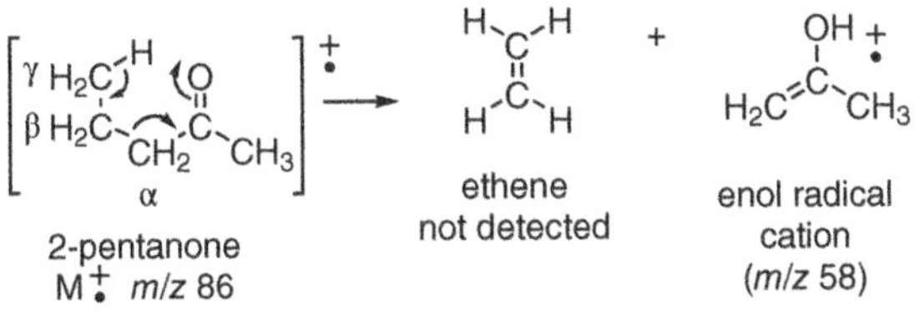

2-pentanone
$M^{+}_{\bullet}$ m/z 86

ethene
not detected

enol radical
cation
(m/z 58)

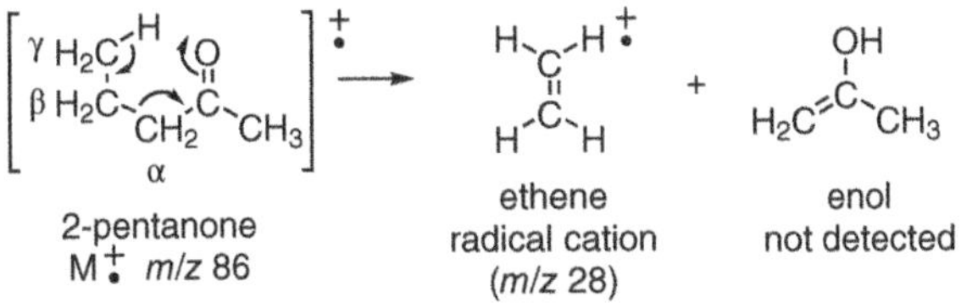

2-pentanone
$M^{+}_{\bullet}$ m/z 86

ethene
radical cation
(m/z 28)

enol
not detected

5. Methyl *t*-butyl ether

m/z	Fragment

31

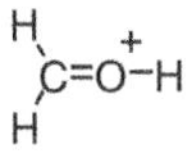

$$\underset{H}{\overset{\displaystyle H}{\diagdown}}C=\overset{+}{O}\text{-}H$$

57

$$+\underset{\displaystyle CH_3}{\overset{\displaystyle CH_3}{\underset{}{C}}}\text{-}CH_3$$

73

$$\overset{+}{O}\text{-}\underset{\displaystyle CH_3}{\overset{\displaystyle CH_3}{C}}\text{-}CH_3 \qquad CH_3\text{-}\overset{+}{O}=C\underset{CH_3}{\overset{CH_3}{\diagdown}} \qquad \text{(base peak)}$$

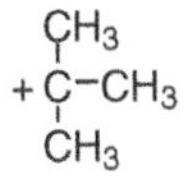

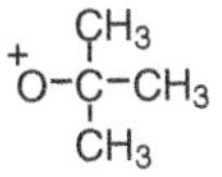

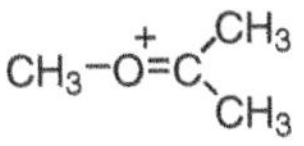

88

$$CH_3\text{-}O\text{-}\underset{\displaystyle CH_3}{\overset{\displaystyle CH_3}{C}}\text{-}CH_3 \overset{+}{\underset{\bullet}{}} \qquad \text{(molecular ion)}$$

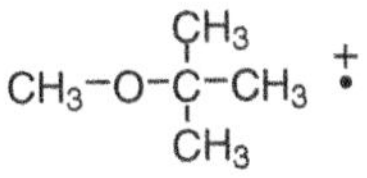

6. 2-Bromopropane

m/z	Fragment	
43	$CH_3-\overset{+}{\underset{H}{C}}-CH_3$	(base peak)
79	$^{79}Br^+$	
81	$^{81}Br^+$	
107	$CH_3-\overset{+}{\underset{H}{C}}-\overset{79}{Br}$	
109	$CH_3-\overset{+}{\underset{H}{C}}-\overset{81}{Br}$	
122	$\underset{\overset{|}{\underset{Br}{79}}}{CH_3-\underset{}{\overset{H}{C}}-CH_3}\,\overset{+}{\cdot}$	(molecular ion)
124	$\underset{\overset{|}{\underset{Br}{81}}}{CH_3-\underset{}{\overset{H}{C}}-CH_3}\,\overset{+}{\cdot}$	(molecular ion)

7. 2-Propanamine

m/z	Fragment	
43	$CH_3-\overset{+}{\underset{H}{C}}-CH_3$	
44	$CH_3-\overset{+}{\underset{H}{C}}=\overset{+}{N}H_2$	(base peak)
59	$\underset{\overset{|}{NH_2}}{CH_3-\underset{}{C}-CH_3}\quad\overset{+}{\cdot}$	(molecular ion)

8. Propanoic acid

m/z	Fragment	
29	CH_3CH_2+	
45	$^+O\equiv C-OH$	
57	$CH_3-CH_2-C\equiv O^+$	
59	$^+CH_2-C\overset{O}{\underset{OH}{\diagdown}}$	
74	$CH_3-CH_2-C\overset{O}{\underset{OH}{\diagdown}}\,\overset{+}{\cdot}$	(molecular ion) (base peak)

Chapter 10

Chemical Tests for Functional Groups

1. An anhydride reacts with aniline to give an anilide and an anilium salt. The sodium hydroxide liberates the aniline from the anilinium salt.

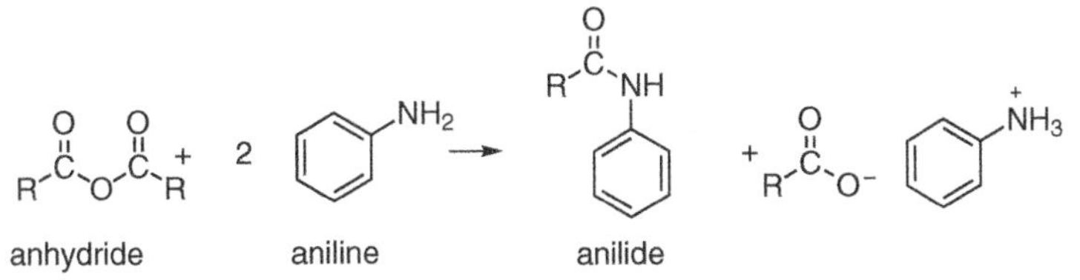

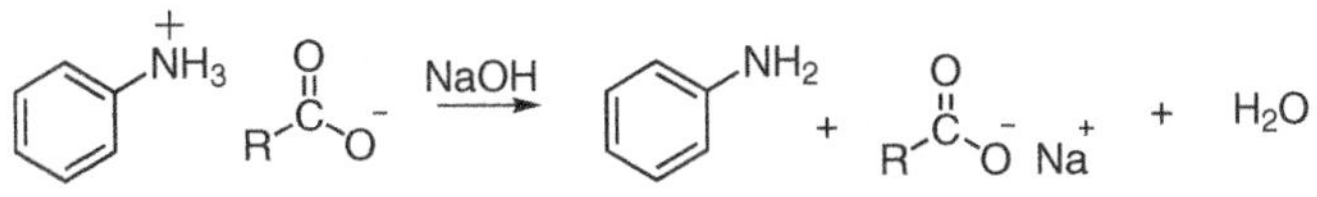

An acid halide reacts with aniline to give an anilide and an anilium salt. The sodium hydroxide liberates the aniline from the anilinium salt.

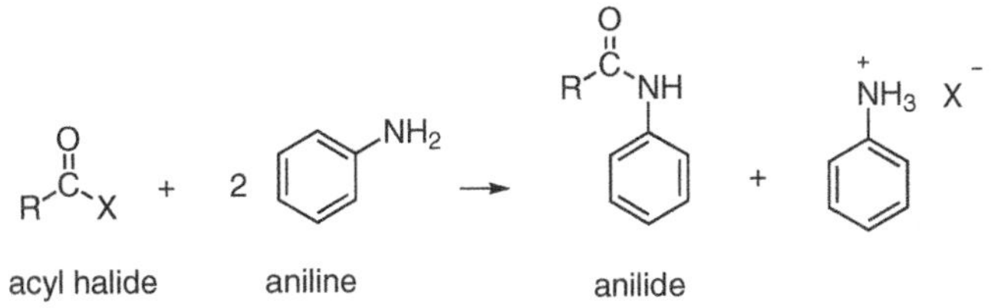

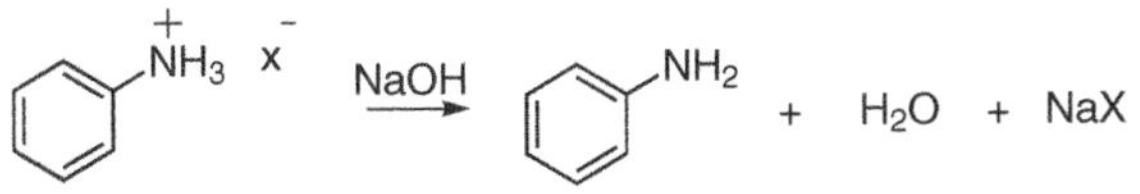

Student Solutions Manual to Accompany The Systematic Identification of Organic Compounds, Ninth Edition. Christine K. F. Hermann, Terence C. Morrill, Ralph L. Shriner, and Reynold C. Fuson.
© 2023 John Wiley & Sons, Inc. Published 2023 by John Wiley & Sons, Inc.

2. Propanoyl chloride reacts with hydroxylamine to form the hydroxamic acid.

$$CH_3CH_2-\overset{\overset{\displaystyle O}{\|}}{C}-Cl \quad + \quad H_2NOH \quad \longrightarrow \quad CH_3CH_2-\overset{\overset{\displaystyle O}{\|}}{C}-NHOH \quad + \quad HCl$$

propanoyl chloride   hydroxylamine   hydroxamic acid

Propanoyl anhydride reacts with hydroxylamine to yield the hydroxamic acid and propanoic acid.

$$CH_3CH_2-\overset{\overset{\displaystyle O}{\|}}{C}-O-\overset{\overset{\displaystyle O}{\|}}{C}-CH_2CH_3 \quad + \quad H_2NOH \quad \longrightarrow \quad CH_3CH_2-\overset{\overset{\displaystyle O}{\|}}{C}-OH \quad + \quad CH_3CH_2-\overset{\overset{\displaystyle O}{\|}}{C}-NHOH$$

propanoic anhydride   hydroxylamine   propanoic acid   hydroxamic acid

The hydroxamic acid is treated with iron chloride to give the ferric hydroxamate complex.

$$3 \;\; CH_3CH_2-\overset{\overset{\displaystyle O}{\|}}{C}-NHOH \quad + \quad FeCl_3 \quad \longrightarrow \quad CH_3CH_2(-\overset{\overset{\displaystyle O}{\|}}{C}-NHO)_3Fe \quad + \quad 3\;HCl$$

hydroxamic   ferric hydroxamate
acid   complex
              (burgundy or magenta color)

To distinguish between the propanoyl chloride and propanoic anhydride, use a test that detects the presence of chlorine. The silver nitrate test (Experiment 35) or the sodium iodide test (Experiment 36) detects the presence of chlorine.

3. Acetic anhydride is hydrolyzed to form acetic acid.

$$CH_3-\overset{\overset{\displaystyle O}{\|}}{C}-O-\overset{\overset{\displaystyle O}{\|}}{C}-CH_3 \quad + \quad H_2O \quad \longrightarrow \quad 2 \;\; CH_3-\overset{\overset{\displaystyle O}{\|}}{C}-OH \quad + \quad heat$$

acetic anhydride   acetic acid

Acetic anhydride undergoes a reaction with ethanol and hydroxide to produce propanoic acid and ethyl propanoate.

$$CH_3CH_2-\overset{\overset{\displaystyle O}{\|}}{C}-O-\overset{\overset{\displaystyle O}{\|}}{C}-CH_2CH_3 \quad + \quad CH_3CH_2OH \quad \overset{NaOH}{\longrightarrow}$$

acetic anhydride   ethanol

$$CH_3CH_2-\overset{\overset{\displaystyle O}{\|}}{C}-OH \quad + \quad CH_3CH_2-\overset{\overset{\displaystyle O}{\|}}{C}-OCH_2CH_3$$

propanoic acid   ethyl propanoate

N-Phenylpropanamide is produced from the reaction of acetic anhydride with aniline.

$$CH_3CH_2-\overset{\overset{\displaystyle O}{\|}}{C}-O-\overset{\overset{\displaystyle O}{\|}}{C}-CH_2CH_3 \quad + \quad 2 \;\; \text{(aniline, } C_6H_5NH_2\text{)} \quad \longrightarrow$$

acetic anhydride   aniline

$$CH_3CH_2-\overset{\overset{\displaystyle O}{\|}}{C}-NH-C_6H_5 \quad + \quad CH_3CH_2-\overset{\overset{\displaystyle O}{\|}}{C}-O^- \quad C_6H_5\overset{+}{N}H_3$$

N-phenylpropanamide

4. Acetic anhydride and hydroxylamine react via the following mechanism.

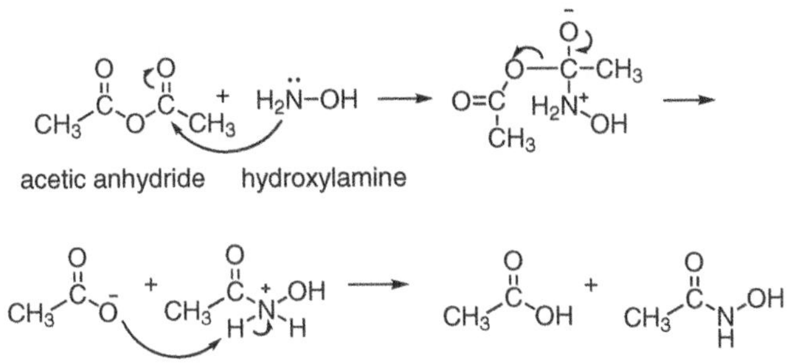

acetic anhydride   hydroxylamine

Acetyl chloride reacts with hydroxylamine in the following mechanism.

acetyl chloride hydroxylamine

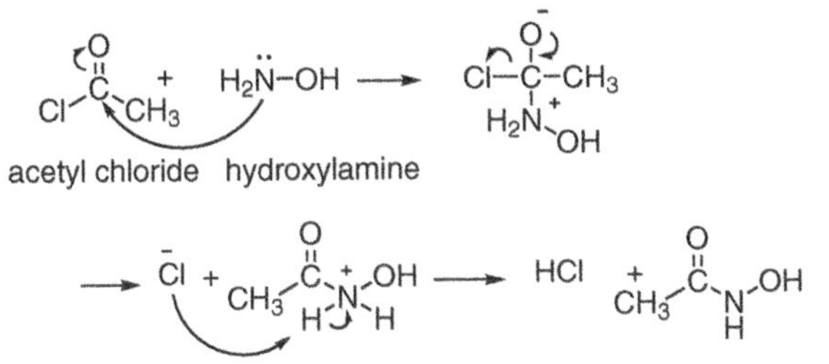

    The acyl halides are more reactive than the anhydrides because the lone pairs on the central oxygen on anhydrides are shared by two carbonyl groups, thereby forming other resonance structures. The contribution of the other resonance structures reduces the positive charge on the carbonyl carbon, making it less reactive than the acyl halides.

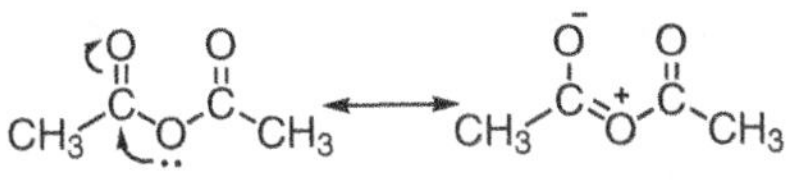

5. Acetyl chloride reacts with water to form acetic acid.

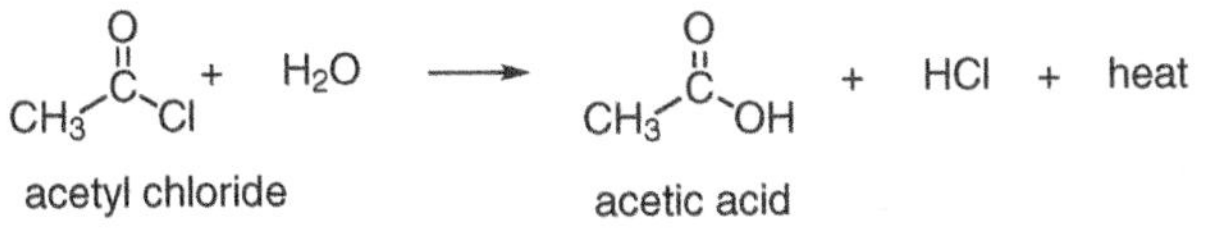

acetyl chloride   acetic acid

Ethyl acetate is formed from acetyl chloride reacting with ethanol.

acetyl chloride   ethyl acetate

Acetyl chloride reacts with aniline to form acetanilide.

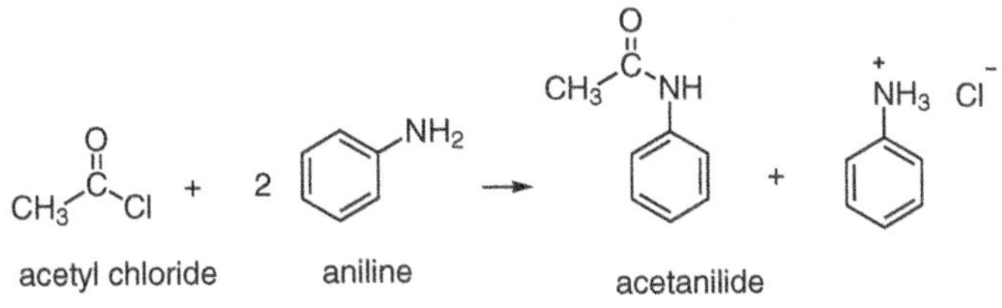

acetyl chloride   aniline   acetanilide

6. Sodium is used in testing neutral compounds. Therefore, it will react very little or not at all with phenol, benzoic acid, oximes, nitromethane, and benzenesulfonamide.

   The test is never used with these compounds for the following reasons. Benzoic acid is more acidic than phenol. Phenol and benzenesulfonamide have approximately the same acidity and are both more acidic than alcohols. Oximes, like most bases, are basic. Nitromethane exists in tautomeric equilibrium with the corresponding nitronic acids through the carbanion.

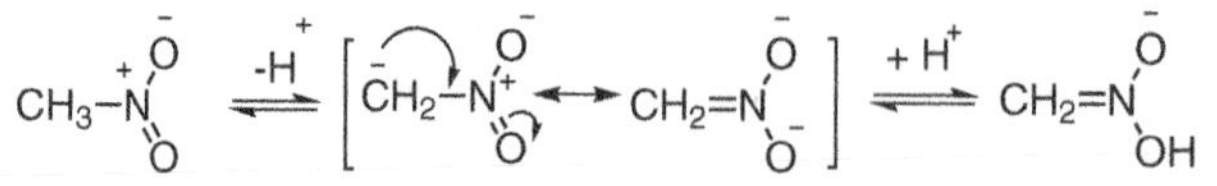

7. If sodium comes into contact with a large amount of water, a dangerous explosion may occur, due to the formation of hydrogen, a flammable gas. Additionally, the sodium must be cleaned prior to use by carefully scraping off the outer oxide layer. Any utensils should be placed in alcohol to decompose any traces of sodium.

8. The mechanism of acetyl chloride with an alcohol is illustrated below.

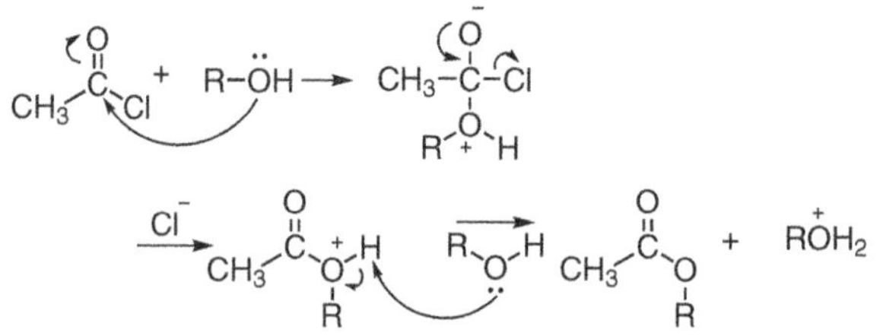

   As seen in the mechanism, a smaller alcohol would react faster than a larger alcohol. Therefore, the trend is primary > secondary > tertiary.

   A competing reaction for tertiary alcohols is the formation of the alkyl chloride.

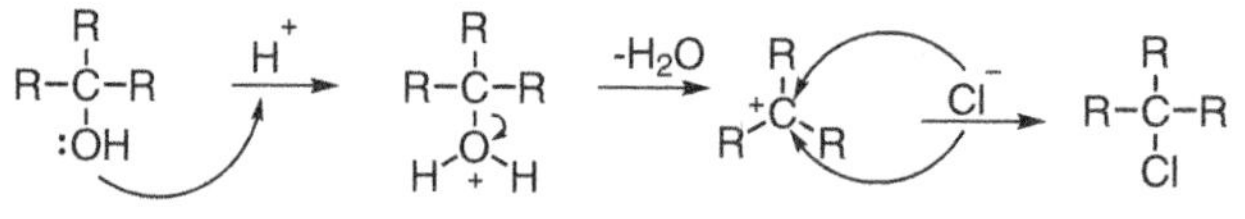

9. Primary alcohols do not react perceptibly.

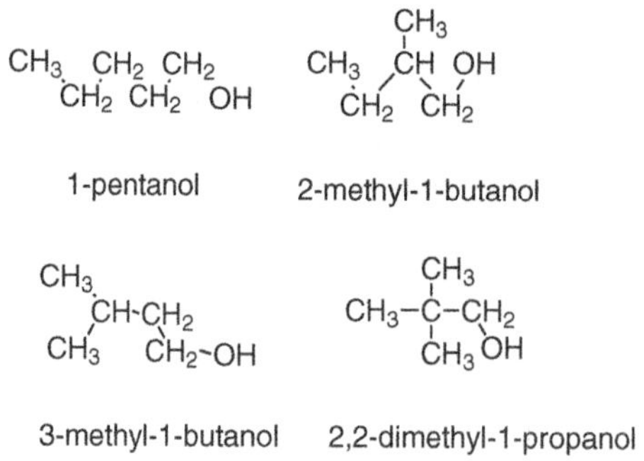

1-pentanol   2-methyl-1-butanol

3-methyl-1-butanol   2,2-dimethyl-1-propanol

Secondary alcohols are between the primary and tertiary alcohols in reactivity.

$$\underset{\text{2-pentanol}}{CH_3\text{-}CH_2\text{-}CH_2\text{-}\underset{\underset{OH}{|}}{CH}\text{-}CH_3} \quad + \quad HCl \quad \xrightarrow{ZnCl_2} \quad \underset{\text{2-chloropentane}}{CH_3\text{-}CH_2\text{-}CH_2\text{-}\underset{\underset{Cl}{|}}{CH}\text{-}CH_3} \quad + \quad H_2O$$

$$\underset{\text{3-pentanol}}{CH_3\text{-}\underset{\underset{CH_2}{|}}{\overset{\overset{OH}{|}}{CH}}\text{-}CH_3} \quad + \quad HCl \quad \xrightarrow{ZnCl_2} \quad \underset{\text{3-chloropentane}}{CH_3\text{-}\underset{\underset{CH_2}{|}}{\overset{\overset{Cl}{|}}{CH}}\text{-}CH_3} \quad + \quad H_2O$$

$$\underset{\text{3-methyl-2-butanol}}{CH_3\text{-}CH\text{-}CH\text{-}CH_3} \quad + \quad HCl \quad \xrightarrow{ZnCl_2} \quad \underset{\text{2-chloro-3-methylbutane}}{CH_3\text{-}CH\text{-}CH\text{-}CH_3} \quad + \quad H_2O$$

The tertiary alcohol reacts the fastest.

$$\underset{\text{2-methyl-2-butanol}}{CH_3\text{-}\underset{\underset{OH}{|}}{\overset{\overset{CH_3}{|}}{C}}\text{-}CH_2\text{-}CH_3} \quad + \quad HCl \quad \xrightarrow{ZnCl_2} \quad \underset{\text{2-chloro-2-methylbutane}}{CH_3\text{-}\underset{\underset{Cl}{|}}{\overset{\overset{CH_3}{|}}{C}}\text{-}CH_2\text{-}CH_3} \quad + \quad H_2O$$

10. Listed below are the relative stabilities of the carbocations.

3° benzylic ≈ 3° allylic > 2° benzylic ≈ 2° allylic ≈ 3° alkyl > 1°
benzylic ≈ 1° allylic ≈ 2° alkyl > 1° alkyl > vinylic ≈ aryl

Allyl alcohol will produce a 1° allylic carbocation, which is roughly equivalent to a 2° alkyl carbocation. The 1° allylic carbocation will react faster than a 1° propyl carbocation, produced from 1-propanol.

allyl alcohol → 1° allylic cation

1-propanol → 1° propyl carbocation

Benzyl alcohol will produce a 1° benzylic carbocation, which is roughly equivalent to a 2° alkyl carbocation. The 1° benzylic carbocation will react faster than a 1° pentyl carbocation, produced from 1-pentanol.

benzyl alcohol → 1° benzylic carbocation

1-pentanol → 1° pentyl cation

11. Butanal and 1-butanol will give a positive test with Jones reagent.

$$3\ \ CH_3CH_2CH_2\overset{O}{\underset{}{C}}H \ \ +\ 2\ CrO_3 \ +\ 3\ H_2SO_4 \ \longrightarrow$$

butanal

$$3\ \ CH_3CH_2CH_2\overset{O}{\underset{}{C}}OH \ \ +\ 3\ H_2O \ +\ Cr_2(SO_4)_3$$

butanoic acid

$$3\ CH_3CH_2CH_2CH_2OH \ \ +\ \ 4\ CrO_3 \ +\ 6\ H_2SO_4 \ \longrightarrow$$

1-butanol

$$3\ \ CH_3CH_2CH_2\overset{O}{\underset{}{C}}OH \ \ +\ 9\ H_2O \ +\ 2\ Cr_2(SO_4)_3$$

butanoic acid

12. Acetyl chloride reacts with 1-butanol, but not butanal.

$$CH_3CH_2CH_2CH_2OH \ \ +\ \ CH_3\overset{O}{\underset{}{C}}Cl \ \longrightarrow \ CH_3\overset{O}{\underset{}{C}}OCH_2CH_2CH_2CH_3 \ \ +\ HCl\ (g)$$

1-butanol   acetyl chloride   butyl acetate

$$CH_3CH_2CH_2\overset{O}{\underset{}{C}}H \ \ +\ \ CH_3\overset{O}{\underset{}{C}}Cl \ \longrightarrow \ \text{no reaction}$$

butanal   acetyl chloride

2,4-Dinitrophenylhydrazine reacts with butanal, but not 1-butanol.

$$CH_3CH_2CH_2\overset{O}{\underset{}{C}}H \ +\ \text{(2,4-dinitrophenylhydrazine)} \ \xrightarrow{\ H_2SO_4\ }$$

butanal   2,4-dinitrophenylhydrazine

2,4-dinitrophenylhydrazone
of butanal

$$CH_3CH_2CH_2CH_2OH \ \ +\ \text{(2,4-dinitrophenylhydrazine)} \ \xrightarrow{\ H_2SO_4\ } \ \text{no reaction}$$

13. 1-Butanol (1° alcohol) and 2-butanol (2° alcohol) react with Jones oxidation. 2-Methyl-2-propanol (3° alcohol) does not react with Jones oxidation.

$$3\ CH_3CH_2CH_2CH_2OH \ \ +\ \ 4\ CrO_3 \ +\ 6\ H_2SO_4 \ \longrightarrow$$

1-butanol   (orange-red)

$$3\ \ CH_3CH_2CH_2\overset{O}{\underset{}{C}}OH \ \ +\ 9\ H_2O \ +\ 2\ Cr_2(SO_4)_3$$

butanoic acid   (intense blue to green)

$$3 \quad \underset{\underset{OH}{|}}{CH_3CHCH_2CH_3} \quad + \quad 2\,CrO_3 \quad + \quad 3\,H_2SO_4 \quad \longrightarrow$$

2-butanol   (orange-red)

$$3 \quad \underset{CH_3}{\overset{\overset{\displaystyle O}{\|}}{C}}_{}CH_2CH_3 \quad + \quad 6\,H_2O \quad + \quad Cr_2(SO_4)_3$$

2-butanone   (intense blue to green)

$$3 \quad \underset{\underset{OH}{|}}{\overset{\overset{CH_3}{|}}{CH_3CCH_3}} \quad + \quad CrO_3 \quad + \quad H_2SO_4 \quad \longrightarrow \quad \text{no reaction}$$

2-methyl-2-propanol   (orange-red)

1-Butanol (1° alcohol) does not react with the Lucas reagent. 2-Methyl-2-propanol (3° alcohol) will react faster with the Lucas reagent than 2-butanol (2° alcohol).

$$CH_3CH_2CH_2CH_2OH \quad + \quad HCl \quad \xrightarrow{\;ZnCl_2\;} \quad \text{no reaction}$$

1-butanol

$$\underset{\underset{OH}{|}}{CH_3CHCH_2CH_3} \quad + \quad HCl \quad \xrightarrow{\;ZnCl_2\;} \quad \underset{\underset{Cl}{|}}{CH_3CHCH_2CH_3} \quad + \quad H_2O$$

2-butanol   2-chlorobutane

$$\underset{\underset{OH}{|}}{\overset{\overset{CH_3}{|}}{CH_3CCH_3}} \quad + \quad HCl \quad \xrightarrow{\;ZnCl_2\;} \quad \underset{\underset{Cl}{|}}{\overset{\overset{CH_3}{|}}{CH_3CCH_3}} \quad + \quad H_2O$$

2-methyl-2-propanol   2-chloro-2-
                                          methylpropane

14. For the TCICA test, secondary alcohols react faster than primary alcohols. Tertiary alcohols do not react. The ranking is below.

$$\underset{\underset{OH}{|}}{CH_3CHCH_2CH_3} \quad > \quad CH_3CH_2CH_2CH_2OH \quad > \quad \underset{\underset{OH}{|}}{\overset{\overset{CH_3}{|}}{CH_3CCH_3}}$$

2-butanol   1-butanol   2-methyl-2-propanol
(2° alcohol)   (1° alcohol)   (3° alcohol)

15. Phenols react the quickest. Aliphatic aldehydes react the second fastest, followed by aromatic aldehydes. Ketones do not react within four hours.

phenol   butanal   benzaldehyde   2-butanone
(aromatic alcohol)   (aldehyde)   (aromatic aldehyde)   (ketone)

16. 2,4-dinitrophenylhydrazine reacts with 2-Pentanone and pentanal are to form the 2,4-dinitrophenylhydrazone.

2-pentanone   2,4-dinitrophenylhydrazine

2,4-dinitrophenylhydrazone
of 2-pentanone

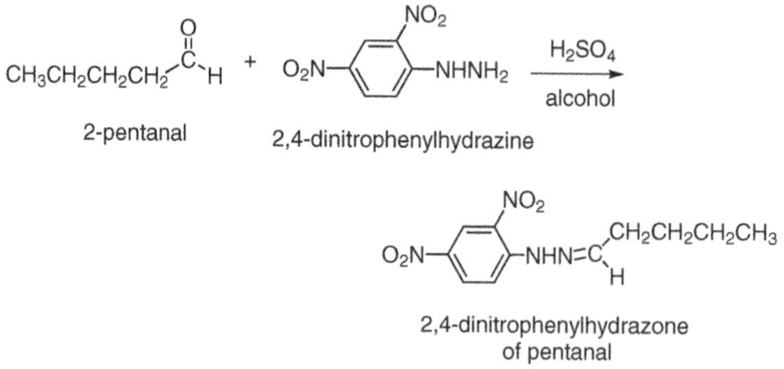

2-pentanal   2,4-dinitrophenylhydrazine

2,4-dinitrophenylhydrazone
of pentanal

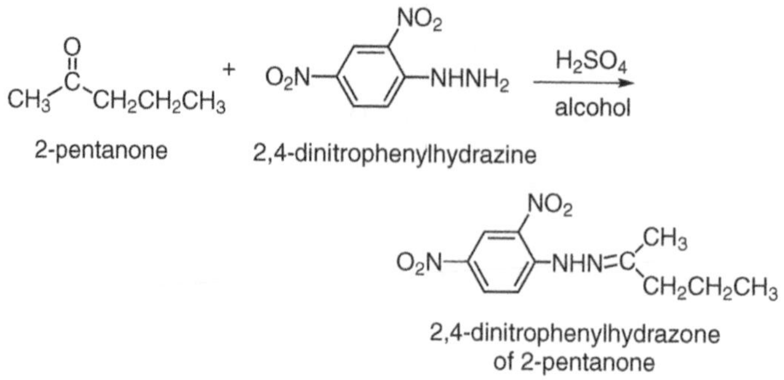

17. 2-Pentanone and pentanal react with hydroxylamine hydrochloride to give the oxime.

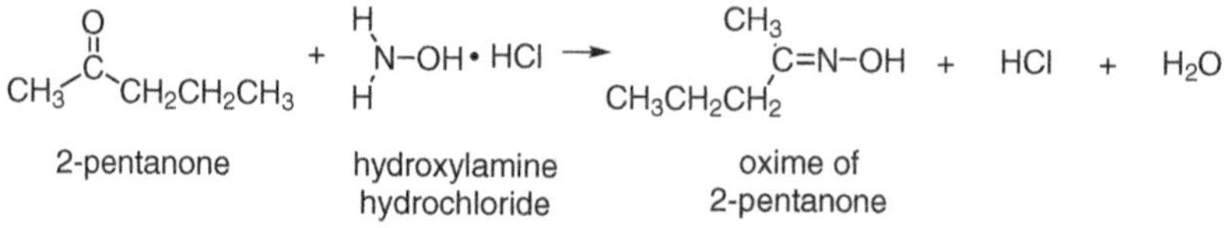

2-pentanone   hydroxylamine
hydrochloride

oxime of
2-pentanone

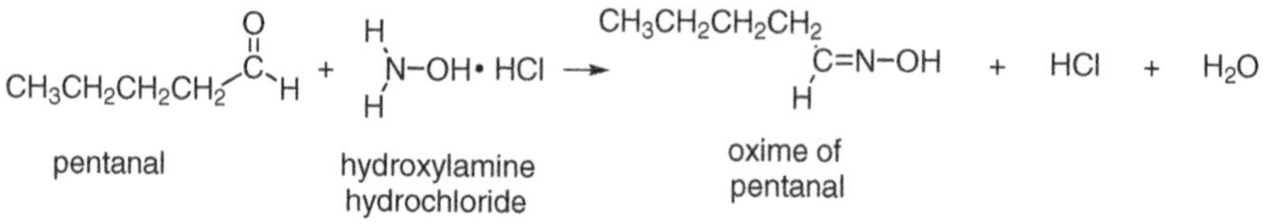

pentanal   hydroxylamine
hydrochloride

oxime of
pentanal

18. The sodium bisulfite test reacts with methyl ketones and cyclic ketones up to cyclooctanone. Because this reaction is very sensitive to steric hindrance, the cyclohexanone would react with sodium bisulfite, whereas the 3-pentanone would not.

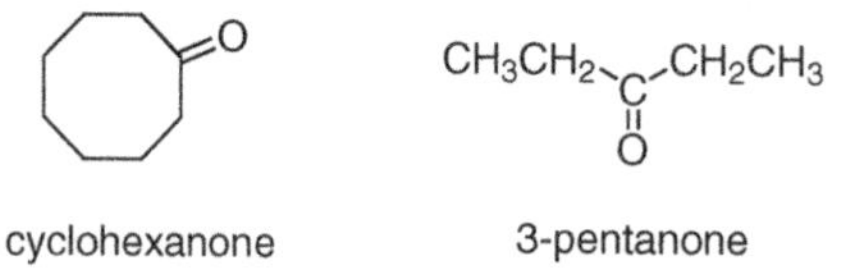

cyclohexanone   3-pentanone

19. Both pinacolone and acetophenone are sterically hindered on one side. The sodium bisulfite reaction does not work with sterically hindered structures.

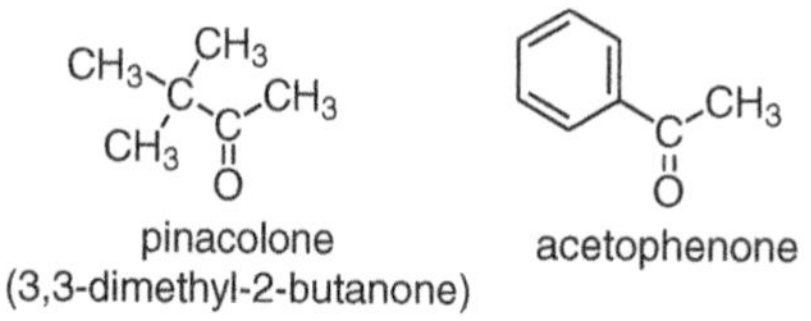

pinacolone
(3,3-dimethyl-2-butanone)

acetophenone

20. Sodium bisulfite will add to an α,β-unsaturated carbonyl compound via a Michael addition. Cinnamaldehyde is an α,β-unsaturated carbonyl compound. A second equivalent of sodium bisulfite would add to the carbonyl.

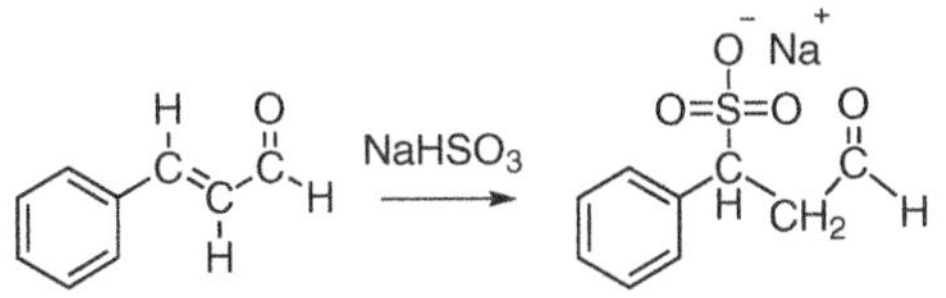

trans-cinnamaldehyde
(trans-3-phenyl-2-propene)

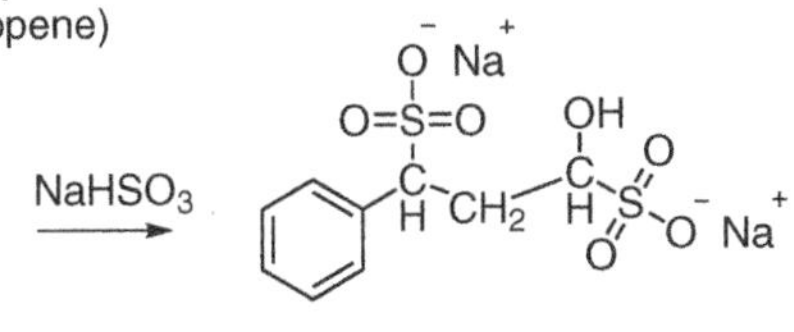

21. The bisulfite addition products that are derived from low molecular weight aldehydes and ketones are soluble in water, but not in ethanol. Therefore, ethanol is used as a solvent. Acetone would be considered a low molecular weight ketone and its bisulfite addition product would be soluble in water.

22. The silver ion could react with a reactive halogen to form a precipitate.

$$Ag^+ + X^- \longrightarrow AgX\ (s)$$

23. The Purpald Test reacts with hexanal, but not 3-hexanone.

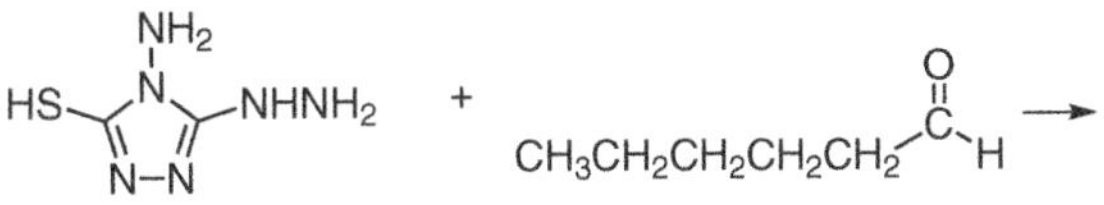

Purpald hexanal

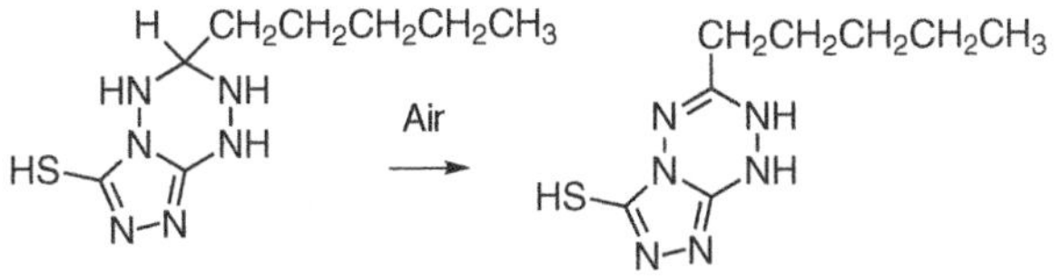

6-mercapto-s-triazolo-
[4,3,b]-s-tetrazine

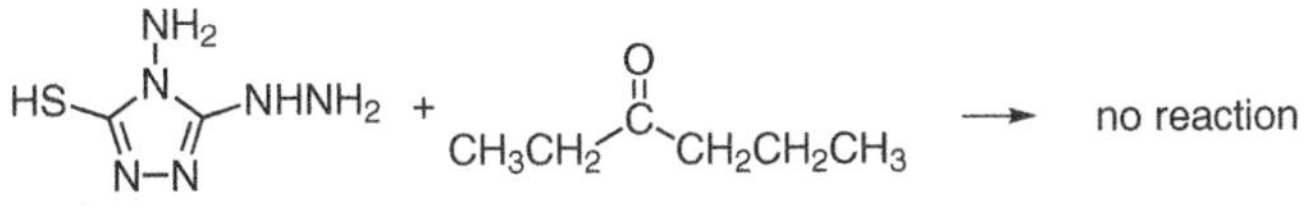

Purpald 3-hexanone

Tollens Test reacts with hexanal, but not 3-hexanone.

$$CH_3CH_2CH_2CH_2CH_2CHO + 2\ Ag(NH_3)_2OH \longrightarrow CH_3CH_2CH_2CH_2CH_2CO_2^-\ NH_4^+ + H_2O + 3\ NH_3 + 2\ Ag(s)$$

hexanal; Tollens reagent; ammonium butyrate

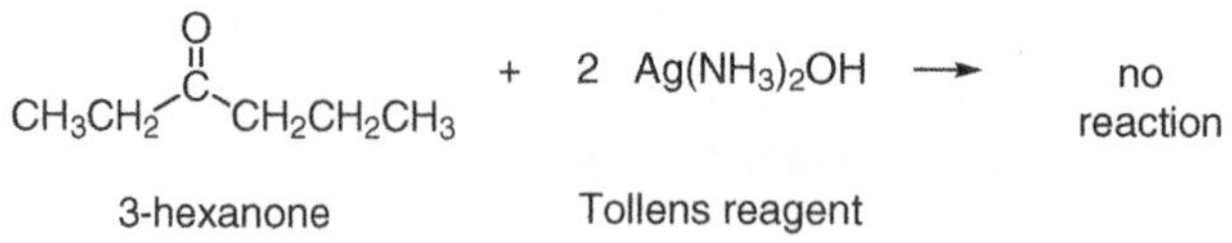

3-hexanone   Tollens reagent

24. Propanamide, *N*-methylpropanamide, and *N,N*-dimethylpropanamide are treated with sodium hydroxide to yield the sodium propanoate and the corresponding amine.

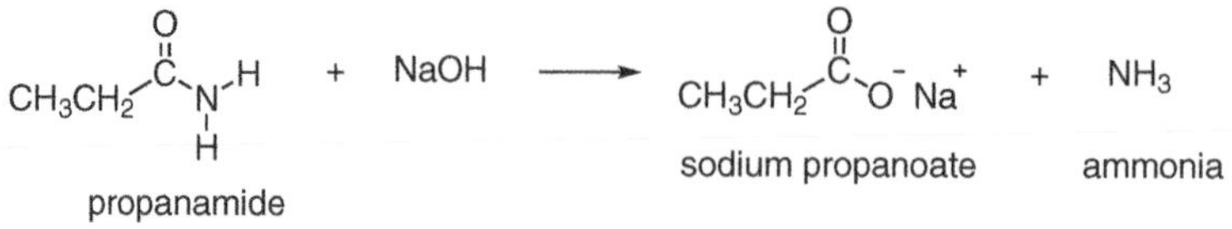

propanamide → sodium propanoate + ammonia

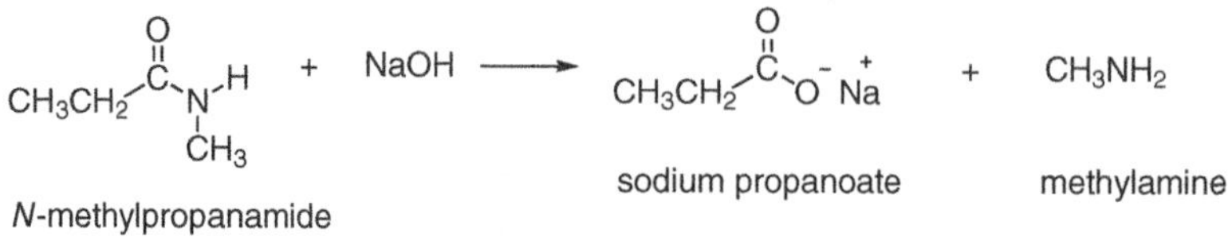

N-methylpropanamide → sodium propanoate + methylamine

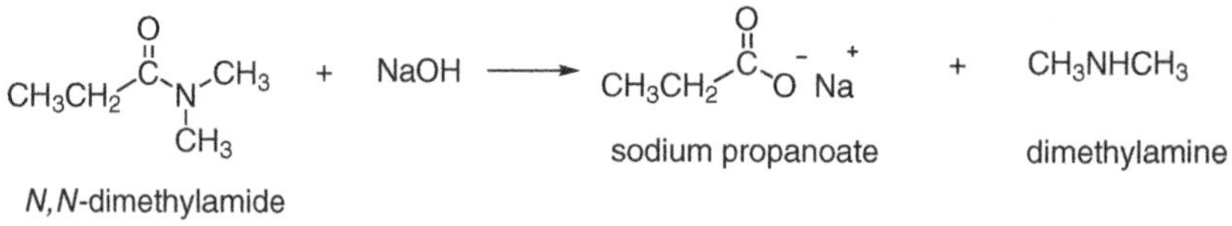

N,N-dimethylamide → sodium propanoate + dimethylamine

25. Propylamine reacts with benzenesulfonyl chloride to yield a soluble salt; acidification results in an insoluble *N*-propylbenzenesulfonamide.

propylamine + benzenesulfonyl chloride → sodium salt of *N*-propylbenzenesulfonamide (soluble)

+ NaCl + 2 H₂O

↓ H⁺

N-propylbenzenesulfonamide (insoluble)

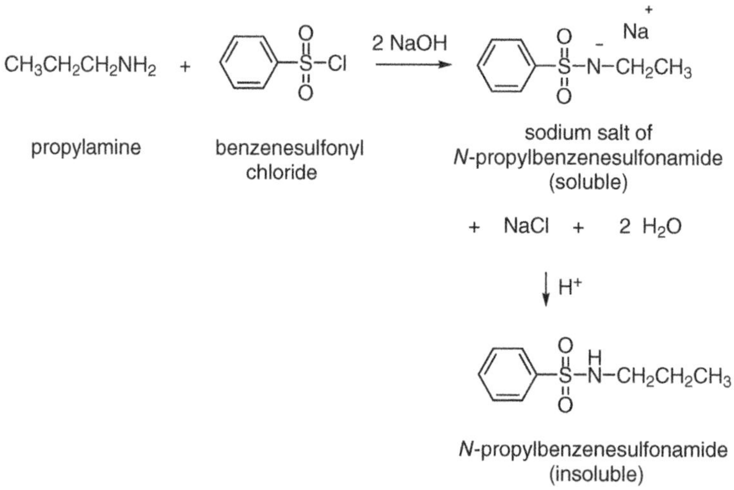

Diethylamine reacts with benzenesulfonyl chloride to give an insoluble *N,N*-diethylbenzenesulfonamide; this product does not dissolve in acid.

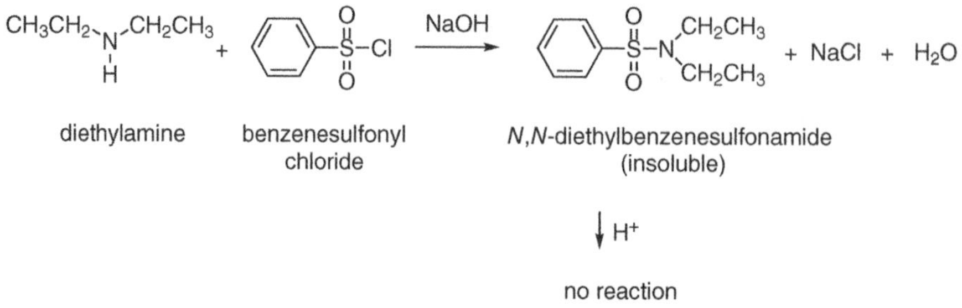

diethylamine + benzenesulfonyl chloride → *N,N*-diethylbenzenesulfonamide (insoluble) + NaCl + H₂O

↓ H⁺

no reaction

Triethylamine forms an intermediate salt with benzenesulfonyl chloride, which breaks apart into the soluble sodium salt of benzenesulfonic acid and the insoluble triethylamine. Acidification results in the soluble triethylammonium salt.

triethylamine   benzenesulfonyl chloride   hydrochloride salt of N,N,N-triethylbenzenesulfonamide

sodium salt of benzenesulfonic acid (soluble)   triethylamine (insoluble)

benzenesulfonic acid   triethylammonium chloride (soluble)

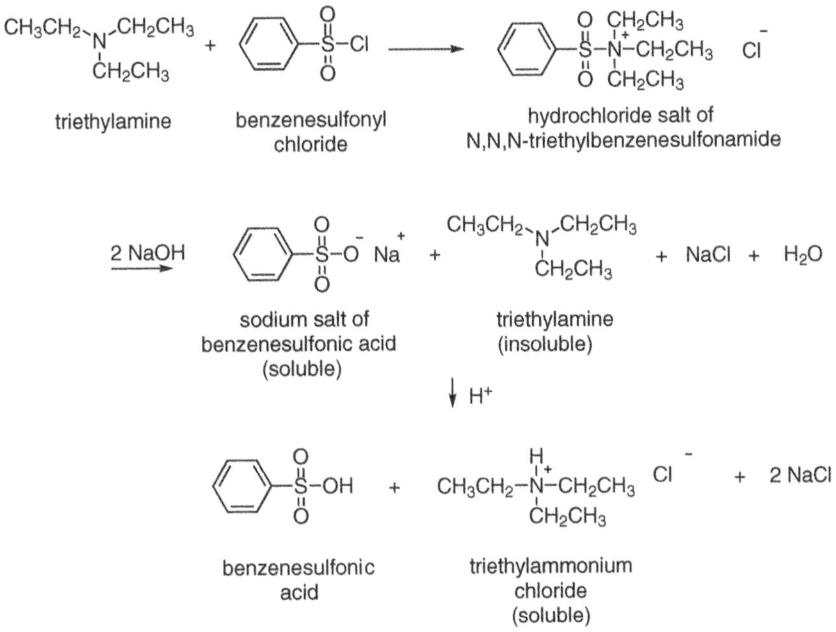

26. Propylamine reacts with nitrous acid to form a diazonium salt, which decomposes spontaneously.

propylamine   nitrous acid   diazonium salt (unstable at 00)

$$N_2\ (g) + CH_3CH_2CH_2OH + CH_3CH_2CH_2Cl + CH_3CH_2CH_2OCH_2CH_2CH_3 + CH_3CH=CH_2$$

Diethylamine reacts with nitrous acid to form a *N,N*-diethyl-*N*-nitrosoamine.

diethylamine   nitrous acid   *N,N*-diethyl-*N*-nitrosoamine (yellow oil or solid)

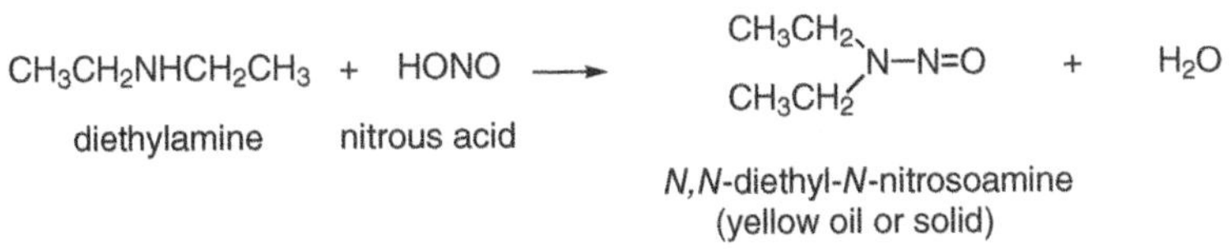

Triethylamine reacts with the acid to form a soluble triethylammonium salt.

triethylamine   triethylamine ammonium salt (soluble)

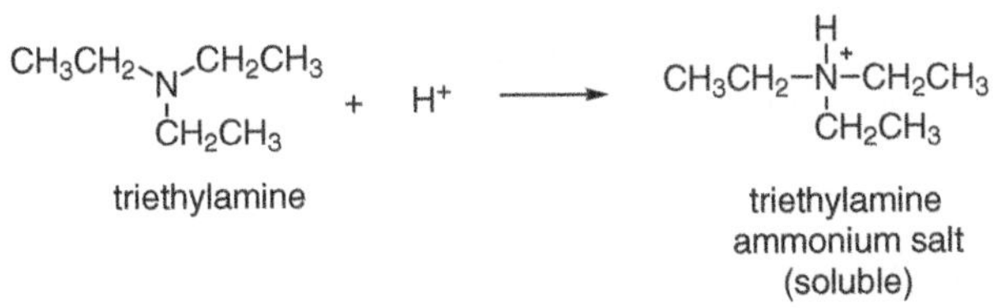

Aniline reacts with nitrous acid to form an aromatic diazonium salt, which loses nitrogen on warming to room temperature to form phenol.

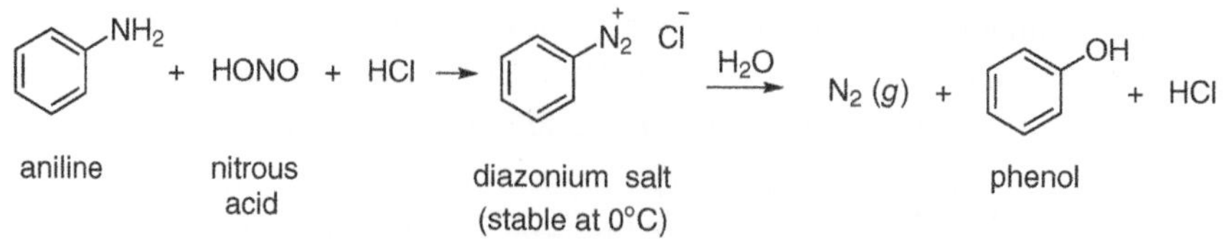

Aniline + HONO + HCl → diazonium salt (stable at 0°C) $\xrightarrow{H_2O}$ N_2 (g) + phenol + HCl

aniline   nitrous acid   diazonium salt (stable at 0°C)   phenol

N-Methylaniline reacts with nitrous acid to form a *N*-methyl-*N*-nitroso-*N*-aniline.

N-methylaniline + HONO (nitrous acid) → *N*-methyl-*N*-nitroso-*N*-aniline (yellow oil or solid) + H_2O

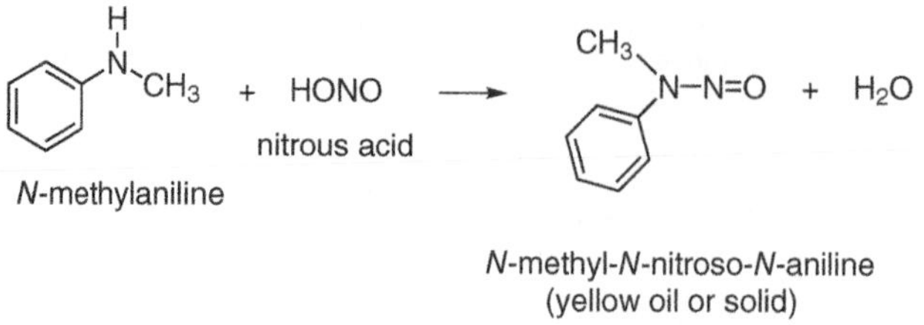

N,*N*-dimethylaniline reacts with nitrous acid to give *N*,*N*-dimethyl-*p*-nitrosoaniline.

N,*N*-dimethylaniline + HONO (nitrous acid) + HCl → Hydrochloride salt of *N*,*N*-dimethyl-*p*-nitrosoaniline (orange color) + H_2O

$\downarrow$ NaOH

N,*N*-dimethyl-*p*-nitrosoaniline (green solid) + NaCl + H_2O

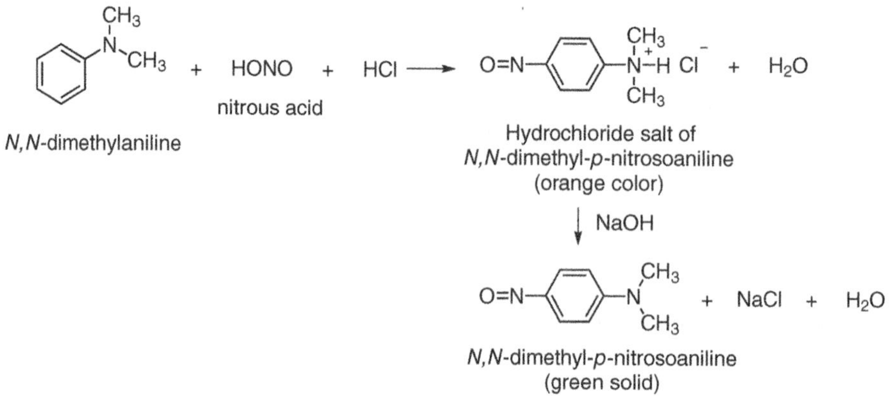

27. Butylamine and diethylamine react with acetyl chloride to form amides. Triethylamine does not react with acetyl chloride because it does not have an active hydrogen.

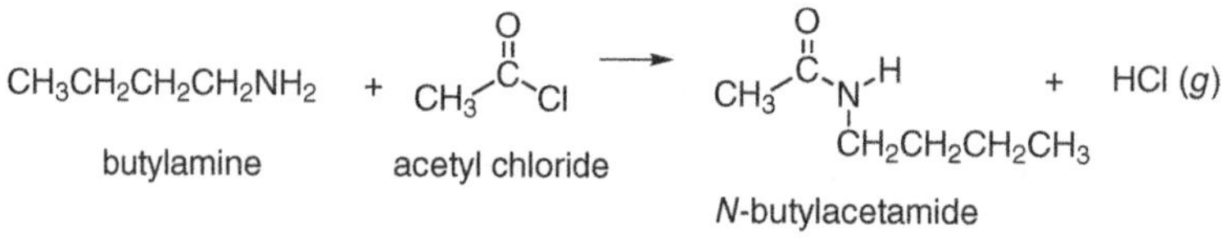

$CH_3CH_2CH_2CH_2NH_2$ + acetyl chloride (CH_3COCl) → *N*-butylacetamide + HCl (g)

butylamine   acetyl chloride   *N*-butylacetamide

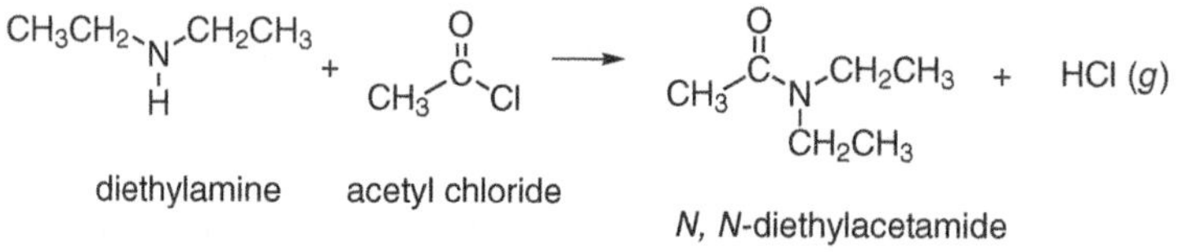

diethylamine (CH_3CH_2)$_2$NH + acetyl chloride (CH_3COCl) → *N*, *N*-diethylacetamide + HCl (g)

diethylamine   acetyl chloride   *N*, *N*-diethylacetamide

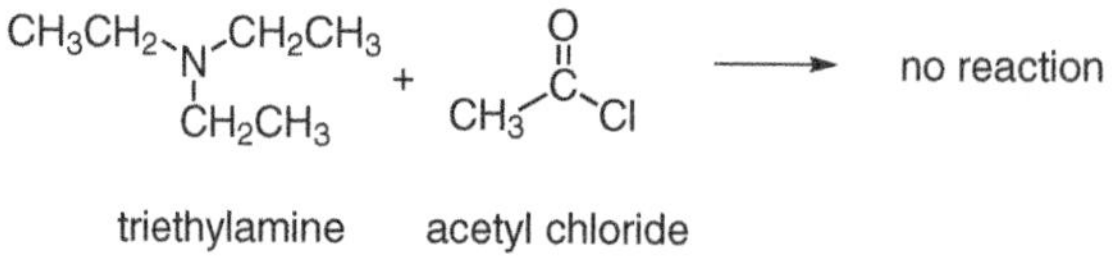

triethylamine (CH_3CH_2)$_3$N + acetyl chloride (CH_3COCl) → no reaction

triethylamine   acetyl chloride

Only diethylamine reacts with carbon disulfide in the presence of ammonium hydroxide.

$CH_3CH_2CH_2CH_2NH_2$ + CS_2 + NH_4OH → no reaction

butylamine   carbon disulfide

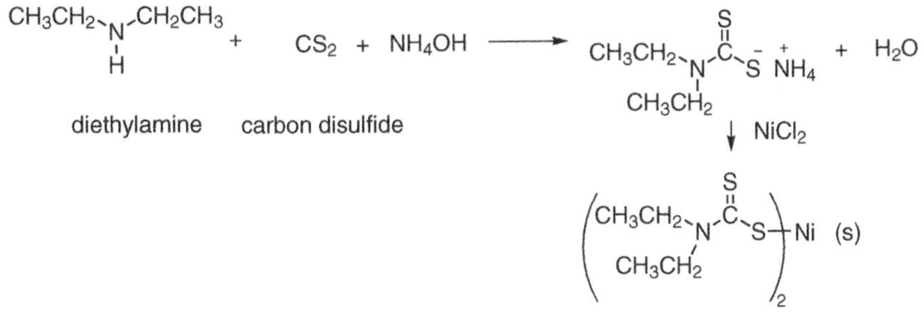

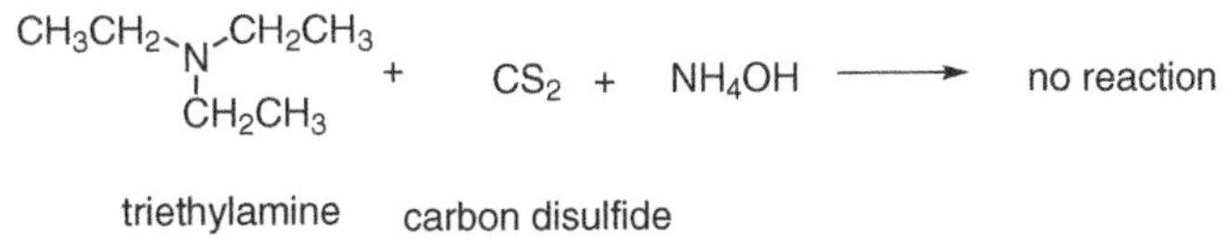

diethylamine   carbon disulfide

triethylamine   carbon disulfide

Butylamine reacts with acetyl chloride, but does not react with nickel chloride, carbon disulfide, and ammonium hydroxide. Diethylamine reacts with both reagents. Triethylamine does not react with any of the reagents.

28. Ninhydrin is the monohydrate of 1,2,3-indanetrione (1,2,3-triketohydrindene).

ninhydrin
(monohydrate of 1,2,3-indanetrion)
or 1,2,3-triketohydrindene)

1,2,3-Indanetrione
(1,2,3-triketohydrindene)

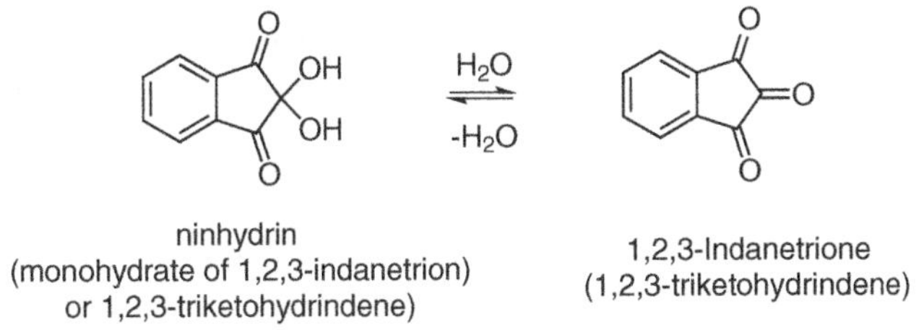

29. D-Fructose is oxidized with periodic acid to formaldehyde, carbon monoxide, and formic acid. D-Glucose is oxidized to formic acid and formaldehyde.

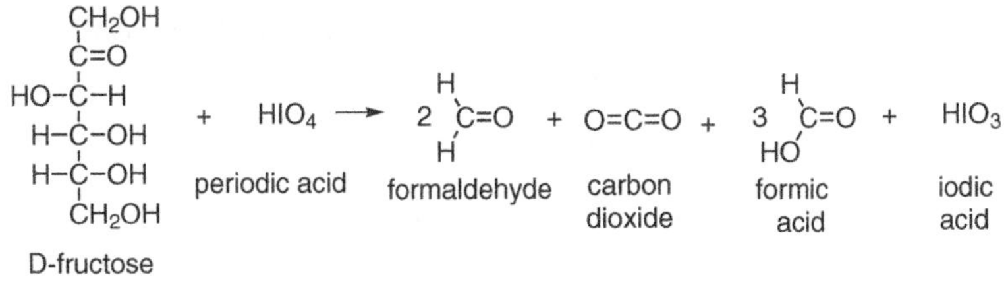

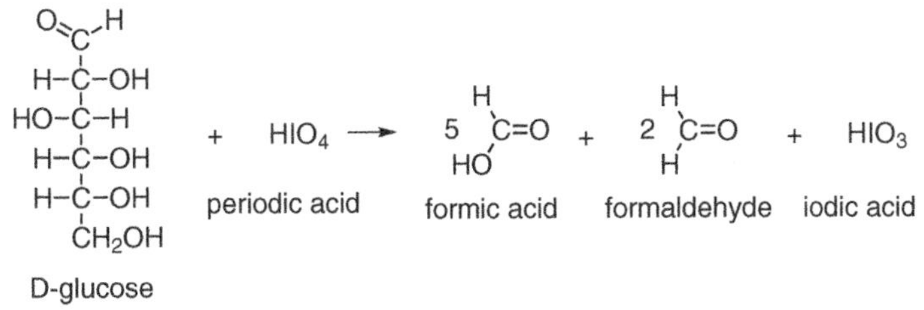

30. Carboxylic acids, due to hydrogen bonding, have higher melting points and boiling points than esters. For example, benzoic acid has a melting point of 122°C, whereas ethyl benzoate has a boiling point of 213°C. If a compound is a liquid, it is more volatile.

31. For ethers in which the groups are either methyl or primary, the cleavage of the ether follows a S_N2 mechanism, as illustrated below. The alkyl halide is formed from the smaller alkyl group.

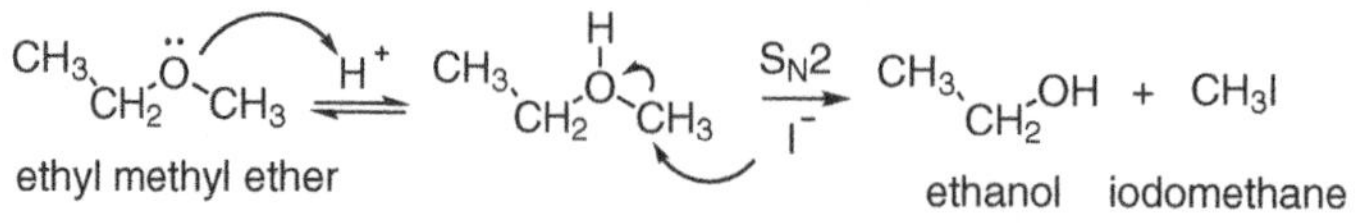

ethyl methyl ether   ethanol   iodomethane

For ethers in which one of the groups is a tertiary alkyl group, the cleavage of the ether follows a S_N1 mechanism. The alkyl halide is formed from the larger alkyl group.

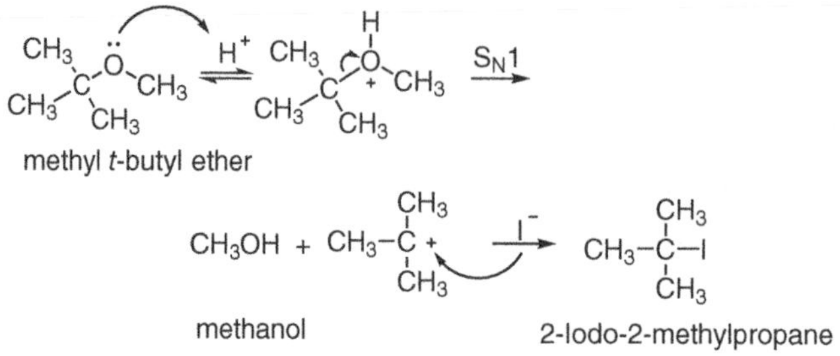

methyl *t*-butyl ether

$$CH_3OH + CH_3-\underset{\underset{CH_3}{|}}{\overset{\overset{CH_3}{|}}{C}} + \quad \xrightarrow{\quad I^- \quad} \quad CH_3-\underset{\underset{CH_3}{|}}{\overset{\overset{CH_3}{|}}{C}}-I$$

methanol   2-Iodo-2-methylpropane

32. Halides that react with silver nitrate follow an S_N1 mechanism. The formation of the carbocation is shown below. The carbocation is stabilized through resonance.

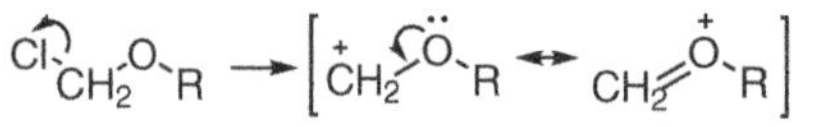

33. Listed below are the relative stabilities of the carbocations.

3° benzylic ≈ 3° allylic > 2° benzylic ≈ 2° allylic ≈ 3° alkyl > 1° benzylic ≈ 1° allylic ≈ 2° alkyl > 1° alkyl > vinylic ≈ aryl

Benzyl chloride forms a 1° benzylic carbocation, which is stabilized through resonance. The 1° benzylic carbocation is more stable than the 1° cyclohexyl carbocation that is formed from cyclohexylmethyl chloride.

benzyl
chloride   1° benzylic carbocation

cyclohexylmethyl   1° cyclohexylmethyl
chloride   carbocation

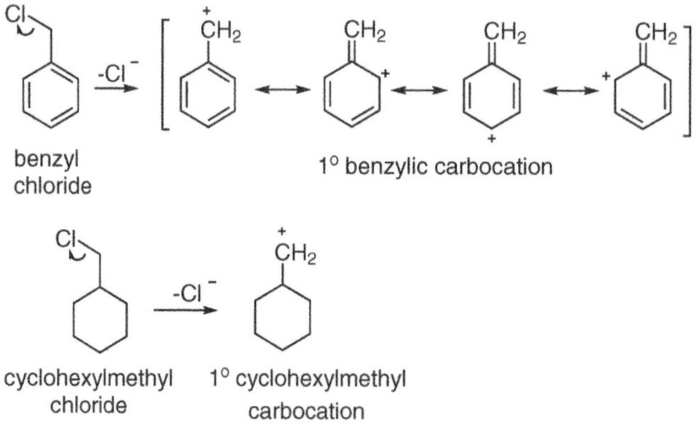

34. The low reactivity of aryl halides and vinyl halides toward ethanolic silver nitrate is attributed to the electronegativity of the sp²-hybridized carbon and to the delocalization of the electrons through resonance. The double bonded carbon is sp²-hybridized and is more electronegative than a sp³-hybridized carbon. It is less likely to donate electrons to a halogen to form a carbocation and a halide ion. In some of the resonance structures listed below, there is a positive charge on the halogen and a negative charge on the carbon. These structures do not favor the placement of a negative charge on the halogen.

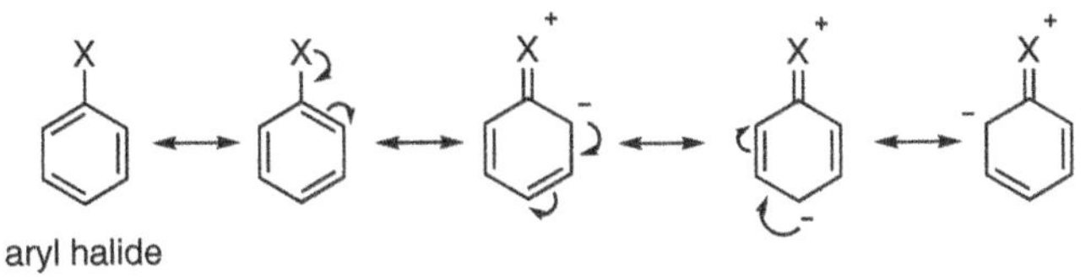

aryl halide

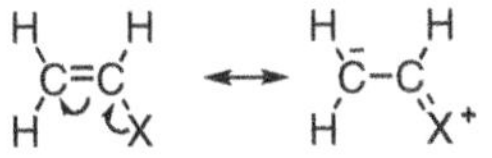

vinyl halide

35. Formation of the tertiary carbocation is sterically inhibited due to the cage-shaped skeleton.

1-chloronorbornane
(1-chlorobicyclo[2.2.1]heptane)

36. 1-Butene is brominated to give 1,2-dibromobutane.

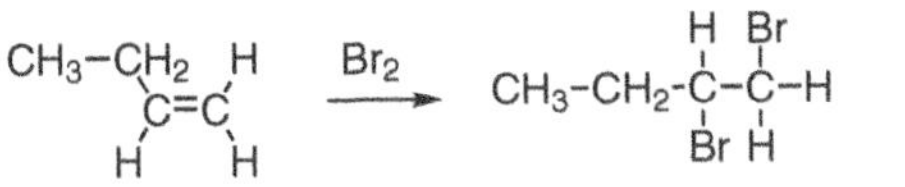

1-butene   1,2-dibromobutane

The bromination of 1-butyne yields 1,1,2,2-tetrabromobutane.

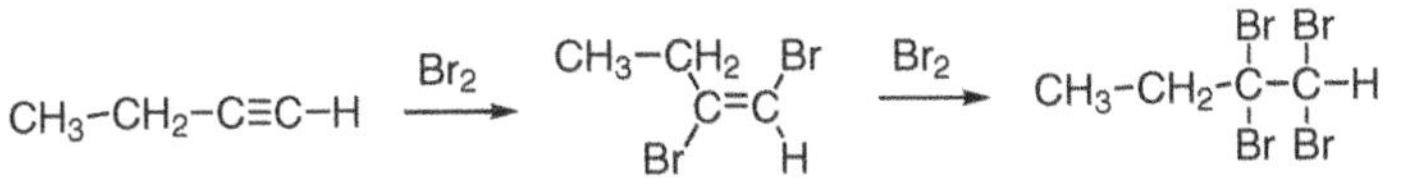

1-butyne   1,2-dibromo-1-butene   1,1,2,2-tetrabromobutane

37. Bromine and potassium permanganate solutions are decolorized in the presence of carbon–carbon double bonds and carbon–carbon triple bonds.

38. The presence of electron-withdrawing groups attached to a doubly bonded carbon may prevent the addition of bromine across the double bond. Potassium permanganate is more versatile in this respect because it is not affected by electron-withdrawing groups.

39. Carbonyl compounds which decolorize bromine solutions usually give a negative potassium permanganate test. Alcohols and compounds such as benzaldehyde and formaldehyde decolorize potassium permanganate solutions, but not bromine solutions. Therefore, to confirm the identity of a carbon-carbon double bond or carbon-carbon triple bond, both tests should be used.

40. 1-Butene is oxidized to 1,2-butanediol with potassium permanganate.

$$3 \quad \underset{H}{\overset{CH_3CH_2}{}}C=C\underset{H}{\overset{H}{}} \quad + \quad 2\ KMnO_4 \quad + \quad 4\ H_2O \longrightarrow$$

1-butene   potassium permanganate
(purple)

$$3 \quad CH_3CH_2\underset{HO}{\overset{H}{\underset{|}{\overset{|}{C}}}}\underset{OH}{\overset{H}{\underset{|}{\overset{|}{C}}}}-H \quad + \quad 2\ KOH \quad + \quad 2\ MnO_2\ (s)$$

1,2-butanediol   manganese dioxide
(brown)

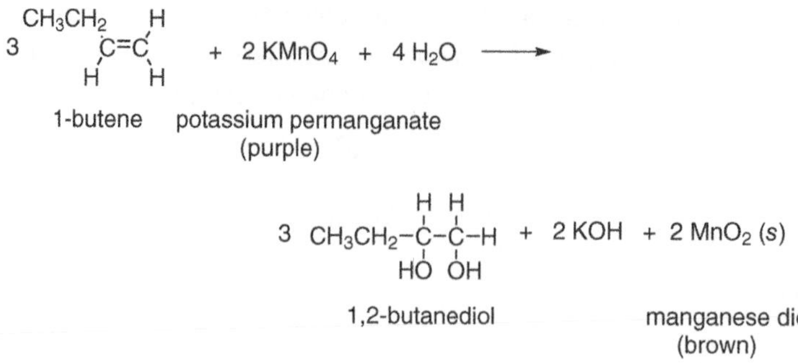

1-Butyne is oxidized to sodium propanoate and sodium formate with potassium permanganate.

$$CH_3CH_2-C{\equiv}C-H \quad + \quad 2\ KMnO_4 \quad + \quad 2\ H_2O \longrightarrow$$

1-butyne   potassium permanganate
(purple)

$$CH_3CH_2\overset{O}{\overset{||}{C}}\overset{-}{O}\ \overset{+}{K} \quad + \quad H\overset{O}{\overset{||}{C}}\overset{-}{O}\ \overset{+}{K} \quad + \quad 2\ H_2O \quad + \quad 2\ MnO_2\ (s)$$

sodium propanoate   sodium formate   manganese dioxide
(brown)

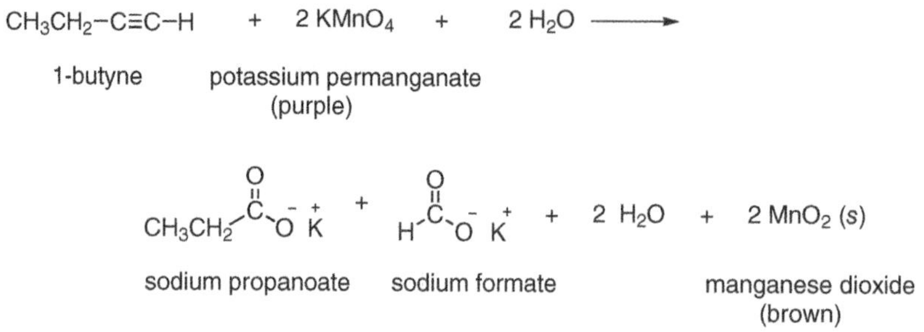

41. 1-Butyne is a terminal alkyne and will react with sodium to produce hydrogen gas and sodium butynide.

$$2\ CH_3CH_2C{\equiv}CH \quad + \quad 2\ Na \longrightarrow 2\ CH_3CH_2C{\equiv}\overset{-}{C}\ \overset{+}{Na} \quad + \quad H_2\ (g)$$

1-butyne   sodium butynide

2-Butyne does not react with sodium.

$$2\ CH_3C{\equiv}CCH_3 \quad + \quad 2\ Na \longrightarrow \text{no reaction}$$

2-butyne

42. Since this reaction is to test for the presence of an aromatic ring, perhaps with a deactivating group, only compounds with the solubility class of I could be used. If the compound contains a strongly activating group such as $-OH$ or $-NH_2$, a violent reaction may occur.

43. The following reactions occur in concentrated sulfuric acid.

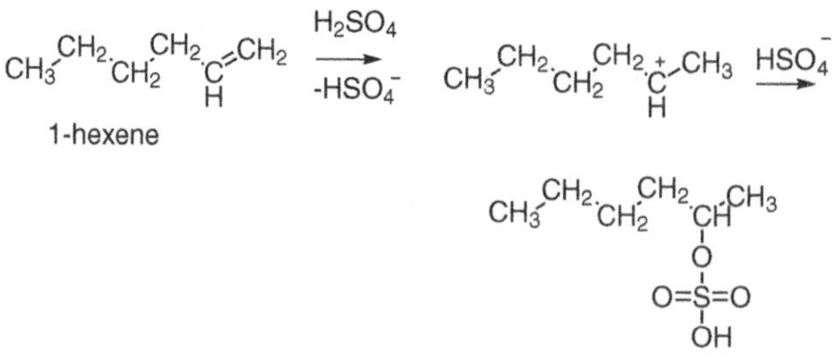

1-hexene

The following reactions occur in fuming sulfuric acid.

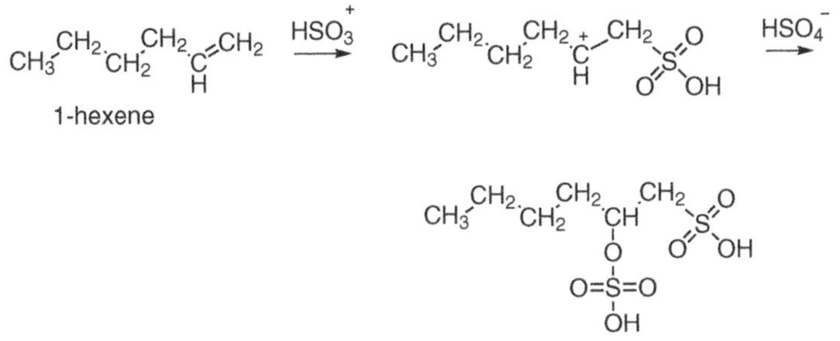

1-hexene

44. To form a tetraarylmethane, a proton transfer would have to occur. Carbon does not have five bonds.

45. Butanenitrile is hydrolyzed to sodium butanoate in the presence of base.

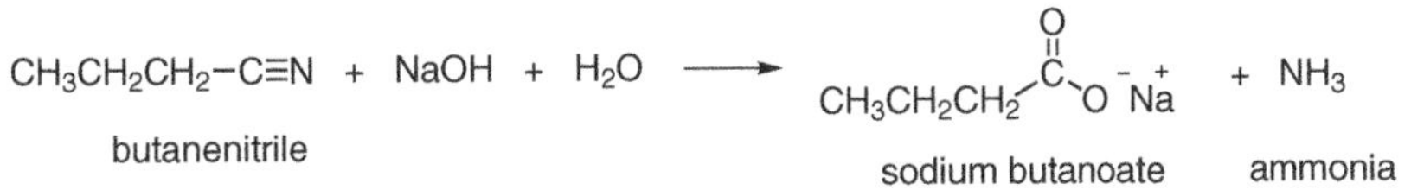

butanenitrile   sodium butanoate   ammonia

46. With sodium hydroxide, benzene will not have a color change. Nitrobenzene will give no color or a very light yellow color. A bluish-purple color develops with 1,3-dinitrobenzene. 1,3,5-Trinitrobenzene will produce a blood-red color.

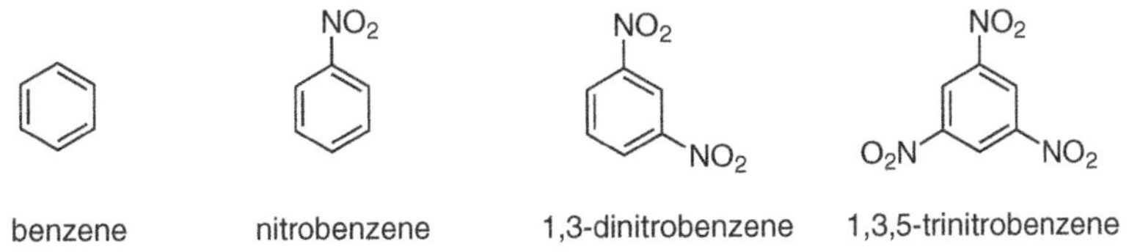

benzene   nitrobenzene   1,3-dinitrobenzene   1,3,5-trinitrobenzene

47. Both hydroxyl groups of 4-hydroxybenzoic acid react with sodium to form the disodium salt.

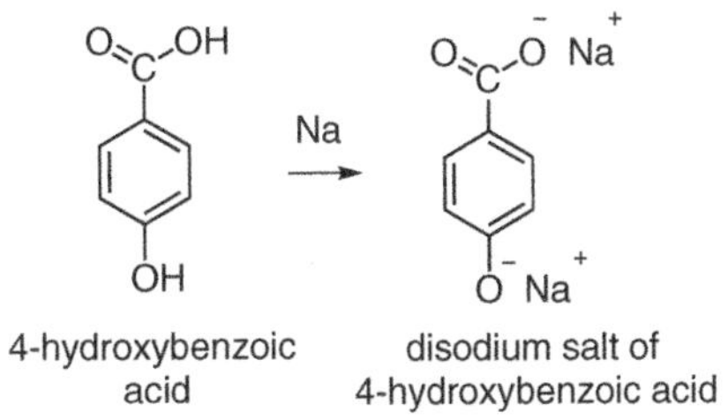

4-hydroxybenzoic   disodium salt of
acid   4-hydroxybenzoic acid

Both hydroxyl groups are also acetylated.

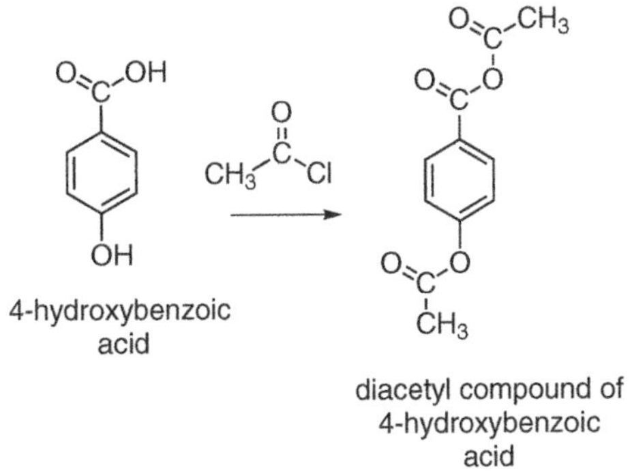

4-hydroxybenzoic
acid

diacetyl compound of
4-hydroxybenzoic
acid

48. 2,4,6-Tribromoaniline will not react with bromine water. The amino group is a strongly activating *ortho*, *para* director and there are no *ortho* or *para* positions open for substitution.

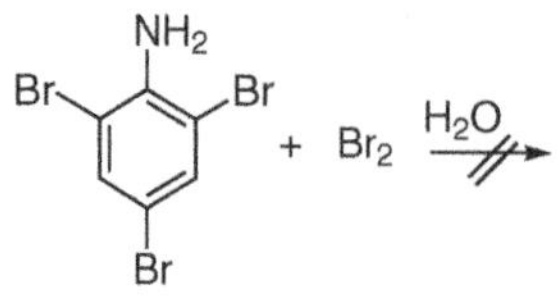

49. Yes, the bromine water is decolorized as the bromine reacts with phenol; one bromine atom substitutes onto the ring and the other bromine atom pairs with the hydrogen that has been substituted. The hydrogen bromide dissolves in the water.

50. The bromine is not hydrolyzed in water. The brominating agent is molecular bromine.

51. The benzene ring attacks the bromine through electrophilic substitution. Electron-donating groups make the ring more reactive. An —O⁻ would donate more electrons to the ring than an —OH. The —OH is a strongly activating group.

Chapter 11

The Preparation of Derivatives

1. The neutralization equivalent is calculated by dividing the molecular weight by the number of —COOH. The calculations are shown below for benzoic acid and phthalic acid.

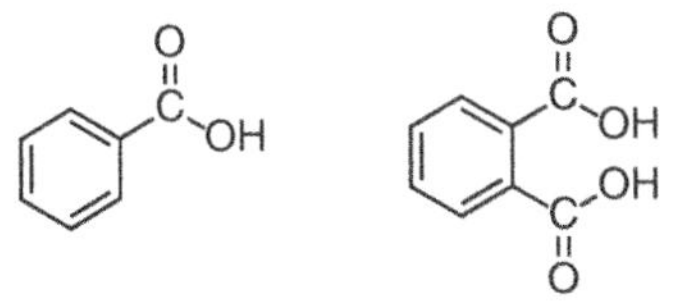

    benzoic acid
    MW = 122

    phthalic acid
    MW = 166

    $$NE = \frac{122}{1} = 122 \qquad NE = \frac{166}{2} = 83$$

2. If the sample is wet, the neutralization equivalent would be too high. The extra water weight would be counted as the weight of the sample and thus would result in the neutralization equivalent being too high.

3. Base does not have any effect on the amine, and thus would not change the neutralization equivalent if the solution was titrated with a base. However, amines can be titrated with hydrochloric acid and the neutralization equivalent determined.

Student Solutions Manual to Accompany The Systematic Identification of Organic Compounds, Ninth Edition.
Christine K. F. Hermann, Terence C. Morrill, Ralph L. Shriner, and Reynold C. Fuson.
© 2023 John Wiley & Sons, Inc. Published 2023 by John Wiley & Sons, Inc.

4. A neutralization equivalent of a phenol is only useful if the K_a value is greater than 1×10^{-6}. Examples are phenols with two electron-withdrawing groups such as dinitrophenols.

5. The neutralization equivalent can only be used on acids which have K_a values greater than 1×10^{-6}.

6. a. Butanoic acid is treated with thionyl chloride, then ammonia to form butanamide.

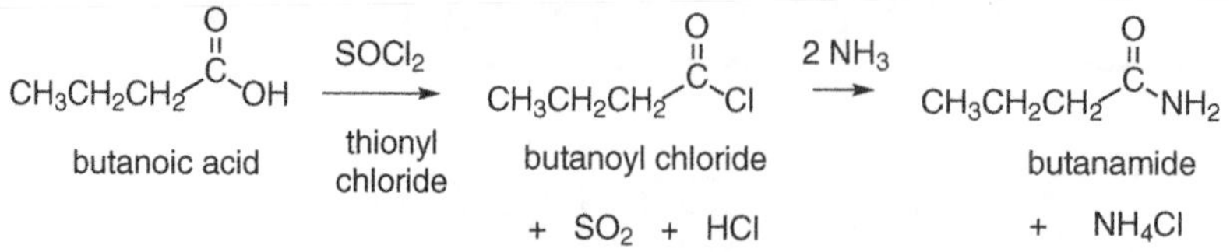

b. The acid halide reacts with 4-toluidine to form the 4-toludide.

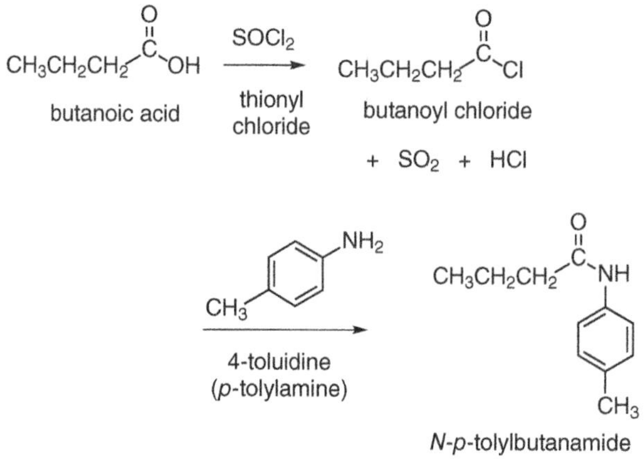

c. Reaction of the acid halide with aniline forms the anilide.

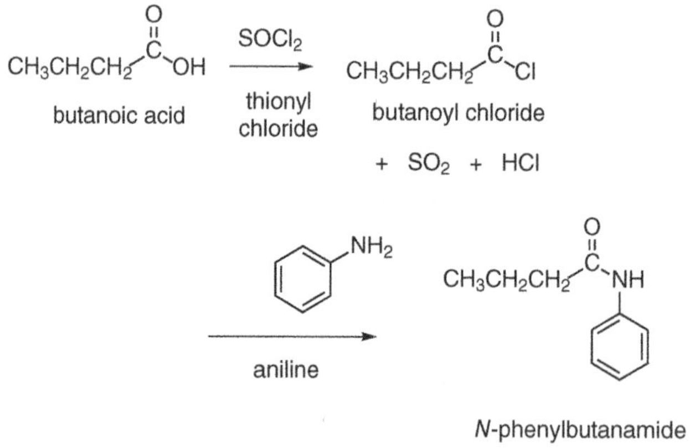

d. The butanoic acid is treated with sodium bicarbonate to form the sodium butanoate, which is then reacted with 4-nitrobenzyl chloride to yield the *N*-nitrobenzyl butanoate.

$$CH_3CH_2CH_2\overset{\overset{O}{\|}}{C}\text{-OH} \xrightarrow{\text{NaHCO}_3} CH_3CH_2CH_2\overset{\overset{O}{\|}}{C}\text{-O}^-\ \text{Na}^+$$

butanoic acid   sodium butanoate

ClCH$_2$—⟨benzene⟩—NO$_2$ →   CH$_3$CH$_2$CH$_2$ CH$_2$—⟨benzene⟩—NO$_2$ + NaCl (with C–O and C=O)

4-nitrobenzyl chloride   *N*-nitrobenzyl butanoate

e. The sodium butanoate reacts with the 4-bromoacyl bromide to produce an ester.

$$CH_3CH_2CH_2\overset{\overset{O}{\|}}{C}\text{-OH} \xrightarrow{\text{NaHCO}_3} CH_3CH_2CH_2\overset{\overset{O}{\|}}{C}\text{-O}^-\ \text{Na}^+$$

butanoic acid   sodium butanoate

BrCH$_2$–C(=O)–⟨benzene⟩–Br   →   C(=O)–⟨benzene⟩–Br + NaBr

4-bromophenacyl bromide

O–CH$_2$, O=C , CH$_2$CH$_2$CH$_3$

2-(4-bromophenyl)-2-oxoethyl butanoate

f. The *S*-benzylthiuronium salt is produced from the reaction of butanoic acid with *S*-benzylthiuronium salt.

$$CH_3CH_2CH_2\overset{\overset{O}{\|}}{C}\text{-OH} \xrightarrow{\text{NaHCO}_3} CH_3CH_2CH_2\overset{\overset{O}{\|}}{C}\text{-O}^-\ \text{Na}^+$$

butanoic acid   sodium butanoate

⟨benzene⟩CH$_2$SC(NH$_2$)$_2$$^+$ Cl$^-$ →   ⟨benzene⟩CH$_2$SCH(NH$_3$)$_2$$^+$ + NaCl

O=C , O$^-$–C–CH$_2$CH$_2$CH$_3$

S-benzylthiuronium salt   *S*-benzylthiuronium salt of butanoic acid

g. Butanoic acid reacts with phenylhydrazine to yield the phenylhydrazide.

CH$_3$CH$_2$CH$_2$–C(=O)–OH + ⟨benzene⟩–NH–NH$_2$ → ⟨benzene⟩–NH–NH–C(=O)–CH$_2$CH$_2$CH$_3$ + H$_2$O

butanoic acid   phenylhydrazine   *N*-phenylhydrazide of butanoic acid

7. a. Acetyl chloride is hydrolyzed to acetic acid.

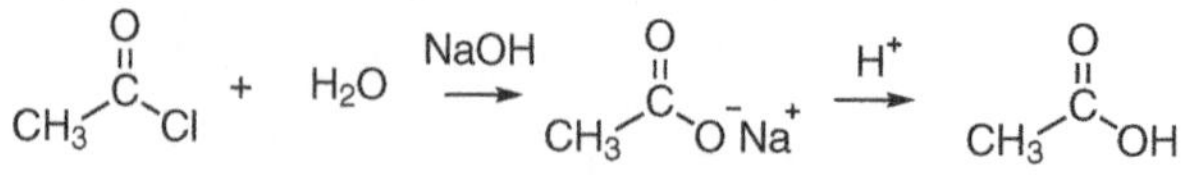

acetyl chloride   sodium acetate   acetic acid

b. An acetamide is formed from the reaction of acetyl chloride with ammonia.

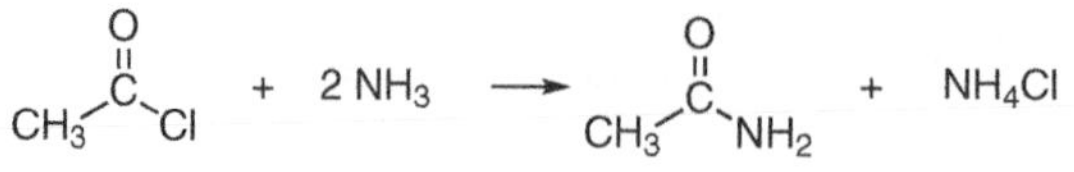

acetyl chloride   acetamide

c. Acetyl chloride reacts with aniline to form an anilide.

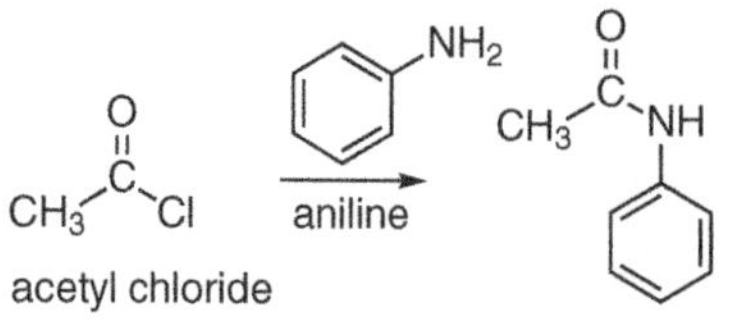

acetyl chloride   aniline   acetanilide

d. Acetyl chloride reacts with 4-toluidine to form a 4-methylacetanilide.

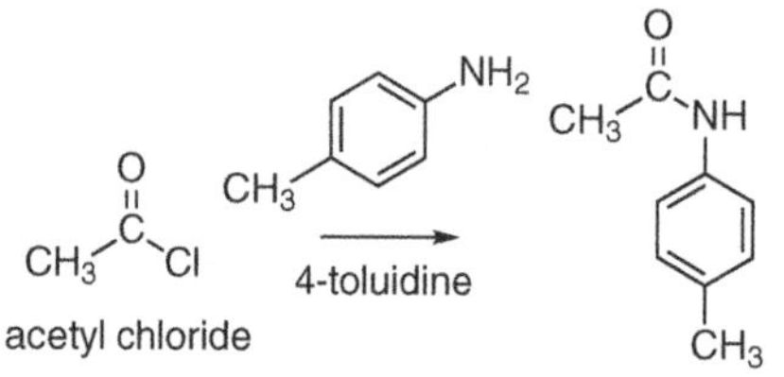

acetyl chloride   4-toluidine   4-methylacetanilide

8. a. Propanoic anhydride is hydrolyzed to propanoic acid.

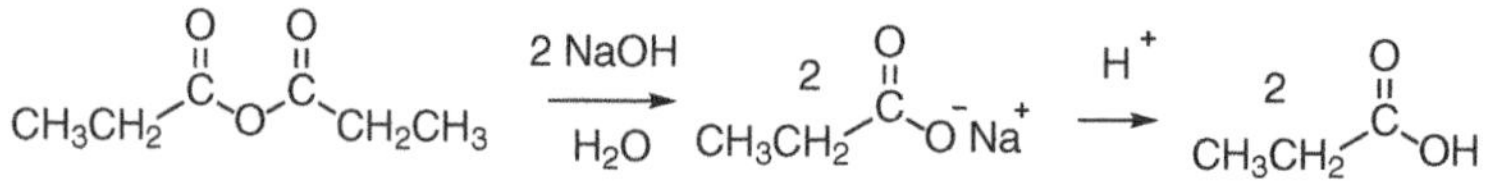

propanoic anhydride   sodium propanoate   propanoic acid

b. Propanoic anhydride is treated with ammonia to form propanamide and propanoic acid.

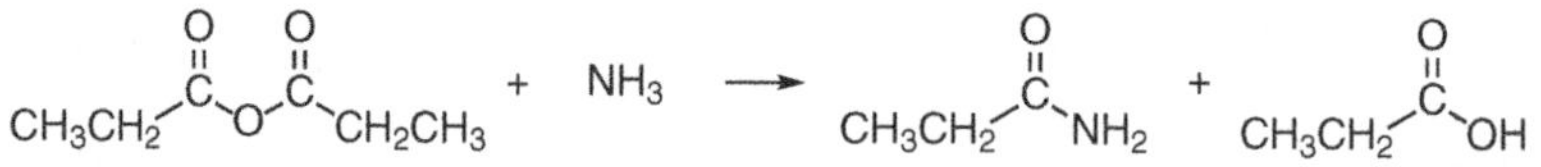

propanoic anhydride   propanamide   propanoic acid

c. Propanoic anhydride reacts with aniline to form the *N*-phenylpropanamide and the anilinium salt.

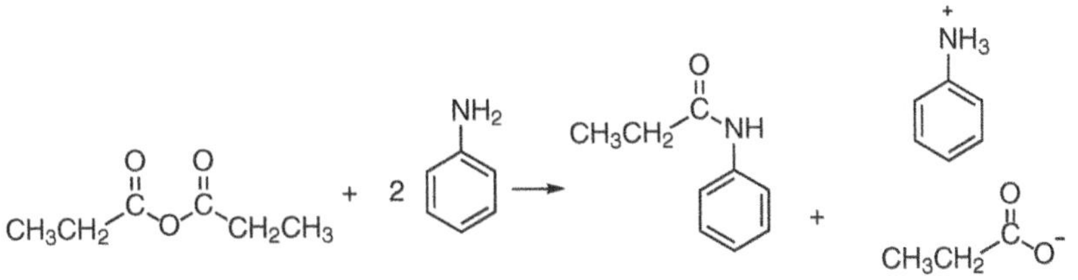

propanoic anhydride   aniline   *N*-phenyl-propanamide

d. The reaction of propanoic anhydride and 4-toluidine yields the 4-toluidide.

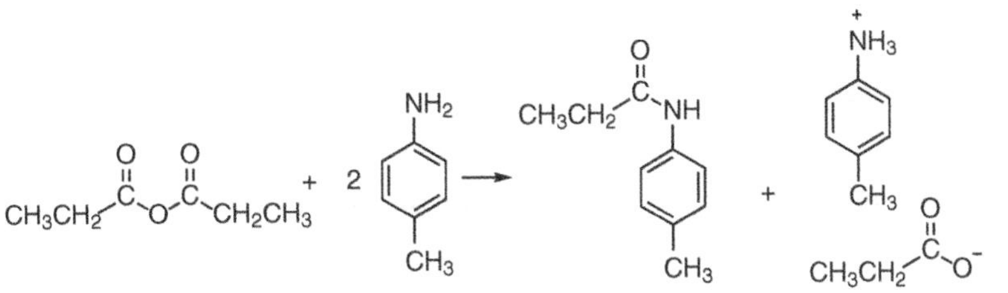

propanoic anhydride   4-toluidine   *N*-(4-tolyl)-propanamide

9. a. 1-Butanol reacts with phenyl isocyanate to make a phenylurethane.

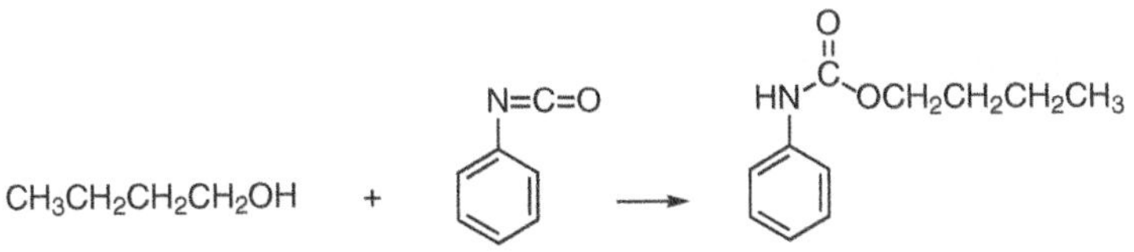

CH₃CH₂CH₂CH₂OH   +

1-butanol   phenyl isocyanate   butyl *N*-phenylcarbamate

b. 1-Butanol reacts with 1-naphthyl isocyanate to yield a 1-naphthylurethane.

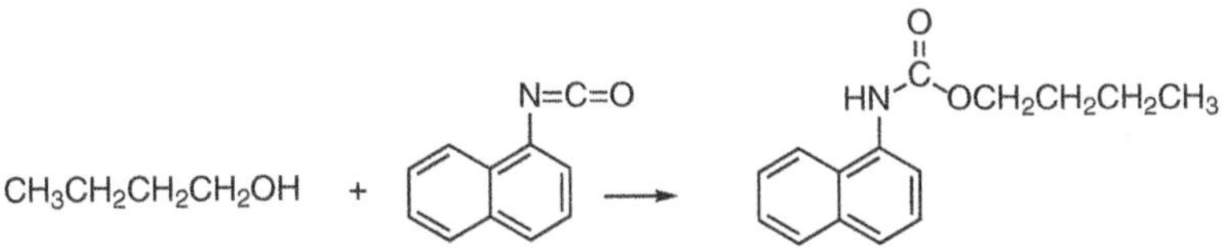

CH₃CH₂CH₂CH₂OH   +

1-butanol   1-naphthyl isocyanate   butyl *N*-naphthylcarbamate

c. 1-Butanol undergoes reaction with 4-nitrobenzoyl chloride to produce butyl 4-nitrobenzoate.

CH₃CH₂CH₂CH₂OH   +   +   HCl

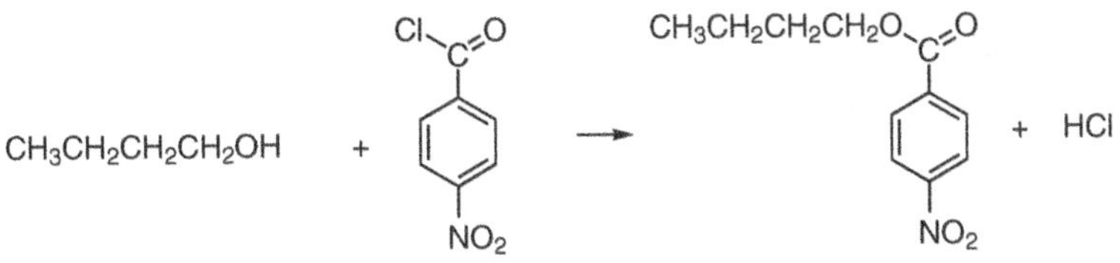

1-butanol   4-nitrobenzoyl chloride   butyl 4-nitrobenzoate

d. Butyl 3,5-dinitrobenzoate is formed from 1-butanol reacting with 3,5-dinitrobenzoyl chloride.

1-butanol 3,5-dinitrobenzoyl chloride butyl 3,5-dinitrobenzoate

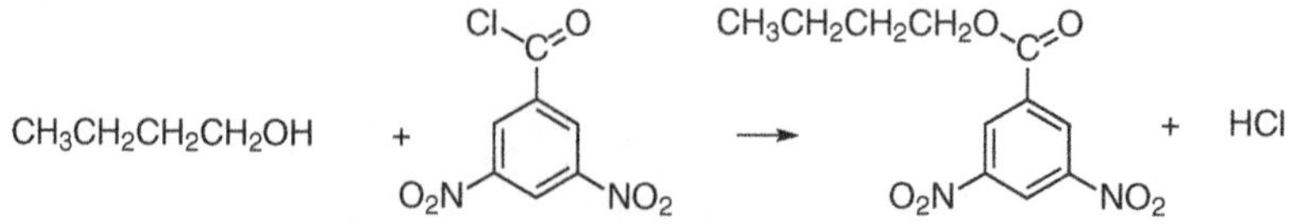

e. 1-Butanol reacts with 3-nitrophthalic anhydride to yield butyl 3-nitrophthalate.

1-butanol 3-nitrophthalic anhydride butyl 3-nitrophthalate

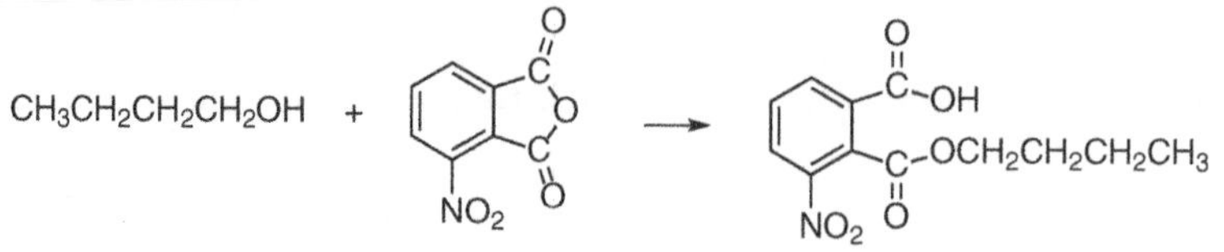

10. a. 2-Butanol reacts with phenyl isocyanate to make a phenylurethane.

2-butanol phenyl isocyanate *sec*-butyl *N*-phenylcarbamate

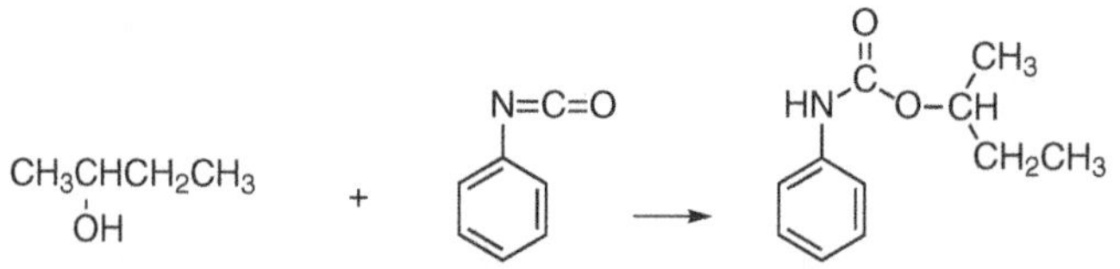

b. 2-Butanol reacts with 1-naphthyl isocyanate to yield a 1-naphthylurethane.

2-butanol 1-naphthyl isocyanate *sec*-butyl *N*-naphthylcarbamate

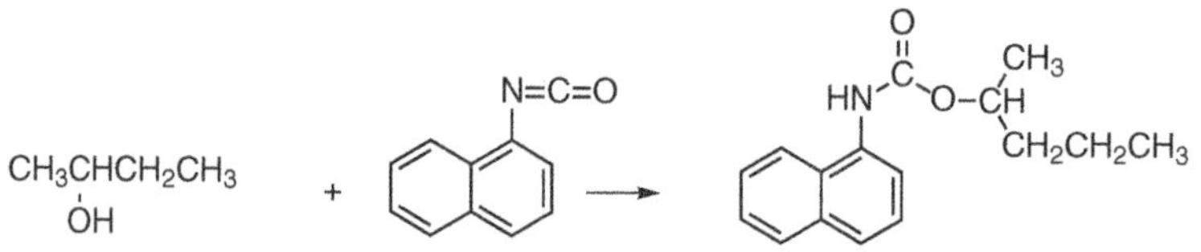

c. 2-Butanol undergoes reaction with 4-nitrobenzoyl chloride to produce *sec*-butyl 4-nitrobenzoate.

2-butanol 4-nitrobenzoyl chloride *sec*-butyl 4-nitrobenzoate

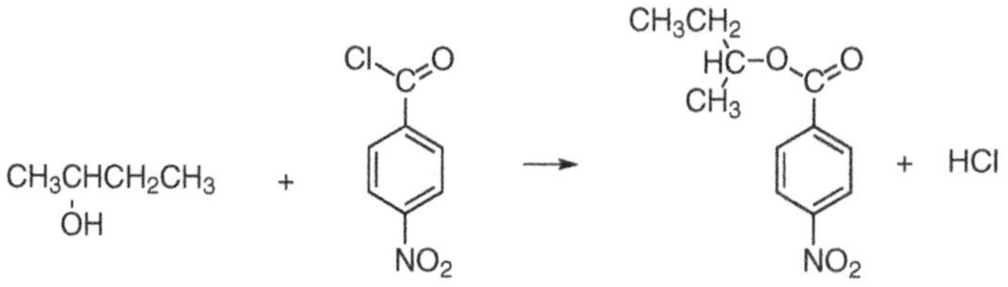

d. *Sec*-butyl 3,5-dinitrobenzoate is formed from 2-butanol reacting with 3,5-dinitro-benzoyl chloride.

CH₃CHCH₂CH₃ | OH
2-butanol

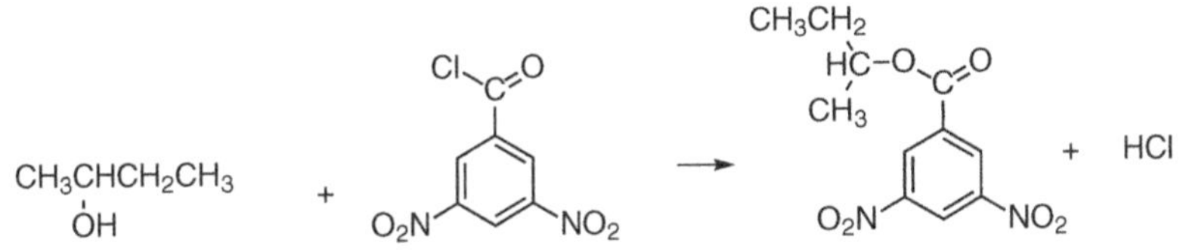

3,5-dinitrobenzoyl chloride

sec-butyl 3,5-dinitrobenzoate

\+ HCl

e. 2-Butanol reacts with 3-nitrophthalic anhydride to yield *sec*-butyl 3-nitrophthalate.

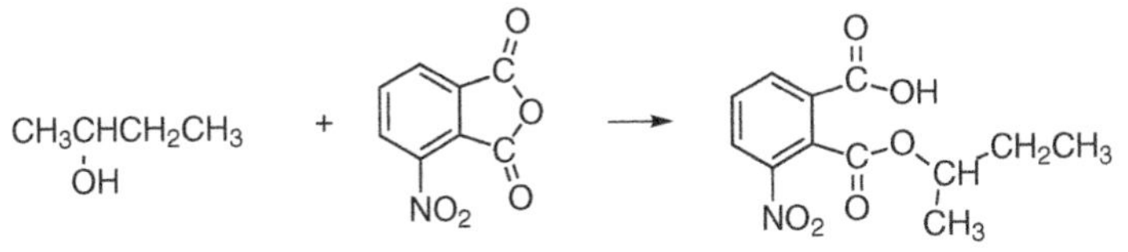

2-butanol

3-nitrophthalic anhydride

sec-butyl 3-nitrophthalate

11. a. 2-Methyl-2-butanol reacts with phenyl isocyanate to yield a phenylurethane.

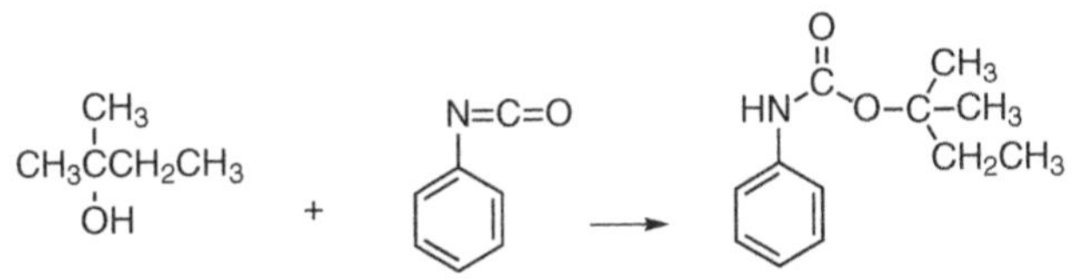

2-methyl-2-butanol

phenyl isocyanate

tert-pentyl *N*-phenylcarbamate

b. 2-Methyl-2-butanol reacts with 1-naphthyl isocyanate to yield a 1-naphthylurethane.

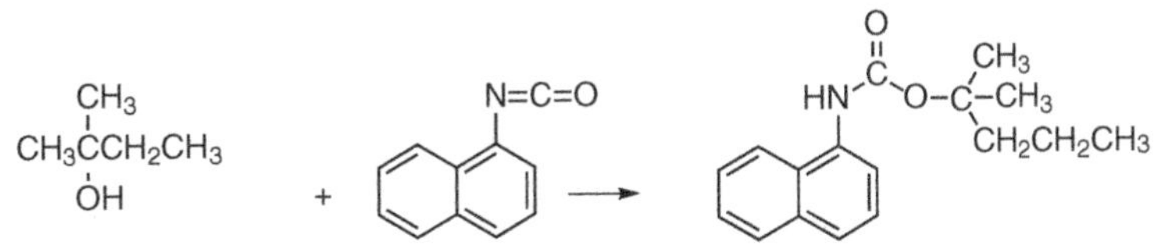

2-methyl-2-butanol

1-naphthyl isocyanate

tert-pentyl *N*-naphthylcarbamate

c. 2-Methyl-2-butanol undergoes reaction with 4-nitrobenzoyl chloride to produce *tert*-pentyl 4-nitrobenzoate.

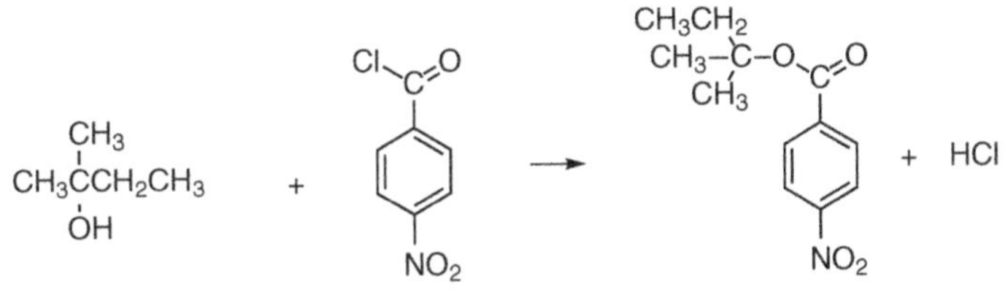

2-methyl-2-butanol

4-nitrobenzoyl chloride

tert-pentyl 4-nitrobenzoate

\+ HCl

d. *Tert*-pentyl 3,5-dinitrobenzoate is formed from 2-methyl-2-butanol reacting with 3,5-dinitrobenzoyl chloride.

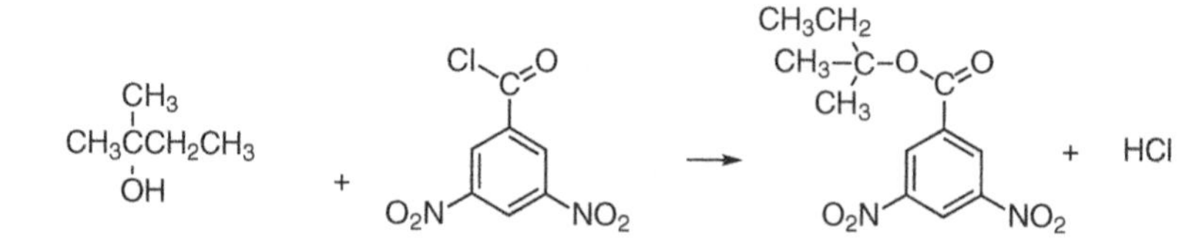

| 2-methyl-2-butanol | 3,5-dinitrobenzoyl chloride | *tert*-pentyl 3,5-dinitrobenzoate |

e. 2-Methyl-2-butanol reacts with 3-nitrophthalic anhydride to yield *tert*-pentyl 3-nitrophthalate.

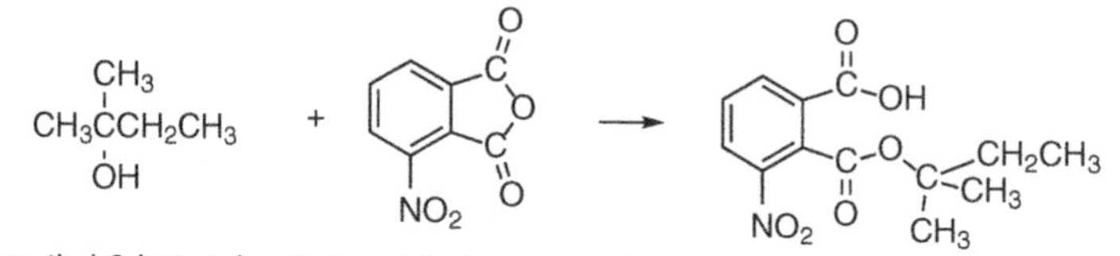

| 2-methyl-2-butanol | 3-nitrophthalic anhydride | *tert*-pentyl 3-nitrophthalate |

12. a. A semicarbazone is formed from the reaction of 2-methylbenzaldehyde with semicarbazide hydrochloride.

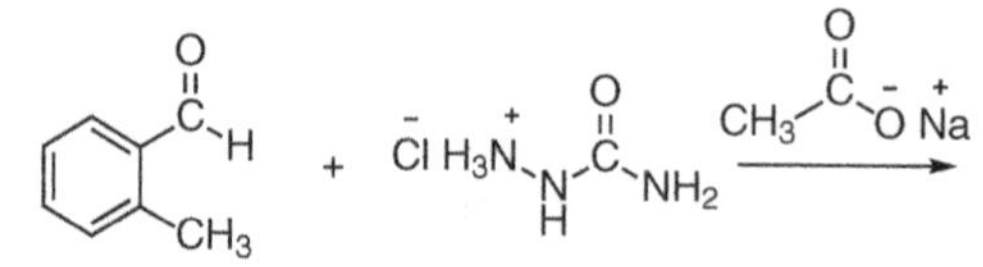

| 2-methylbenzaldehyde | semicarbazide hydrochloride |

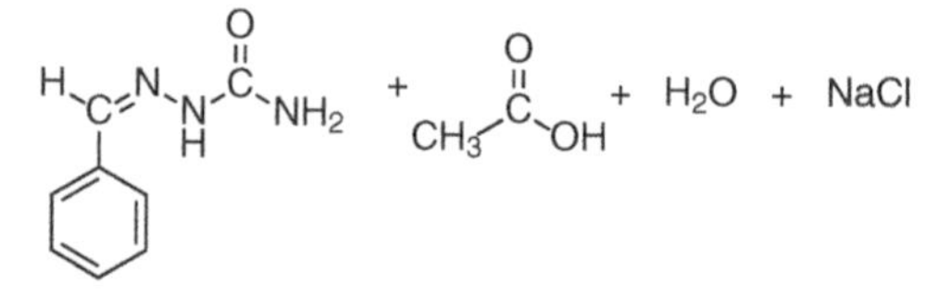

semicarbazone of
2-methylbenzaldehyde

b. The 2,4-dinitrophenylhydrazone is produced from the reaction of 2-methyl-benzaldehyde with 2,4-dinitrophenylhydrazine.

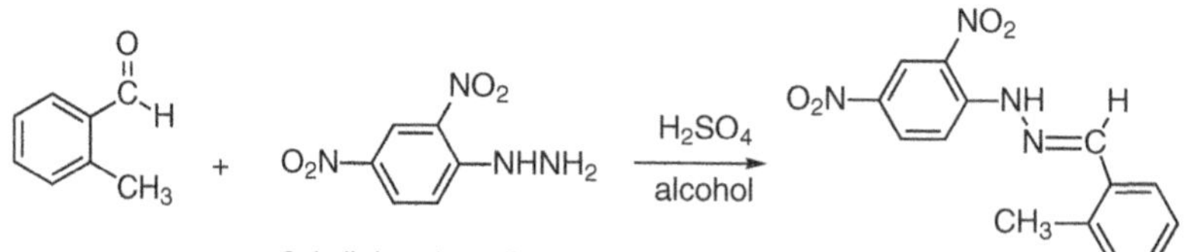

| 2-methyl-benzaldehyde | 2,4-dinitrophenylhydrazine |

2,4-dinitrophenylhydrazone
of 2-methylbenzaldehyde

c. The reaction of 2-methylbenzaldehyde with 4-nitrophenylhydrazine yields the 4-nitrophenylhydrazone.

2-methyl-
benzaldehyde   4-nitrophenylhydrazine

4-nitrophenylhydrazone
of 2-methylbenzaldehyde

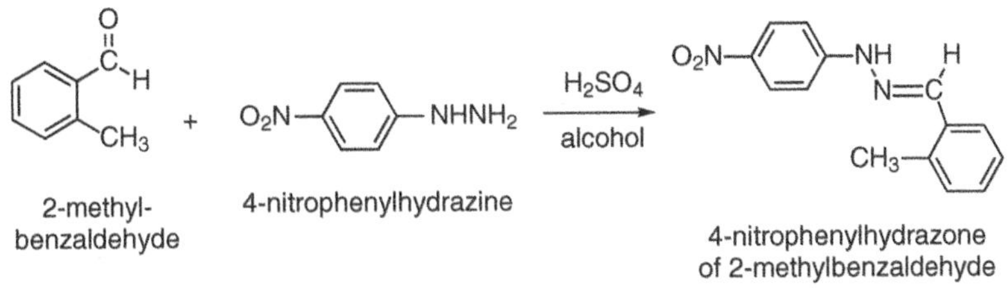

d. Phenylhydrazine reacts with 2-methylbenzaldehyde to form a phenylhydrazone.

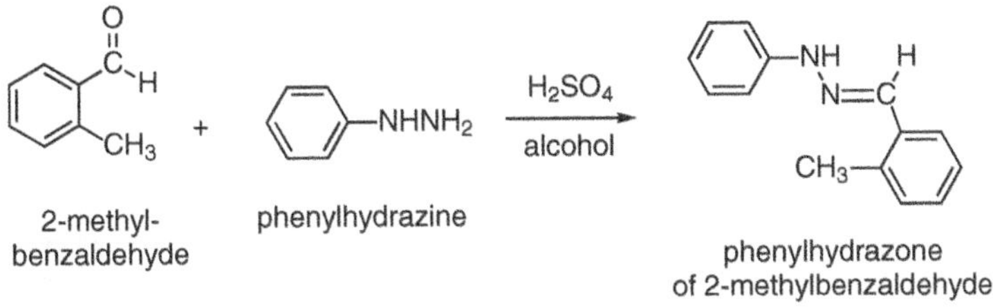

2-methyl-
benzaldehyde   phenylhydrazine

phenylhydrazone
of 2-methylbenzaldehyde

e. Reaction of the 2-methylbenzaldehyde with hydroxylamine hydrochloride yields the oxime.

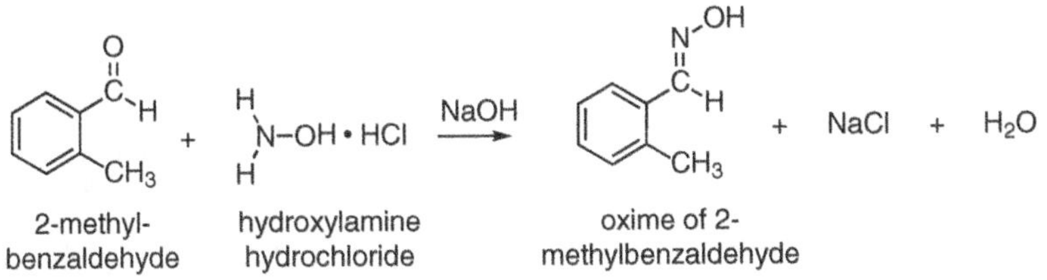

2-methyl-   hydroxylamine   oxime of 2-
benzaldehyde   hydrochloride   methylbenzaldehyde

f. 2-Methylbenzaldehyde reacts with dimedon to give the methone derivative.

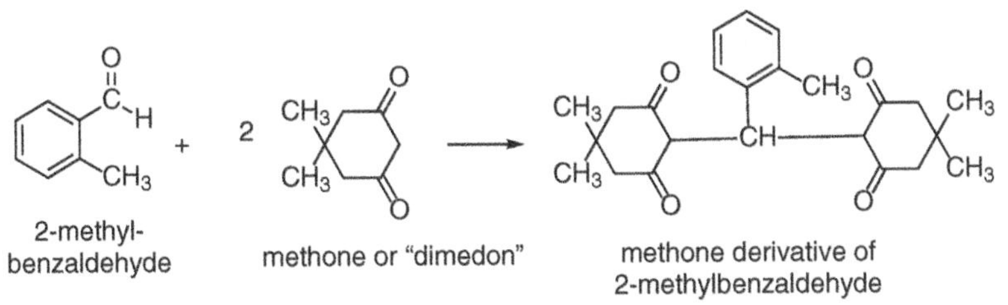

2-methyl-   methone or "dimedon"   methone derivative of
benzaldehyde   2-methylbenzaldehyde

13. a. A semicarbazone is formed from the reaction of 3-pentanone with semicarbazide hydrochloride.

3-pentanone   semicarbazide
                     hydrochloride

semicarbazone of
3-pentanone

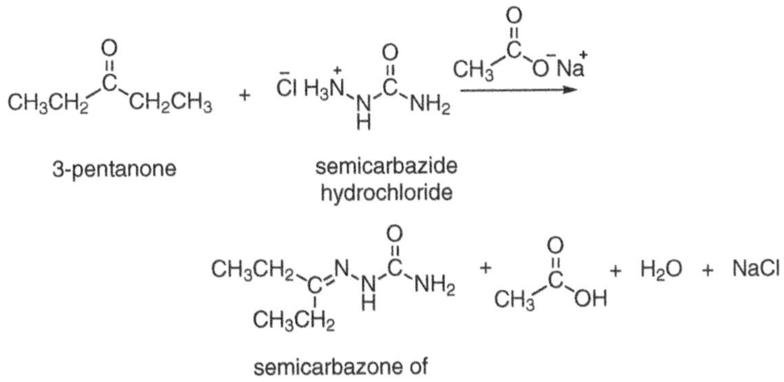

b. The 2,4-dinitrophenylhydrazone is produced from the reaction of 3-pentanone with 2,4-dinitrophenylhydrazine.

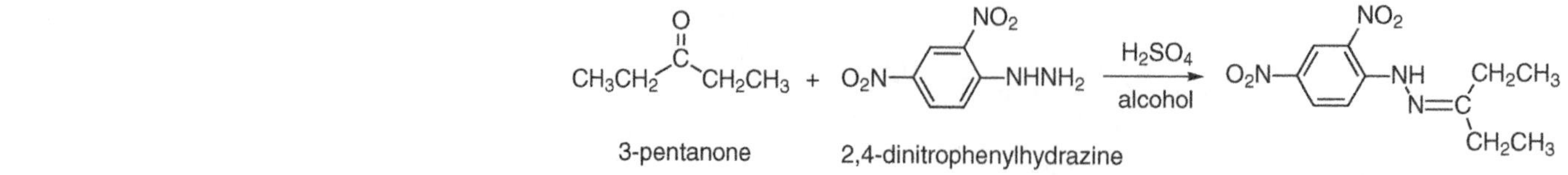

3-pentanone   2,4-dinitrophenylhydrazine

2,4-dinitrophenylhydrazone
of 3-pentanone

c. The reaction of 3-pentanone with 4-nitrophenylhydrazine yields the 4-nitro-phenylhydrazone.

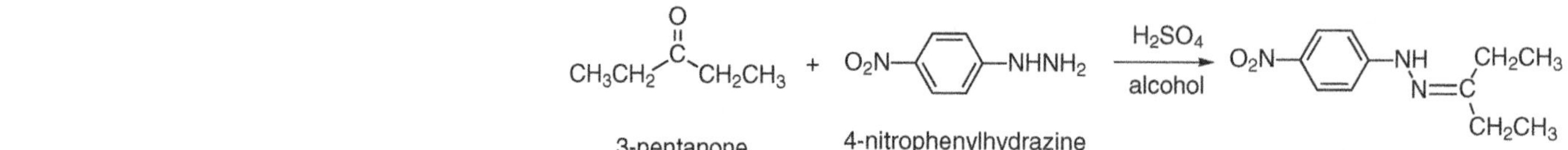

3-pentanone   4-nitrophenylhydrazine

4-nitrophenylhydrazone
of 3-pentanone

d. Phenylhydrazine reacts with 3-pentanone to form a phenylhydrazone.

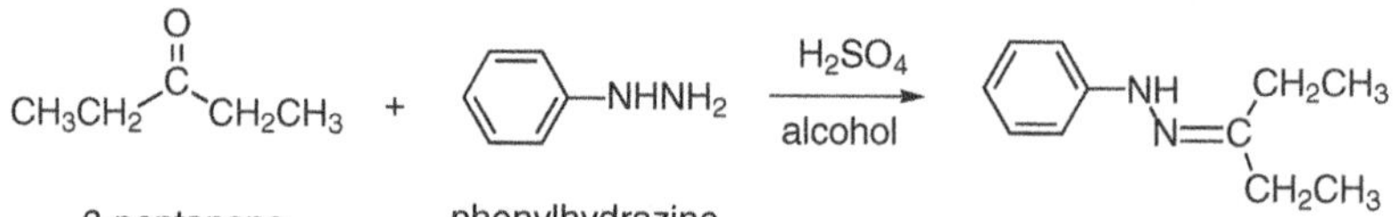

3-pentanone   phenylhydrazine

phenylhydrazone
of 3-pentanone

e. Reaction of the 3-pentanone with hydroxylamine hydrochloride yields the oxime.

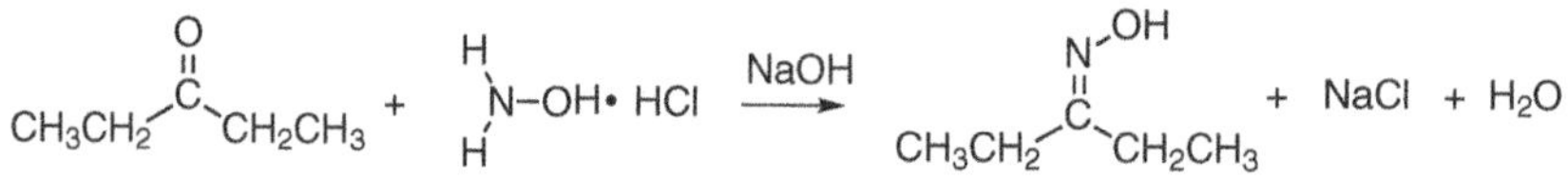

3-pentanone   hydroxylamine
hydrochloride   oxime of 3-pentanone

14. Xanthydrol reacts with hexanamide to form a 9-acylamidoxanthene.

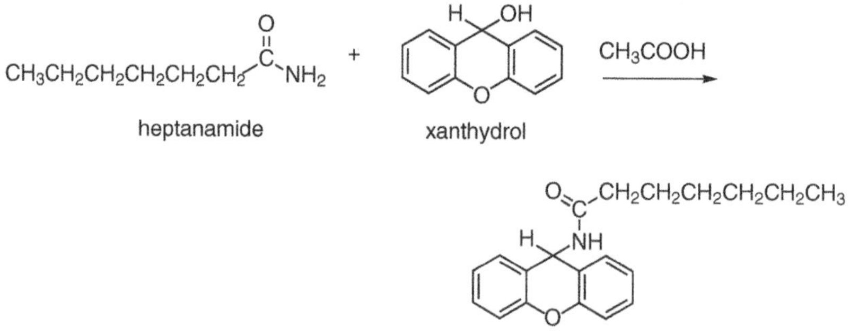

heptanamide   xanthydrol

9-hexanoylamidoxanthene

15. The basic hydrolysis of *N*-methyl-*N*-phenylethanamide gives sodium acetate and *N*-methylaniline.

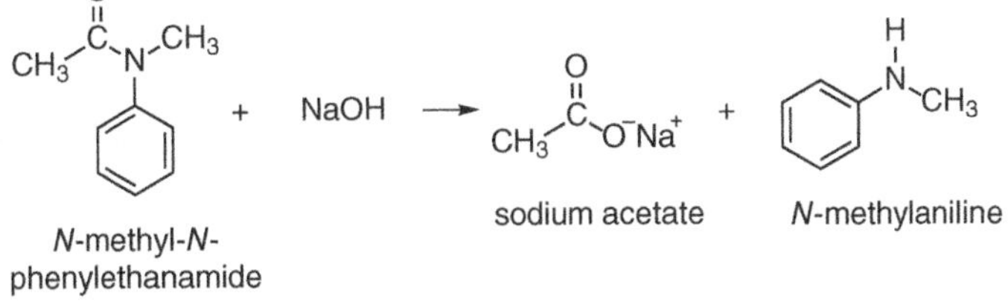

N-methyl-*N*-phenylethanamide + NaOH → sodium acetate + *N*-methylaniline

Sodium acetate reacts with 4-nitrobenzyl chloride to produce 4-nitrobenzyl acetate and with 4-bromophenacyl bromide to form 4-bromophenacyl acetate.

sodium acetate + 4-nitrobenzyl chloride →

4-nitrobenzyl acetate + NaCl

sodium acetate + 4-bromophenacyl bromide →

4-bromophenacyl acetate + NaBr

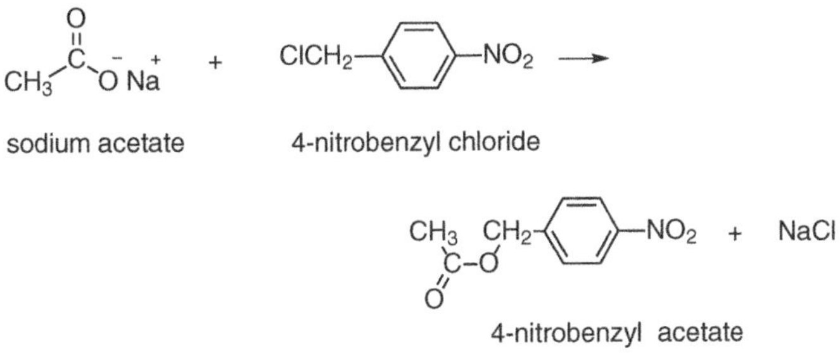

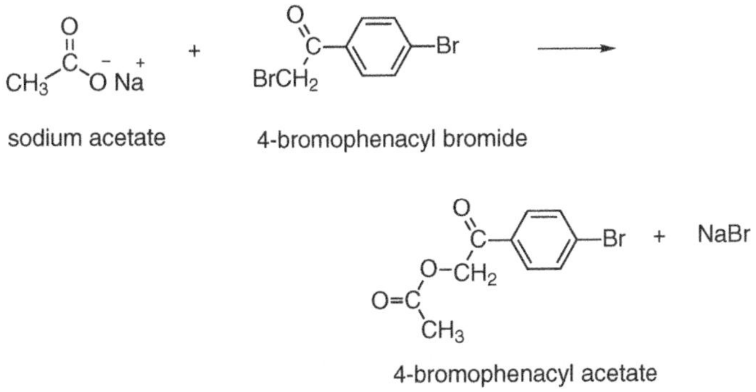

N-Methylaniline reacts with acetic anhydride to form *N*-methyl-*N*-phenylacetamide and with benzoyl chloride to form *N*-methyl-*N*-phenylbenzamide.

N-methylaniline + acetic anhydride → *N*-methyl-*N*-phenylacetamide + acetic acid

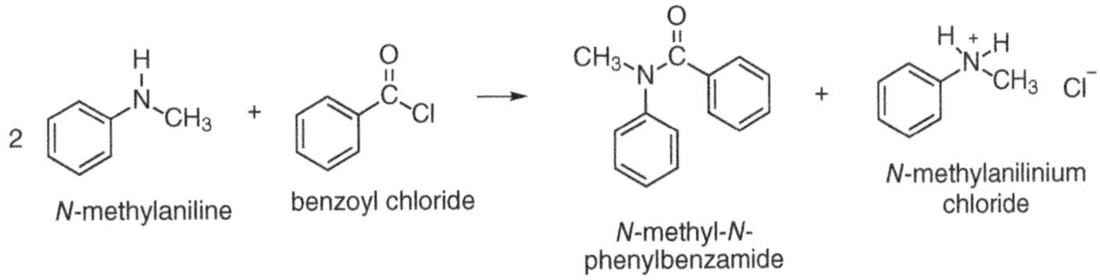

2 *N*-methylaniline + benzoyl chloride → *N*-methyl-*N*-phenylbenzamide + *N*-methylanilinium chloride

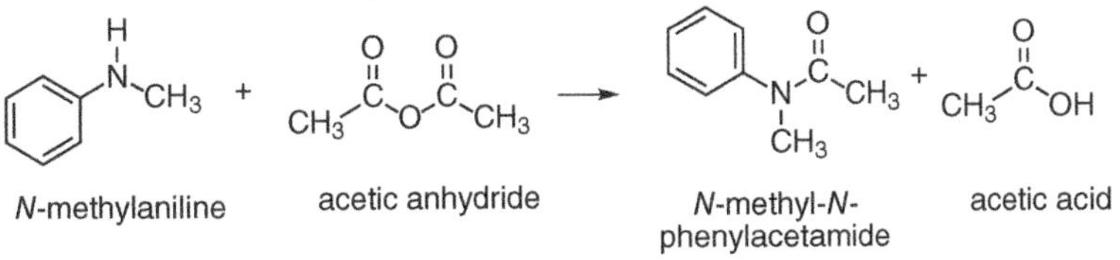

16. *N*-Ethylbutanamide is acidified to produce diethylammonium sulfate, which is then treated with basic to form ethylamine. Butanoic acid is also a product.

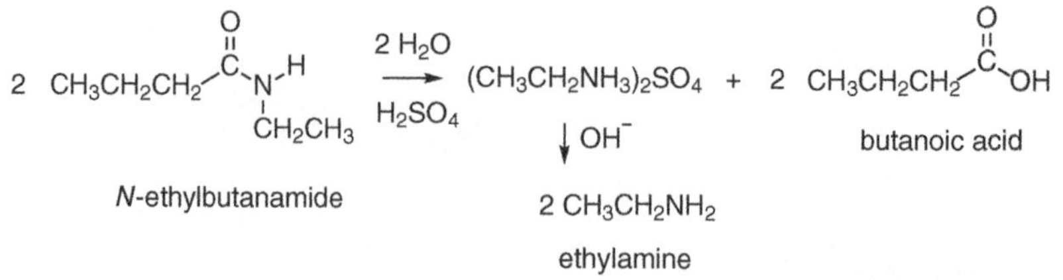

$$2 \ CH_3CH_2CH_2 \overset{O}{\underset{CH_2CH_3}{\overset{\|}{C}}} N\overset{H}{} \xrightarrow[H_2SO_4]{2 \ H_2O} (CH_3CH_2NH_3)_2SO_4 \ + \ 2 \ CH_3CH_2CH_2 \overset{O}{\overset{\|}{C}} OH$$

N-ethylbutanamide   butanoic acid

↓ OH⁻

2 CH₃CH₂NH₂

ethylamine

17. *N*-Methylbenzamide undergoes base hydrolysis to yield sodium benzoate and methylamine. The sodium benzoate with treated with acid to form benzoic acid.

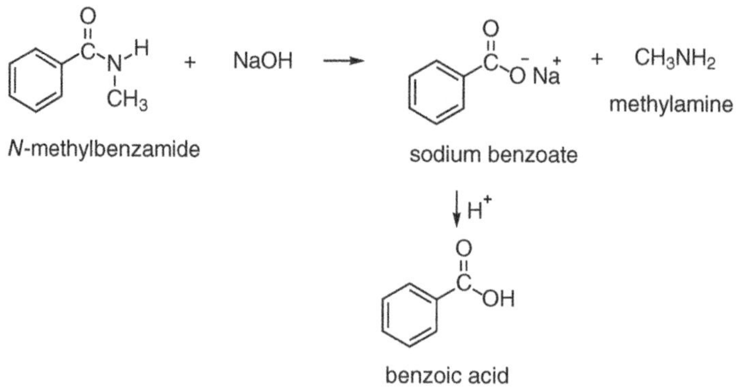

N-methylbenzamide   +   NaOH ⟶   sodium benzoate   +   CH₃NH₂

methylamine

↓ H⁺

benzoic acid

18. a. Cyclohexylamine reacts with acetic anhydride to yield *N*-cyclohexylacetamide.

cyclohexylamine   acetic anhydride   *N*-cyclohexylacetamide   acetic acid

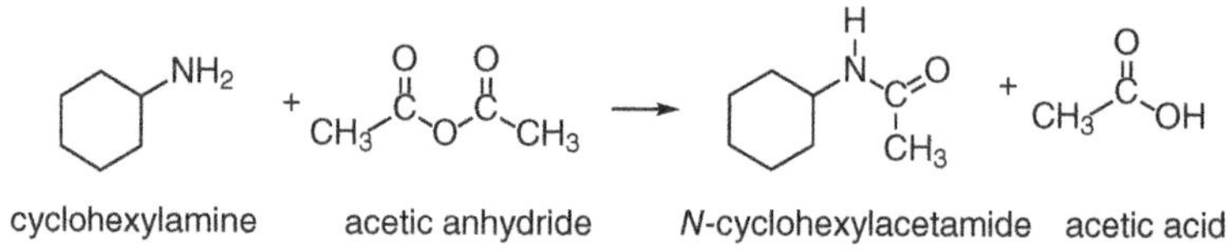

b. Treatment of cyclohexylamine with benzoyl chloride gives *N*-cyclohexylbenzamide.

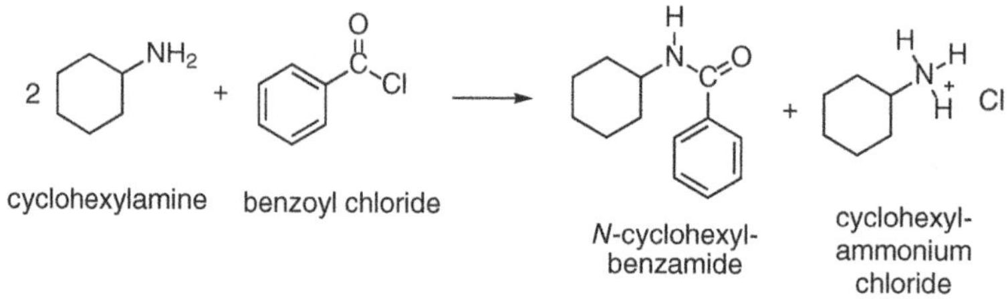

cyclohexylamine   benzoyl chloride   *N*-cyclohexyl-   cyclohexyl-
                                            benzamide   ammonium
                                                               chloride

c. The reaction of cyclohexylamine with benzenesulfonyl chloride yields
N-cyclohexylbenzenesulfonamide.

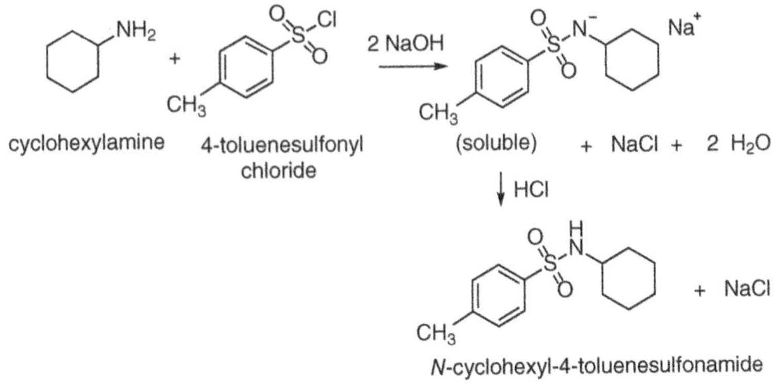

cyclohexylamine · · · 4-toluenesulfonyl chloride · · · (soluble) + NaCl + 2 H$_2$O

↓ HCl

N-cyclohexyl-4-toluenesulfonamide · · · + NaCl

d. 4-Toluenesulfonyl chloride reacts with cyclohexylamine to give *N*-cyclohexyl-4-toluenesulfonamide.

cyclohexylamine · · · 4-toluenesulfonyl chloride · · · (soluble) + NaCl + 2 H$_2$O

↓ HCl

N-cyclohexyl-4-toluenesulfonamide · · · + NaCl

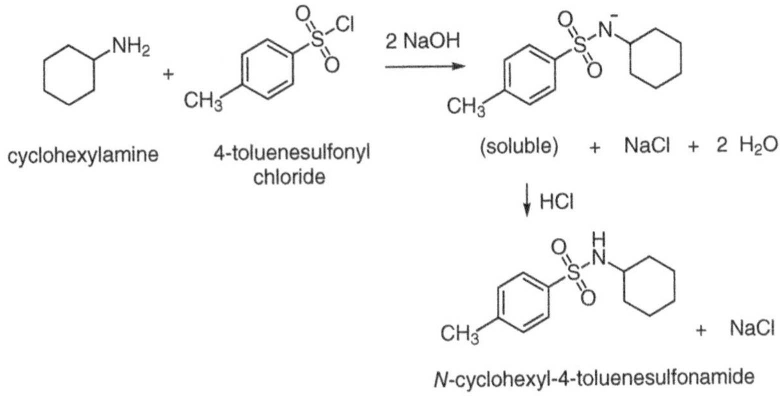

e. Cyclohexylamine reacts with phenyl isocyanate to form 1-cyclohexyl-3-phenylthiourea.

cyclohexylamine · · · phenyl isocyanate · · · 1-cyclohexyl-3-phenylthiourea

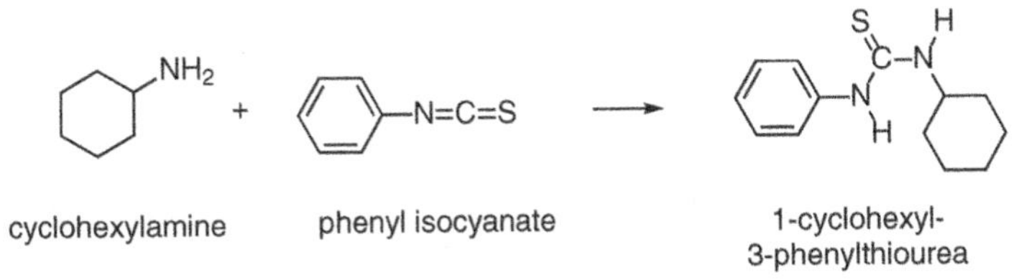

f. Cyclohexylamine undergoes acidification with hydrogen chloride and forms cyclohexylammonium hydrochloride.

cyclohexylamine · · · cyclohexylammonium hydrochloride

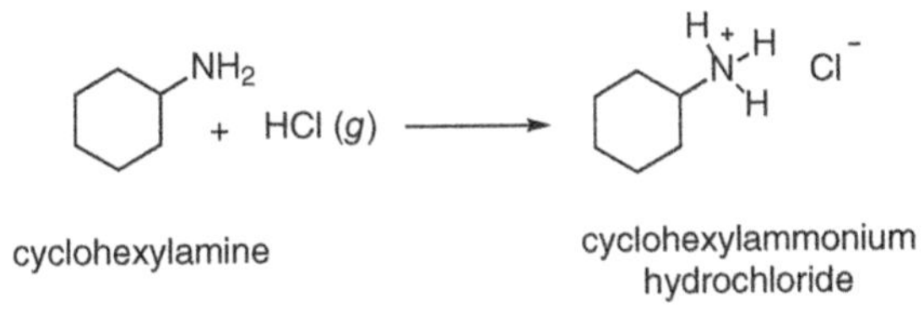

19. a. Diethylamine reacts with acetic anhydride to yield *N,N*-diethylacetamide.

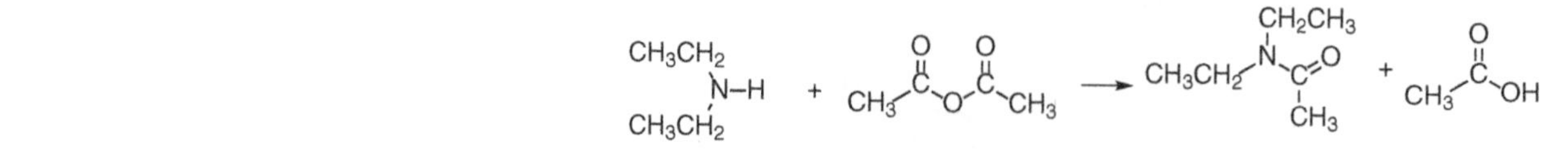

diethylamine · acetic anhydride · *N,N*-diethylacetamide · acetic acid

b. Treatment of diethylamine with benzoyl chloride gives *N,N*-diethylbenzamide.

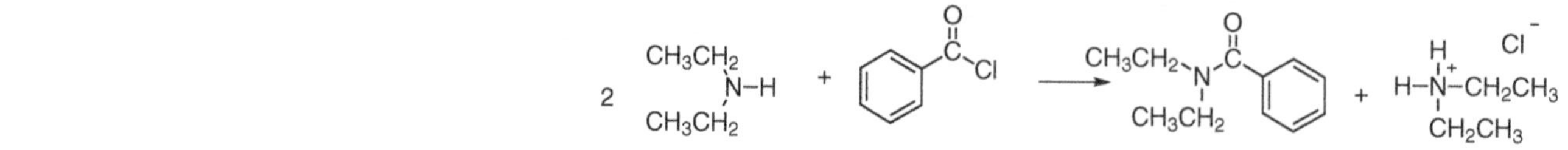

diethylamine · benzoyl chloride · *N,N*-diethyl-benzamide · diethyl ammonium chloride

c. The reaction of diethylamine with benzenesulfonyl chloride yields *N,N*-diethylbenzenesulfonamide.

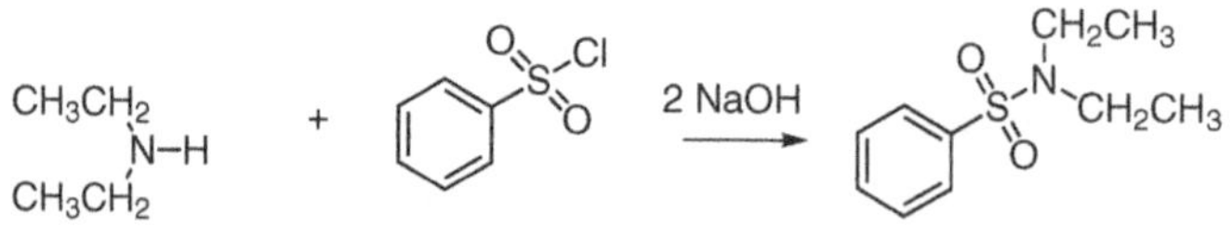

diethylamine · benzenesulfonyl chloride · *N,N*-diethylbenzene-sulfonamide

+ NaCl + H_2O

d. 4-Toluenesulfonyl chloride reacts with diethylamine to give *N,N*-diethyl-4-toluenesulfonamide.

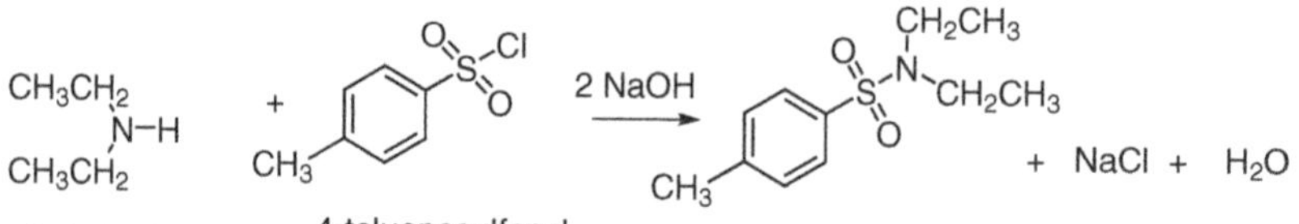

diethylamine · 4-toluenesulfonyl chloride · *N,N*-diethyl-4-toluene-benzene sulfonamide

+ NaCl + H_2O

e. Diethylamine reacts with phenyl isocyanate to form 1,1-diethyl-3-phenylthiourea.

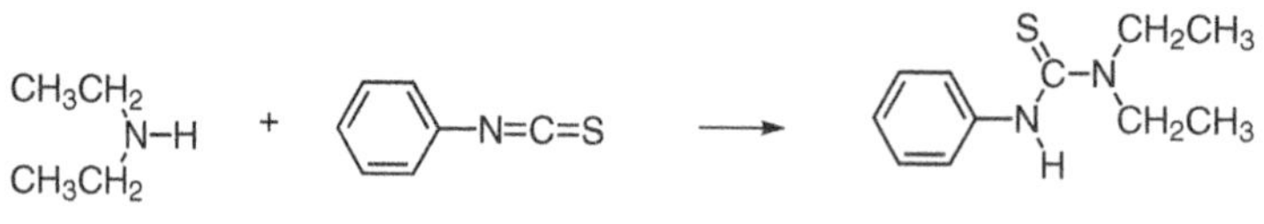

diethylamine · phenyl isocyanate · 1,1-diethyl-3-phenylthiourea

f. Diethylamine undergoes acidification with hydrogen chloride forms diethylammonium hydrochloride.

diethylamine

diethylammonium
hydrochloride

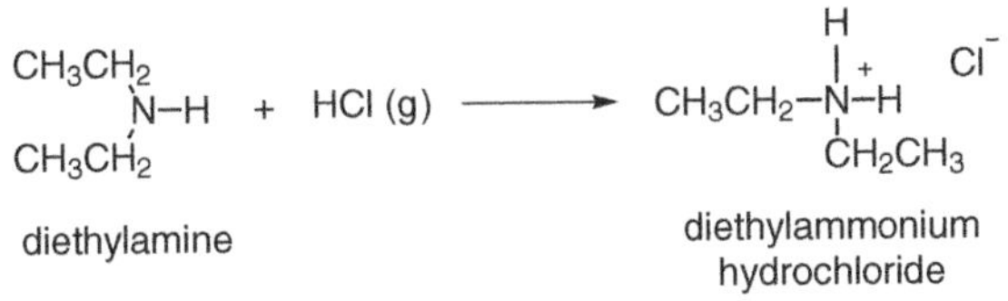

20. a. *N,N*-Dimethylaniline reacts with chloroplatinic acid to yield *N,N*-dimethylanilium
platinate.

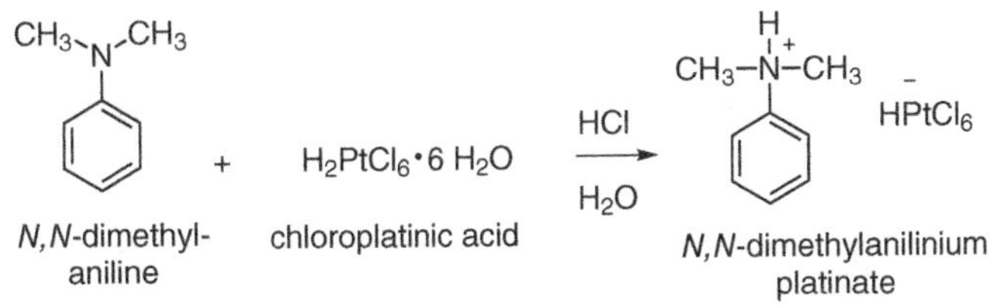

N,N-dimethyl-
aniline

chloroplatinic acid

N,N-dimethylanilinium
platinate

b. The reaction of *N,N*-dimethyl aniline with methyl 4-toluenesulfonate gives the
N,N,N-trimethylanilinium 4-toluenesulfonate.

N,N-dimethyl-
aniline

methyl 4-toluenesulfonate

N,N,N-trimethylanilinium
4-toluenesulfonate

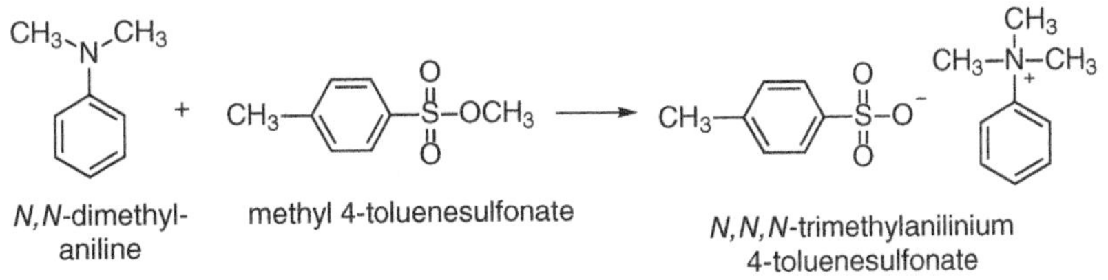

c. Treatment of *N,N*-dimethylaniline with methyl iodide produces the *N,N,N*-
trimethylanilum iodide.

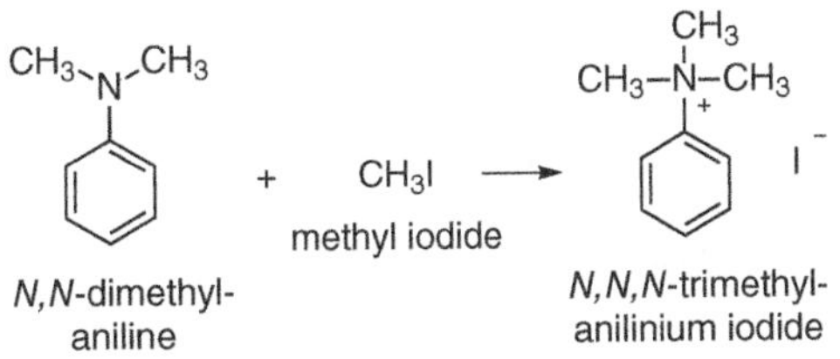

N,N-dimethyl-
aniline

methyl iodide

N,N,N-trimethyl-
anilinium iodide

d. The reaction of *N,N*-dimethylaniline with hydrochloric acid gives the *N,N*-
dimethylanilinium choride.

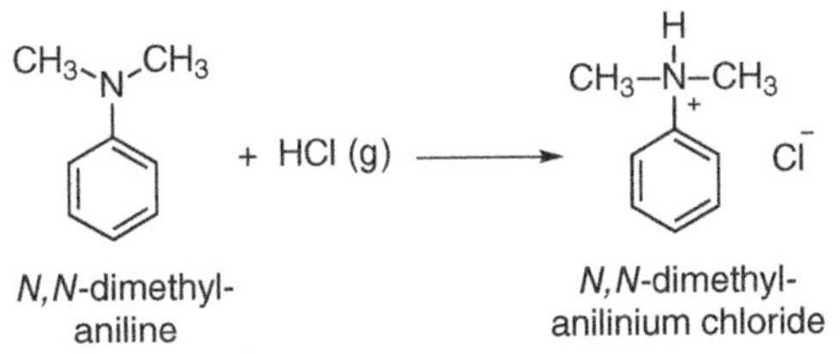

N,N-dimethyl-
aniline

N,N-dimethyl-
anilinium chloride

21. a. In the Hinsberg reaction, 4-toluenesulfonyl chloride reacts with glycine to yield (4-toluenesulfonamide)acetic acid.

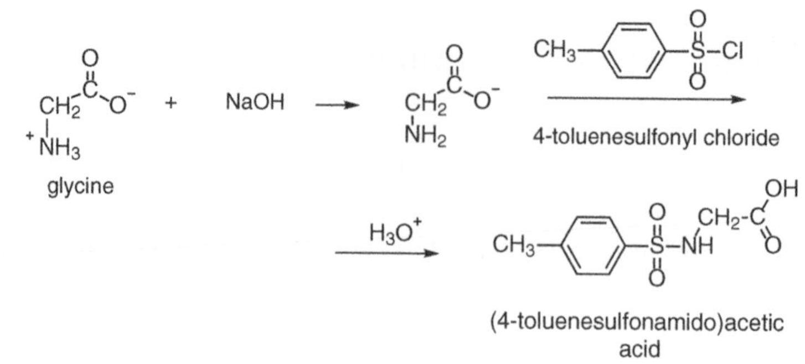

4-toluenesulfonyl chloride

(4-toluenesulfonamido)acetic acid

b. Glycine reacts with phenyl isocyanate to yield (3-phenylureido)acetic acid.

glycine   phenyl isocyanate   (3-phenylureido)acetic acid

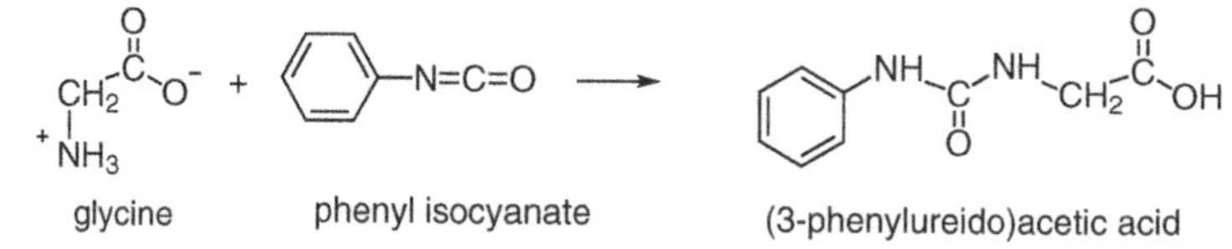

c. Glycine is acetylated with acetic anhydride to produce acetylaminoacetic acid.

glycine   acetic anhydride   acetylaminoacetic acid   acetic acid

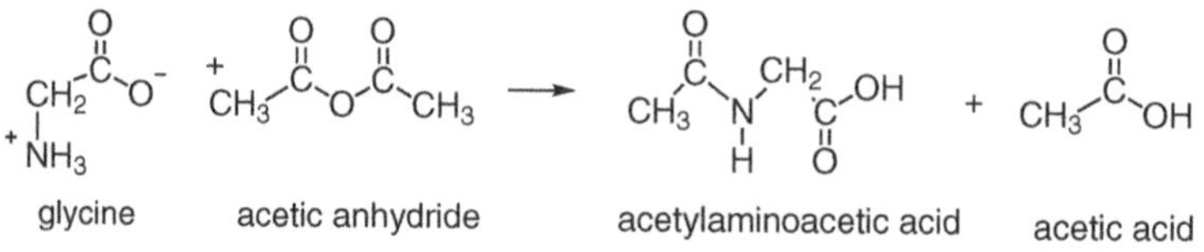

d. Glycine reacts with benzoyl chloride to form benzoylaminoacetic acid.

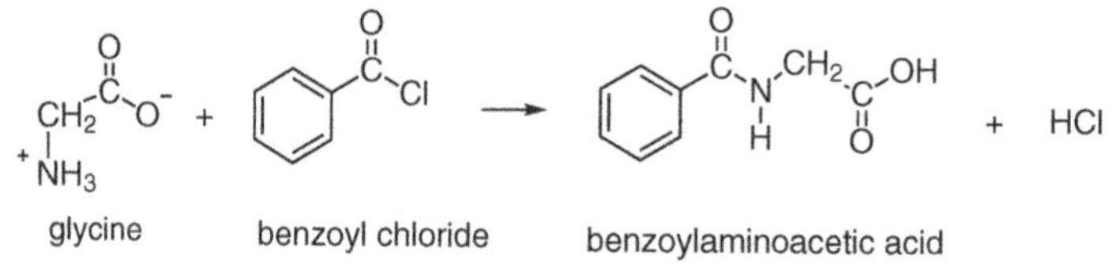

glycine   benzoyl chloride   benzoylaminoacetic acid

e. (3,5-Dinitrobenzamido)acetic acid is produced from the reaction of glycine with 3,5-dinitrobenzoyl chloride.

glycine   3,5-dinitrobenzoyl chloride   (3,5-dinitrobenzamido)acetic acid

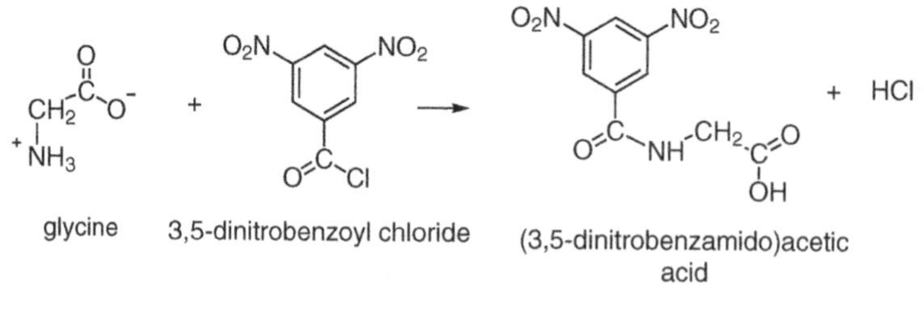

f. The reaction of Sanger's reagent (2,4-dinitrofluorobenzene) with glycine yields the (2,4-dinitrophenylamino)acetic acid.

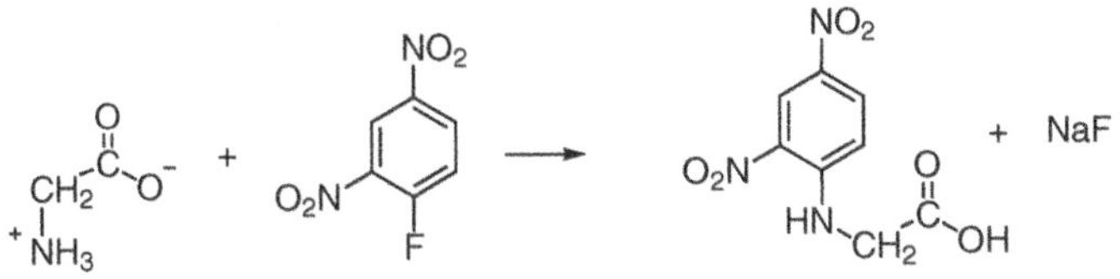

glycine   2,4-dinitrofluorobenzene   (2,4-dinitrophenylamino)-
                   (Sanger's reagent)   acetic acid

22. a. In the Hinsberg reaction, 4-toluenesulfonyl chloride reacts with alanine to yield (*N*-methyl-4-toluenesulfonamide)acetic acid.

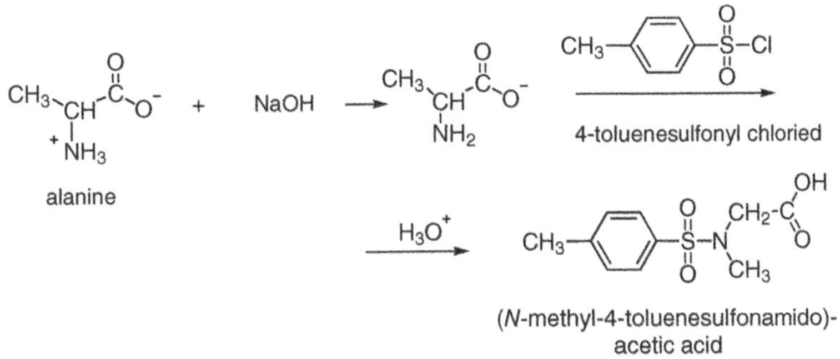

alanine

4-toluenesulfonyl chloried

(*N*-methyl-4-toluenesulfonamido)-
acetic acid

b. Alanine reacts with phenyl isocyanate to yield (1-methyl-3-phenylureido)acetic acid.

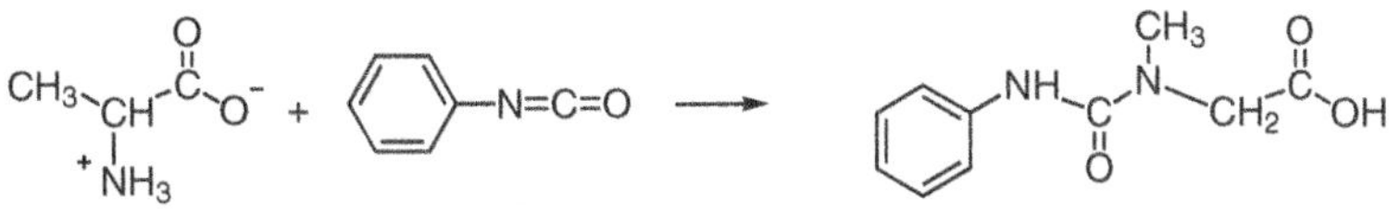

alanine   phenyl isocyanate   (1-methyl-3-phenylureido)-
                                          acetic acid

c. Alanine is acetylated with acetic anhydride to produce acetyl-*N*-methylamino-acetic acid.

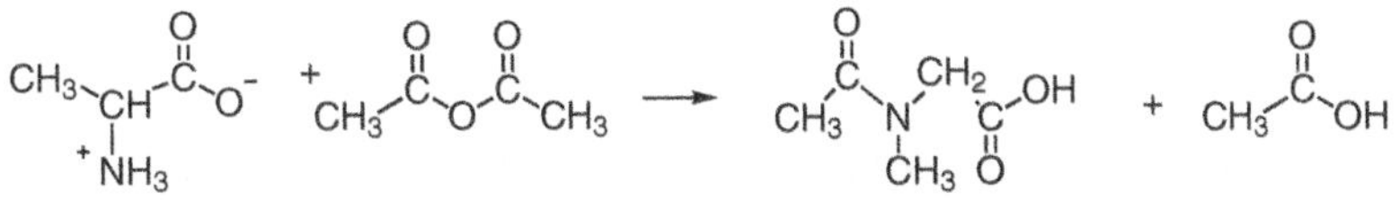

alanine   acetic anhydride   acetyl-*N*-methyl-   acetic acid
                                  aminoacetic acid

d. Alanine reacts with benzoyl chloride to form benzoyl-*N*-methylaminoacetic acid.

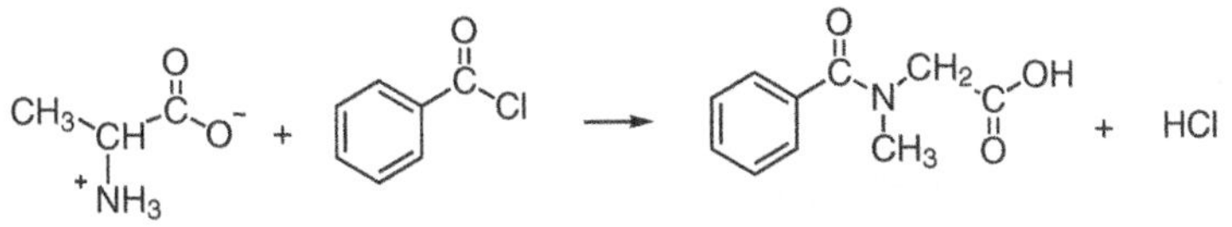

alanine   benzoyl chloride   benzoyl-*N*-methyl-
                                 aminoacetic acid

e. (*N*-Methyl-3,5-dinitrobenzamido)acetic acid is produced from the reaction of alanine with 3,5-dinitrobenzoyl chloride.

alanine   3,5-dinitrobenzoyl chloride   (*N*-methyl-3,5-dinitrobenzamido)-acetic acid

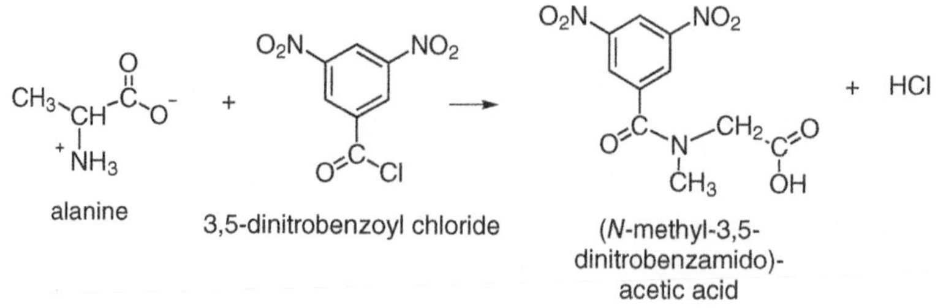

f. The reaction of Sanger's reagent (2,4-dinitrofluorobenzene) with alanine yields the (*N*-methyl-2,4-dinitrophenylamino)acetic acid.

alanine   2,4-dinitrofluorobenzene (Sanger's reagent)   (*N*-methyl-2,4-dinitrophenylamino)-acetic acid

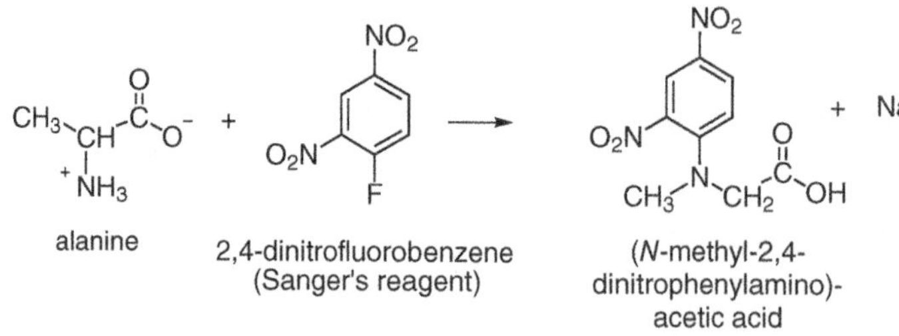

23. a. The osazone is produced from the reaction of D-galactose with excess phenylhydrazine.

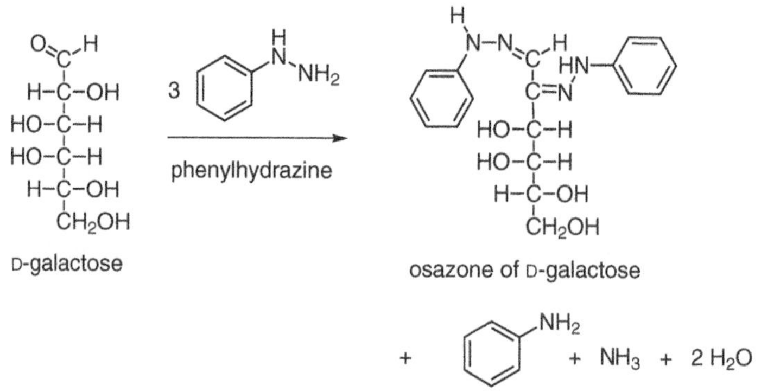

D-galactose   phenylhydrazine   osazone of D-galactose

b. D-Galactose reacts with 4-nitrophenylhydrazine to form the 4-nitrophenylhydrazone.

D-galactose   4-nitrophenylhydrazine   4-nitrophenylhydrazone of D-galactose

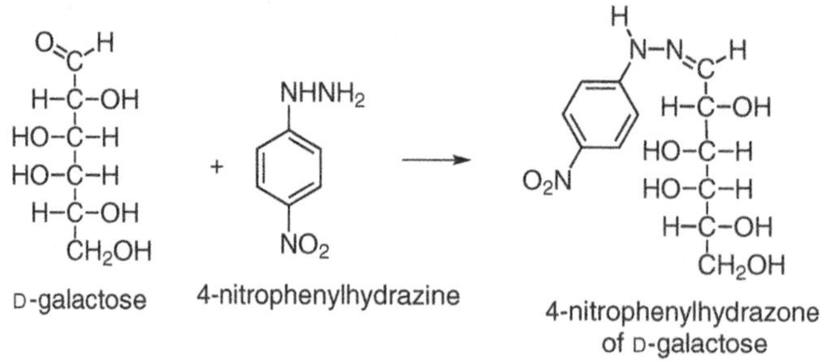

c. The 4-bromophenylhydrazone is formed from the reaction of D-galactose with 4-bromophenylhydrazine.

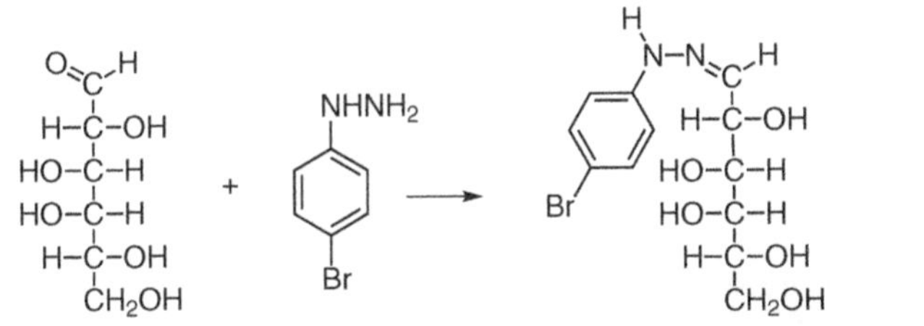

D-galactose   4-bromopheny-   4-bromophenylhydrazone
                hydrazine   of D-galactose

d. All —OHs are acetylated when D-galactose reacts with acetic anhydride.

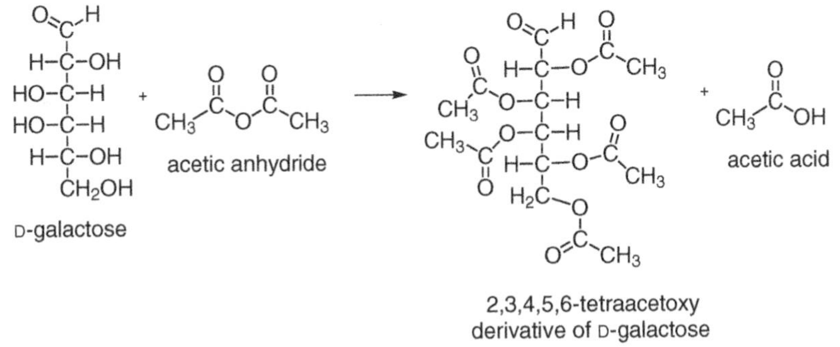

D-galactose   acetic anhydride   acetic acid

2,3,4,5,6-tetraacetoxy
derivative of D-galactose

24. a. The osazone is produced from the reaction of D-fructose with excess phenylhydrazine.

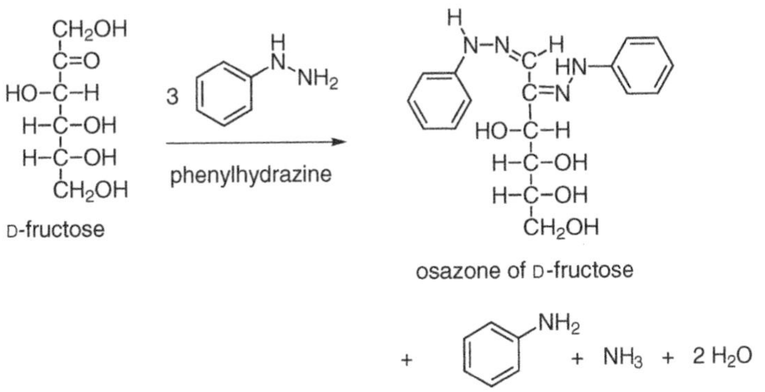

D-fructose   phenylhydrazine   osazone of D-fructose

$+$ $+$ NH_3 + 2 H_2O

b. D-Fructose reacts with 4-nitrophenylhydrazine to form the 4-nitrophenylhydrazone.

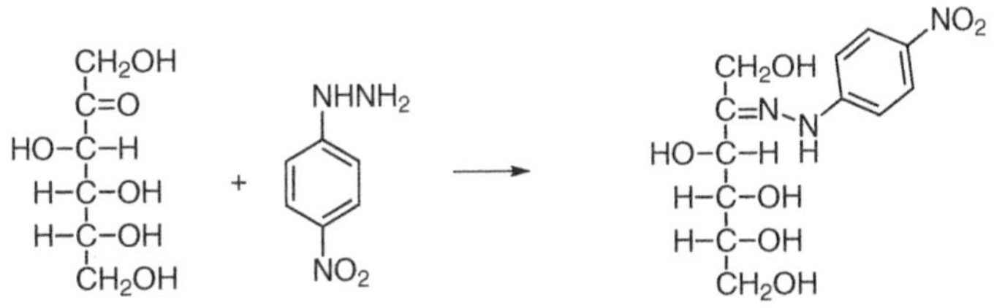

D-fructose   4-nitrophenylhydrazine   4-nitrophenylhydrazone
                                            of D-fructose

c. The 4-bromophenylhydrazone is formed from the reaction of D-fructose with 4-bromophenylhydrazine.

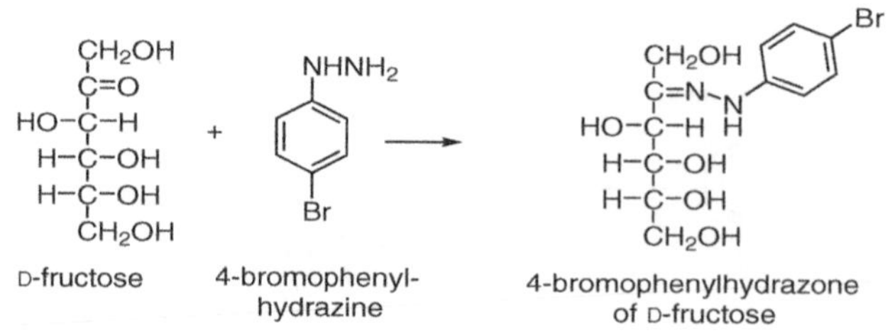

d. All —OHs are acetylated when D-fructose reacts with acetic anhydride.

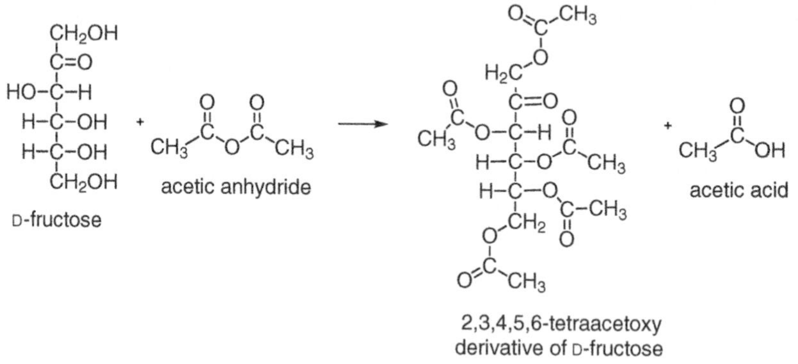

2,3,4,5,6-tetraacetoxy derivative of D-fructose

25. Simple esters, with bp of below 110 °C, would be hydrolyzed easily by method **a**. Esters with bp of 110°–200 °C would require a reflux time of 1–2 hours.

26. **a.** 4-Phenylphenacyl acetate is hydrolyzed to acetic acid and α-hydroxy-4-phenyl-acetophenone.

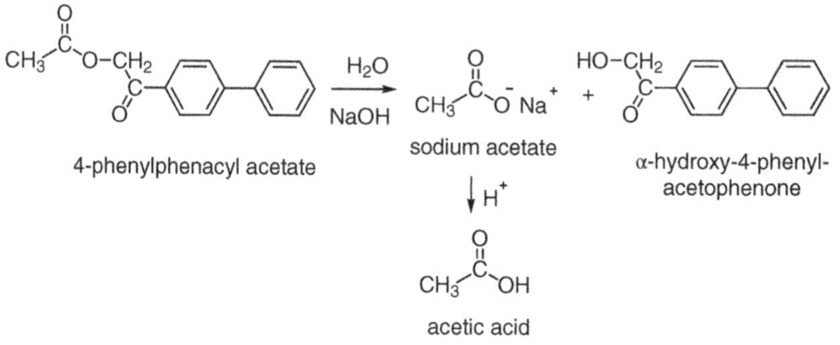

4-phenylphenacyl acetate

acetic acid

b. Ethylene glycol dibenzoate is hydrolyzed to benzoic acid and ethylene glycol.

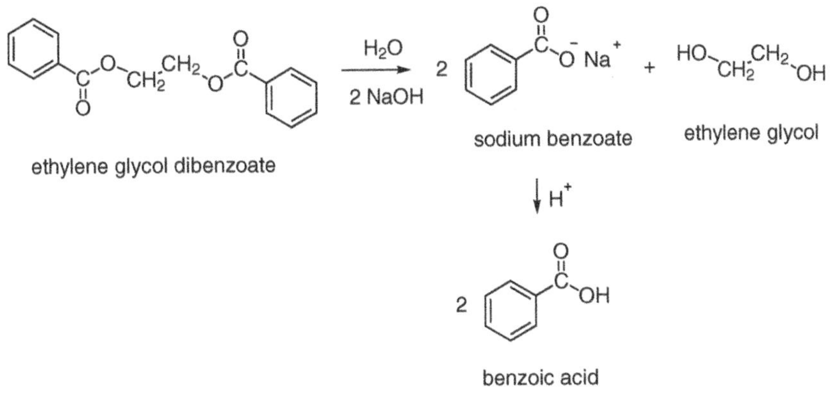

ethylene glycol dibenzoate

sodium benzoate   ethylene glycol

benzoic acid

c. Oxalic acid and 1-butanol are the products from the hydrolysis of dibutyl oxalate.

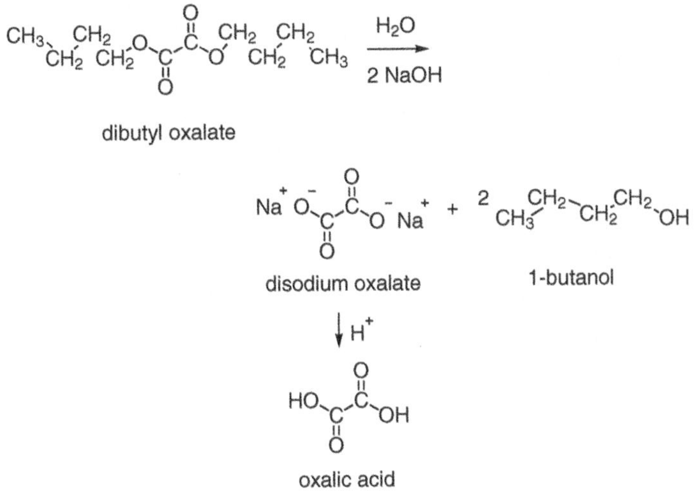

dibutyl oxalate

disodium oxalate   1-butanol

oxalic acid

d. Glycerol triacetate is hydrolyzed to acetic acid and glycerol.

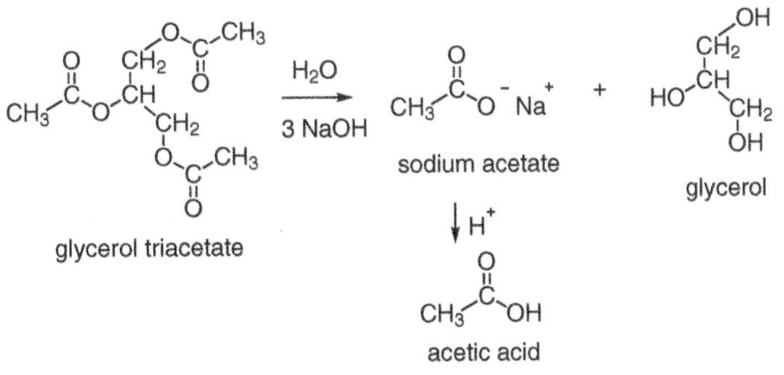

glycerol triacetate

sodium acetate   glycerol

acetic acid

e. Diethyl phthalate is hydrolyzed to phthalic acid and ethanol.

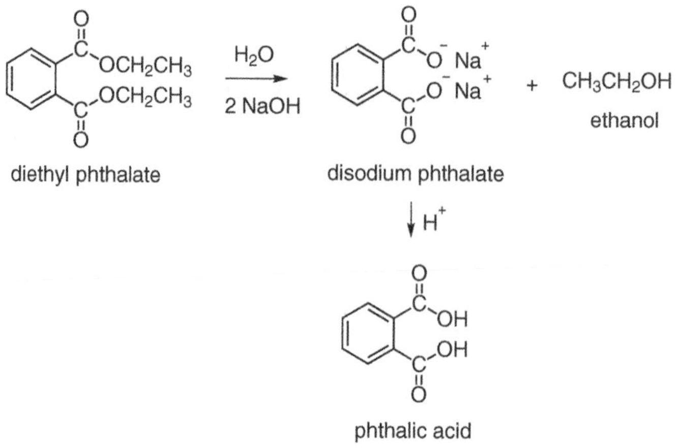

diethyl phthalate   disodium phthalate   + CH_3CH_2OH ethanol

phthalic acid

For **a**, the semicarbazone derivative (mp 146 °C) of α-hydroxy-4-phenylaceto-phenone (mp 86 °C) and the 4-toluidide derivative (mp 153 °C) of acetic acid (bp 118 °C) could be prepared.

For **b**, the phenylurethane derivative (mp 157 °C) of ethylene glycol (bp 197 °C) and the 4-toluidide derivative (mp 158 °C) of benzoic acid (mp 122 °C) could be prepared.

For **c**, the phenylurethane derivative (mp 63 °C) of 1-butanol (bp 118 °C) and the di-4-toluidide derivative (mp 268 °C) of oxalic acid [mp 101 °C (dihydrate), mp 188 °C (anhydrous)] could be prepared.

For **d**, the phenylurethane derivative (mp 180 °C) of glycerol (bp 290 °C) and the 4-toluidide derivative (mp 153 °C) of acetic acid (bp 118 °C) could be prepared.

For **e**, the phenylurethane derivative (mp 52 °C) of ethanol (bp 78 °C) and the dianilide derivative (mp 255 °C) of phthalic acid (mp 208 °C) could be prepared.

27. a. The saponification equivalent of ethyl acetoacetate is 65.

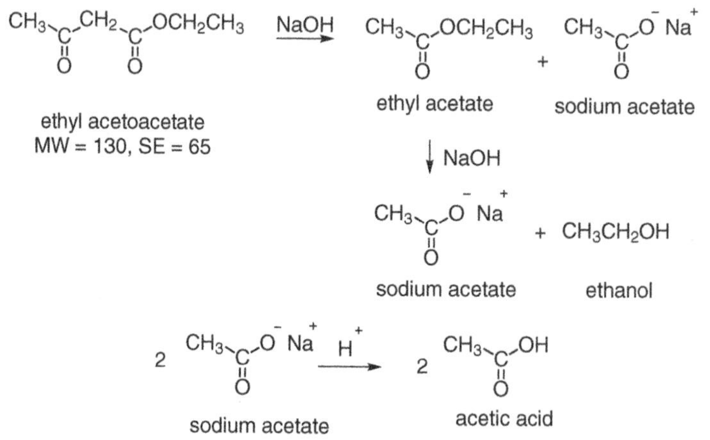

ethyl acetoacetate
MW = 130, SE = 65

ethyl acetate   sodium acetate

sodium acetate   ethanol

sodium acetate   acetic acid

b. The saponification equivalent of ethyl hydrogen phthalate is 92.

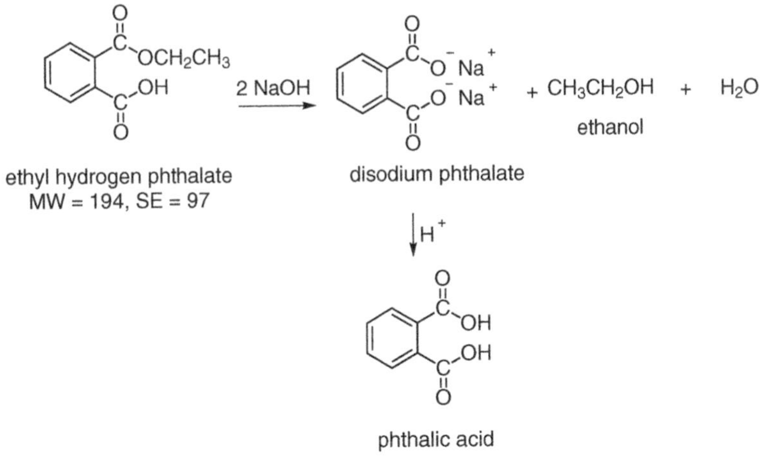

2 NaOH
+ CH_3CH_2OH + H_2O
ethanol
ethyl hydrogen phthalate
MW = 194, SE = 97
disodium phthalate
H^+
phthalic acid

c. The saponification equivalent of diethyl propanedioate is 80.

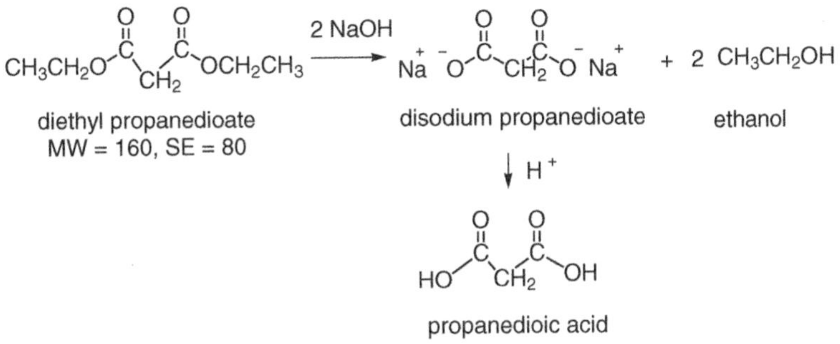

2 NaOH
+ 2 CH_3CH_2OH
diethyl propanedioate
MW = 160, SE = 80
disodium propanedioate
ethanol
H^+
propanedioic acid

d. The saponification equivalent of ethyl cyanoacetate is 56.5.

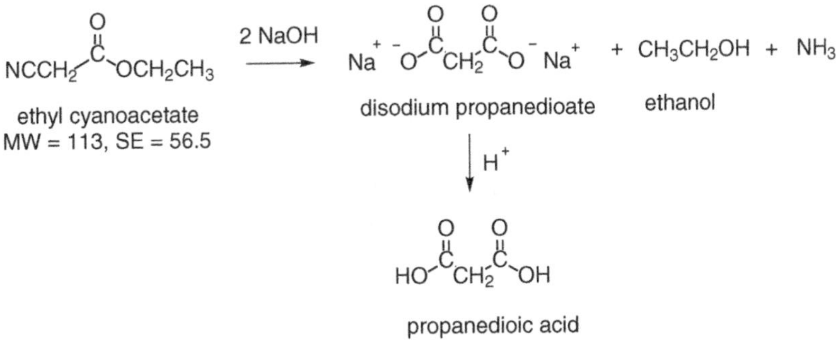

2 NaOH
+ CH_3CH_2OH + NH_3
ethyl cyanoacetate
MW = 113, SE = 56.5
disodium propanedioate
ethanol
H^+
propanedioic acid

e. The saponification equivalent of dibutyl phthalate is 139.

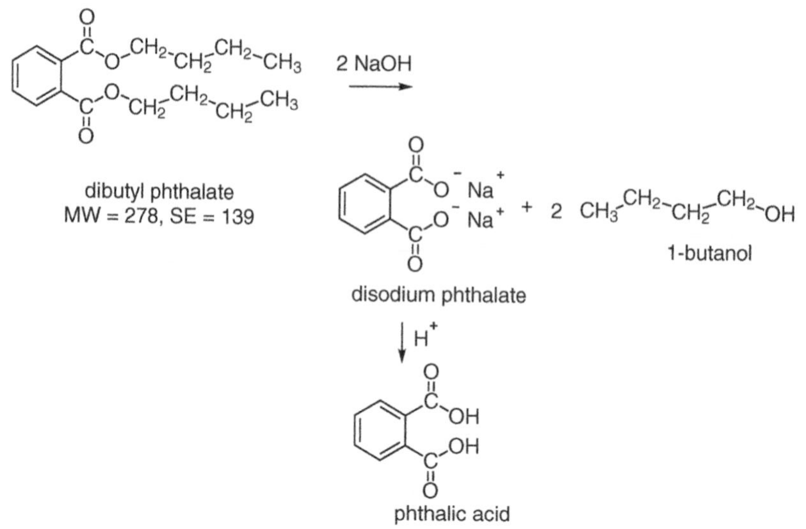

dibutyl phthalate
MW = 278, SE = 139

disodium phthalate

1-butanol

phthalic acid

28. When benzaldehyde is treated with sodium hydroxide, a Cannizzaro reaction occurs to form sodium benzoate and benzyl alcohol. Since only half of the benzaldehyde molecules become a carboxylic acid, the saponification equivalent is half of the molecular weight.

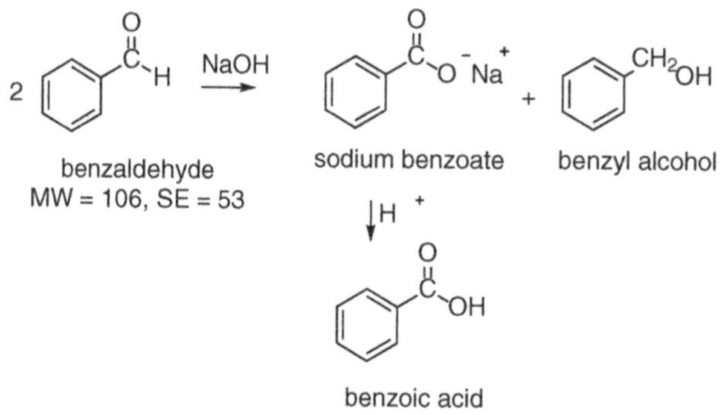

benzaldehyde
MW = 106, SE = 53

sodium benzoate   benzyl alcohol

benzoic acid

29. If an ester is partially hydrolyzed, the saponification equivalent would be unreliable since the value would be between the molecular weight of the carboxylic acid and the molecular weight of the ester.

30. The calculations are shown below.

$$\text{moles of ethyl bromide} = \frac{2.5\,g}{102\,g/mol} = 0.023 \text{ moles of ethyl bromide}$$

$$\text{moles of magnesium} = \frac{0.5\,g}{24.3\,g/mol} = 0.021 \text{ moles of magnesium}$$

The magnesium is the limiting reagent.

31. a. Propanamide can be synthesized by treating methyl propanoate with ammonia.

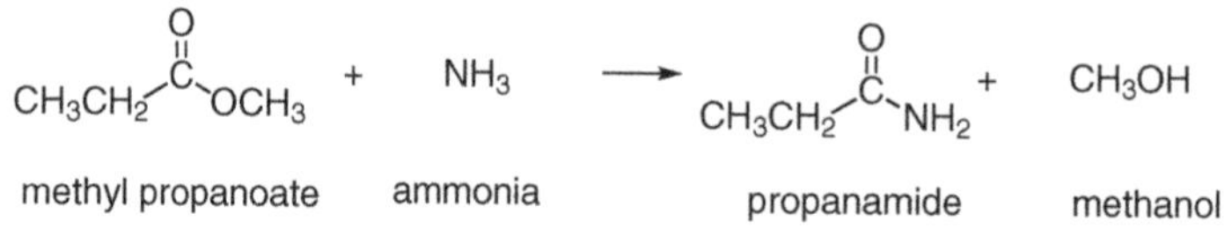

methyl propanoate   ammonia   propanamide   methanol

b. Methyl propanoate can be converted to the 4-toluidide derivative using 4-toludine.

$$CH_3CH_2Br \quad + \quad Mg \quad \longrightarrow \quad CH_3CH_2MgBr$$

ethyl bromide   magnesium   ethylmagnesium bromide

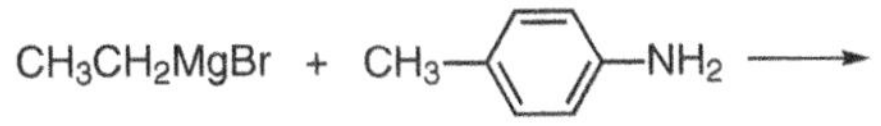

$$CH_3CH_2MgBr \quad + \quad CH_3-\!\!\!\langle\text{benzene ring}\rangle\!\!\!-NH_2 \quad \longrightarrow$$

ethylmagnesium   4-toluidine
bromide

$$CH_3-\!\!\!\langle\text{benzene ring}\rangle\!\!\!-NHMgBr \quad + \quad CH_3CH_3$$

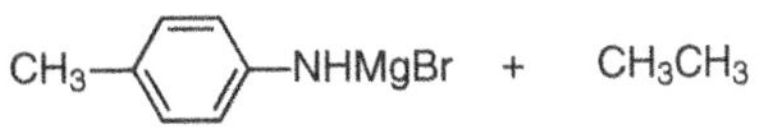

4-toluidinomagnesium bromide

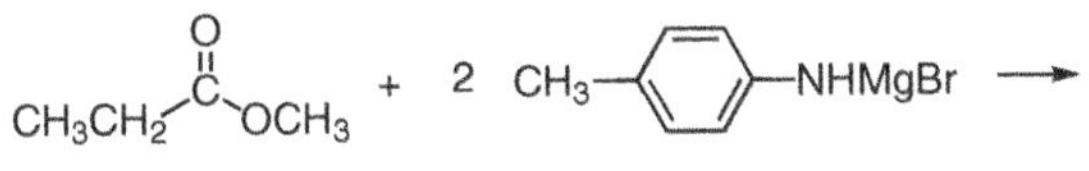

methyl propanoate   4-toludinomagnesium
bromide

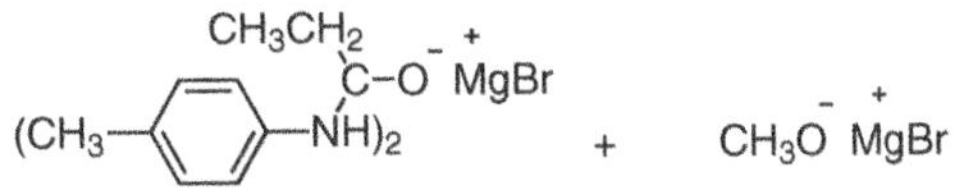

1,1-di-(4-toluidino)-1-propoxy-
magnesium bromide

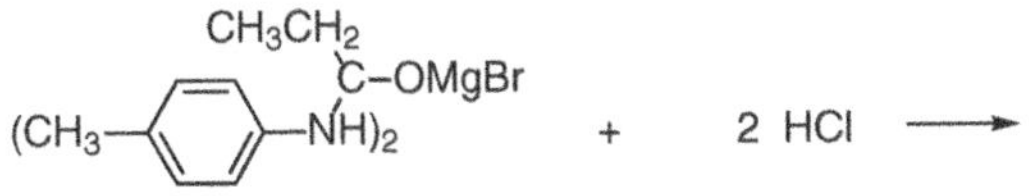

$$CH_3O^-\ MgBr^+$$

1,1-di-(4-toluidino)-1-propoxy-
magnesium bromide

+   2 HCl $\longrightarrow$

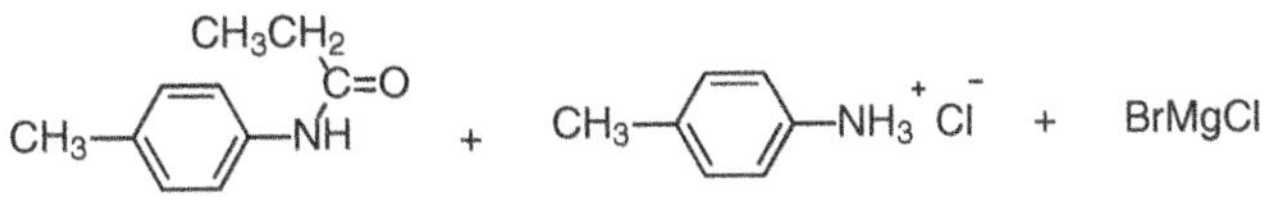

N-4-tolylpropanamide   4-toluidinium chloride   BrMgCl

c. Methyl propanoate reacts with 3,5-dinitrobenzoic acid to give the 3,5-dinitrobenzoate.

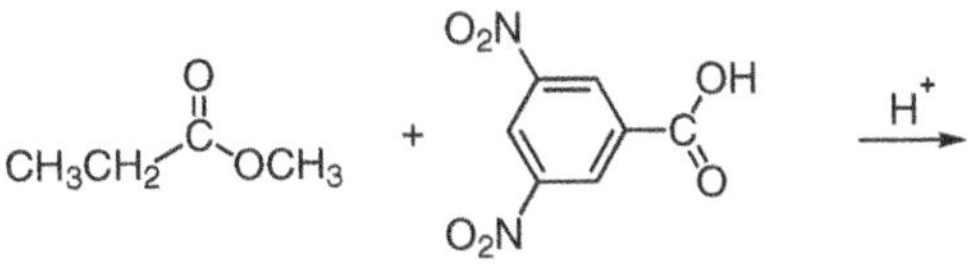

methyl propanoate   3,5-dinitrobenzoic acid

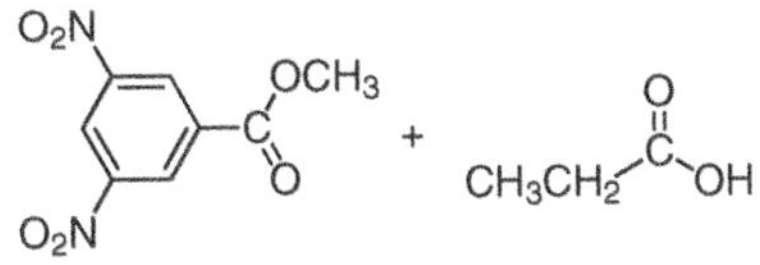

methyl 3,5-   propanoic acid
dinitrobenzamide

d. The N-benzylamide is formed from the reaction of methyl propanoate with benzylamine.

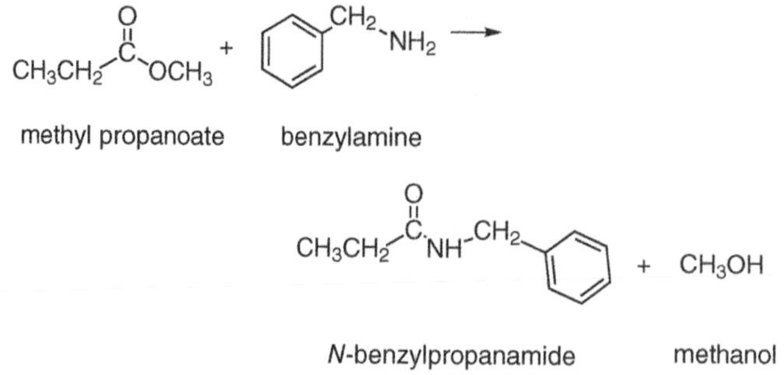

methyl propanoate   benzylamine

N-benzylpropanamide   methanol

e. Methyl propanoate reacts with hydrazine to yield the hydrazide.

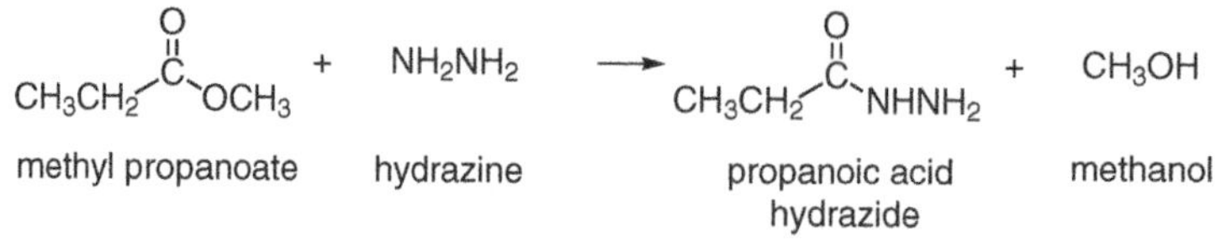

methyl propanoate   hydrazine   propanoic acid   methanol
                                   hydrazide

32. a. Benzamide can be synthesized by treating ethyl benzoate with ammonia.

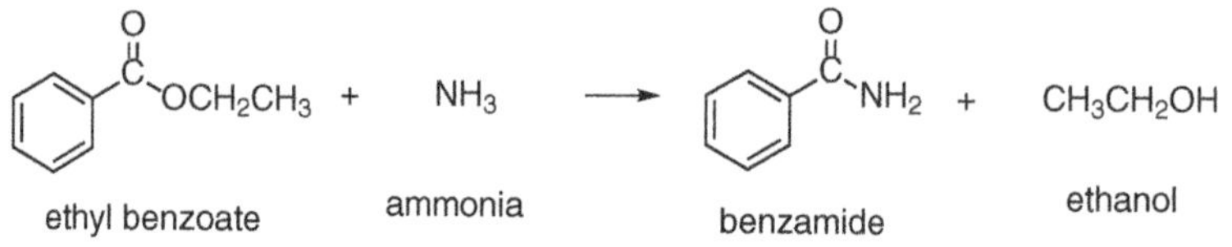

ethyl benzoate   ammonia   benzamide   ethanol

b. Ethyl benzoate can be converted to the 4-toluidide derivative using 4-toludine.

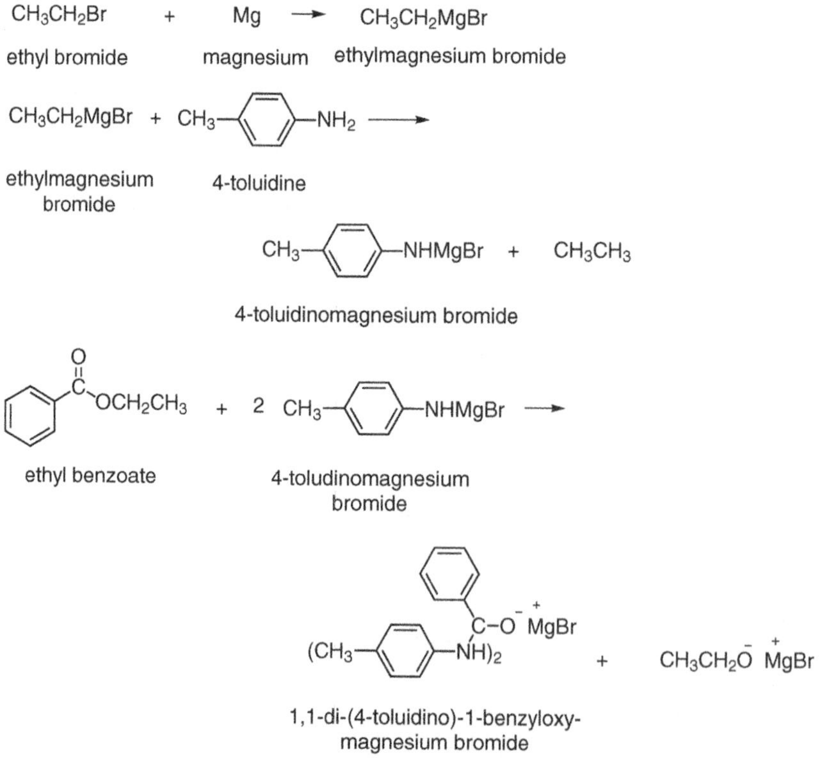

$$CH_3CH_2Br \quad + \quad Mg \longrightarrow CH_3CH_2MgBr$$

ethyl bromide   magnesium   ethylmagnesium bromide

ethylmagnesium   4-toluidine
bromide

4-toluidinomagnesium bromide

ethyl benzoate   4-toludinomagnesium
                             bromide

1,1-di-(4-toluidino)-1-benzyloxy-
magnesium bromide

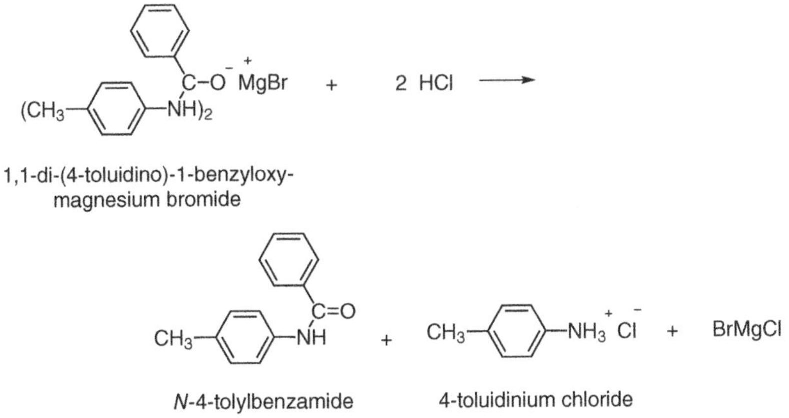

1,1-di-(4-toluidino)-1-benzyloxy-
magnesium bromide

N-4-tolylbenzamide   4-toluidinium chloride

c. Ethyl benzoate reacts with 3,5-dinitrobenzoic acid to give the 3,5-dinitrobenzoate.

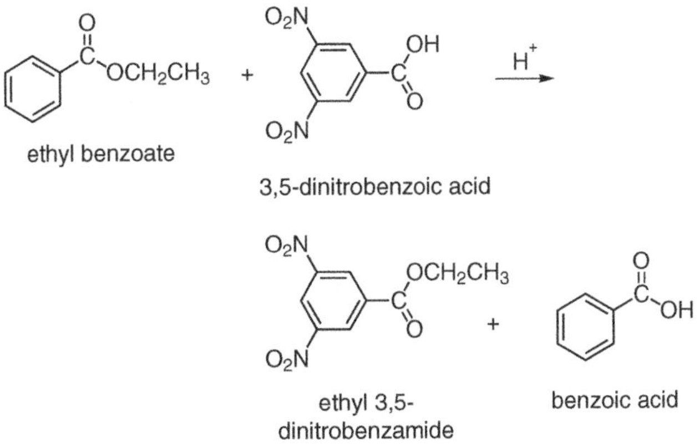

ethyl benzoate

3,5-dinitrobenzoic acid

ethyl 3,5-   benzoic acid
dinitrobenzamide

d. The N-benzylamide is formed from the reaction of ethyl benzoate with benzylamine.

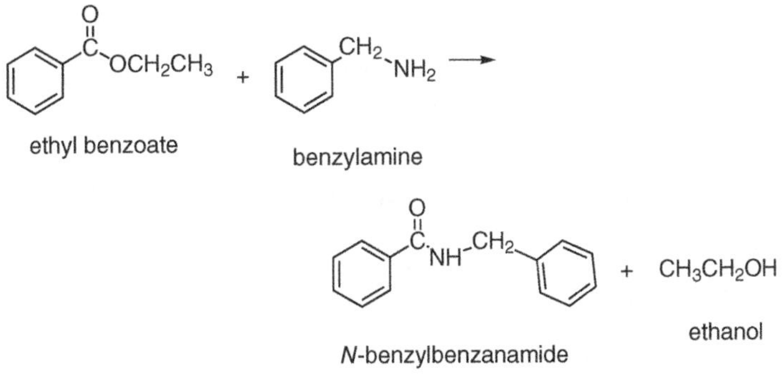

ethyl benzoate   benzylamine

+   CH₃CH₂OH

ethanol

N-benzylbenzanamide

e. Ethyl benzoate reacts with hydrazine to yield the hydrazide.

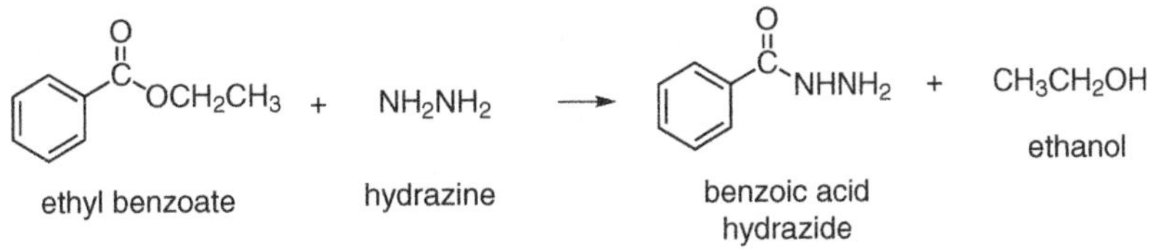

ethyl benzoate   hydrazine   benzoic acid   ethanol
                                     hydrazide

33. If an unsymmetrical ether was treated with 3,5-dinitrobenzoyl chloride, then two 3,5-dinitrobenzoates would be formed. An example is shown below with methyl ethyl ether.

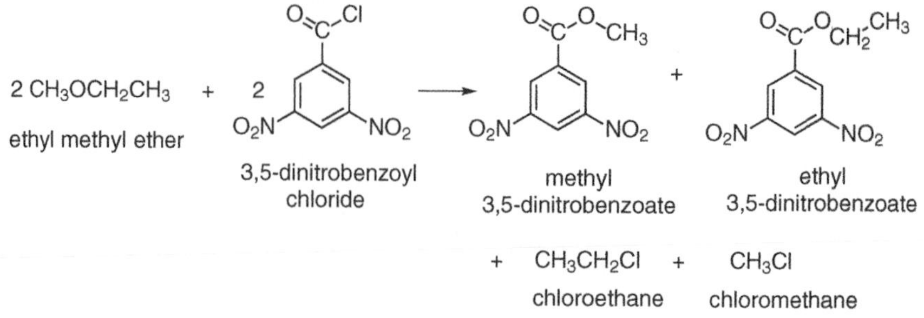

2 $CH_3OCH_2CH_3$ + 2 [3,5-dinitrobenzoyl chloride] → [methyl 3,5-dinitrobenzoate] + [ethyl 3,5-dinitrobenzoate]

ethyl methyl ether

3,5-dinitrobenzoyl chloride

methyl 3,5-dinitrobenzoate

ethyl 3,5-dinitrobenzoate

+ CH_3CH_2Cl + CH_3Cl

chloroethane   chloromethane

34. The reaction of dipropyl ether with 3,5-dinitrobenzoyl chloride yields propyl 3,5-dinitrobenzoate.

$CH_3CH_2CH_2OCH_2CH_2CH_3$ + 2 [3,5-dinitrobenzoyl chloride] → [propyl 3,5-dinitrobenzoate] + $CH_3CH_2CH_2Cl$

dipropyl ether

3,5-dinitrobenzoyl chloride

propyl 3,5-dinitrobenzoate

1-chloropropane

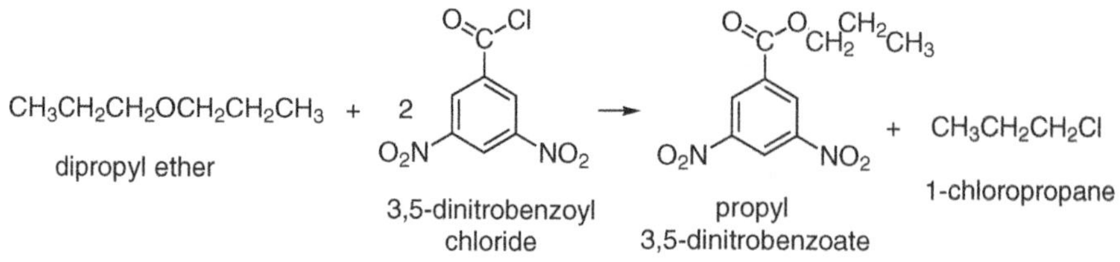

35. a. 1,2-Dimethoxybenzene reacts with chlorosulfonic acid to form the intermediate 3,4-dimethoxybenzenesulfonyl chloride, which reacts with ammonium carbonate or ammonium hydroxide to give the 3,4-dimethoxybenzenesulfonamide.

1,2-dimethoxy-benzene   —2 $ClSO_3H$, chlorosulfonic acid→   3,4-dimethoxy-benzenesulfonyl chloride   —$(NH_4)_2CO_3$ or NH_4OH→   3,4-dimethoxybenzene-sulfonamide

+ HCl + H_2SO_4

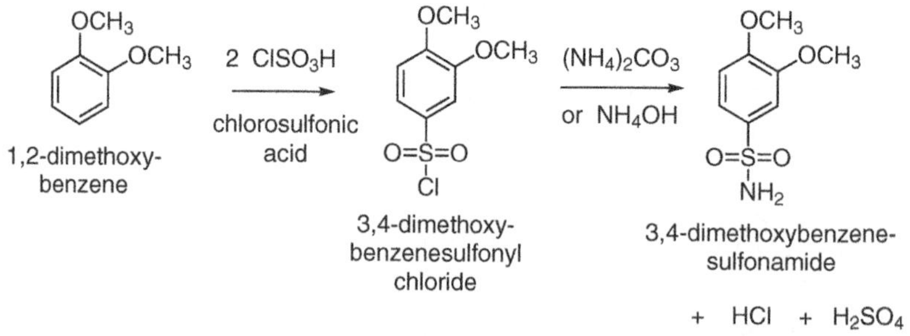

b. 1,2-Dimethoxybenzene is nitrated to give 1,2-dimethoxy-4-nitrobenzene.

1,2-dimethoxy-benzene   —HNO_3, H_2SO_4→   1,2-dimethoxy-4-nitrobenzene

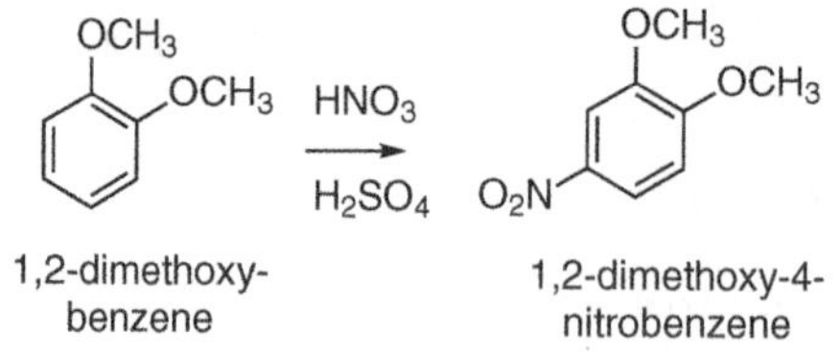

c. Bromination of 1,2-dimethoxybenzene produces 4,5-dibromo-1,2-dimethoxybenzene.

1,2-dimethoxybenzene   —Br_2→   4,5-dibromo-1,2-dimethoxybenzene

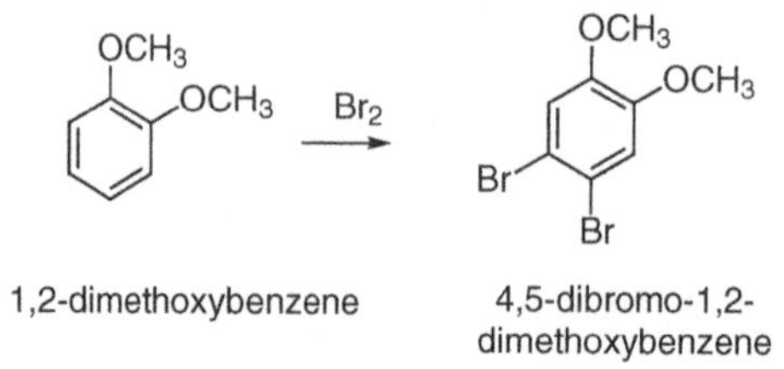

36. Potassium iodide is added to alkyl chlorides to speed up the reaction. This is not needed for vicinal dihalides. A base added to a vicinal dihalide may result in the formation of an alkyne.

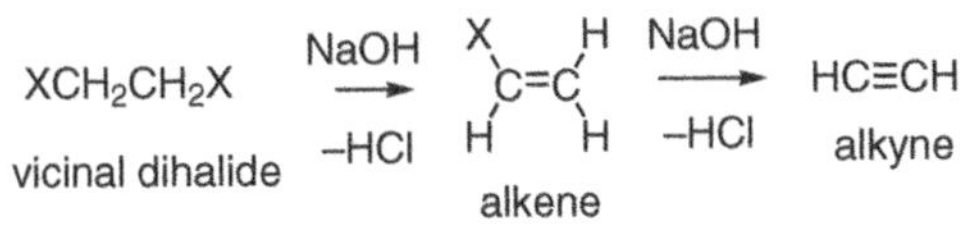

37. **a.** The Grignard reagent is prepared from the reaction of 1-bromo-2-methylpropane with magnesium. This Grignard reagent reacts with phenyl isocyanate to make the 3-methyl-*N*-phenylbutanamide.

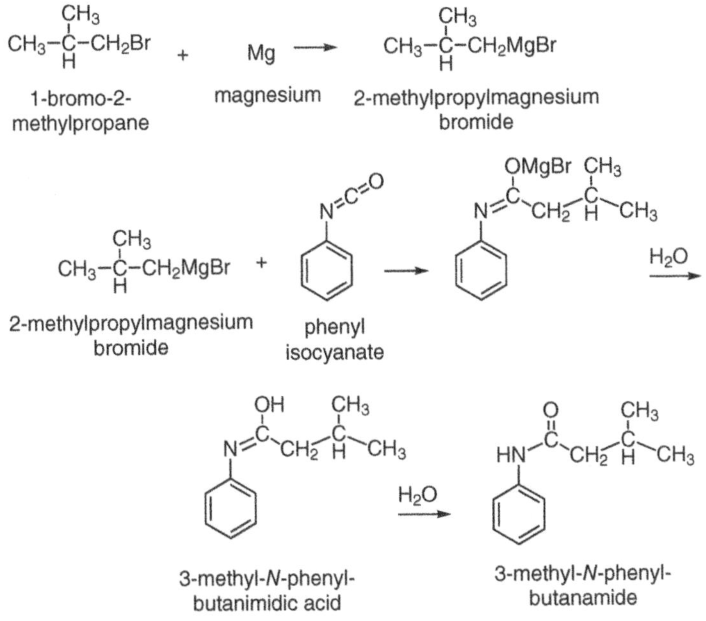

b. 2-Methylpropylmagnesium bromide is made from the reaction of 1-bromo-2-methylpropane with magnesium. This Grignard reagent undergoes a reaction with naphthyl isocyanate to yield the 3-methyl-*N*-naphthalen-1-ylbutanamide.

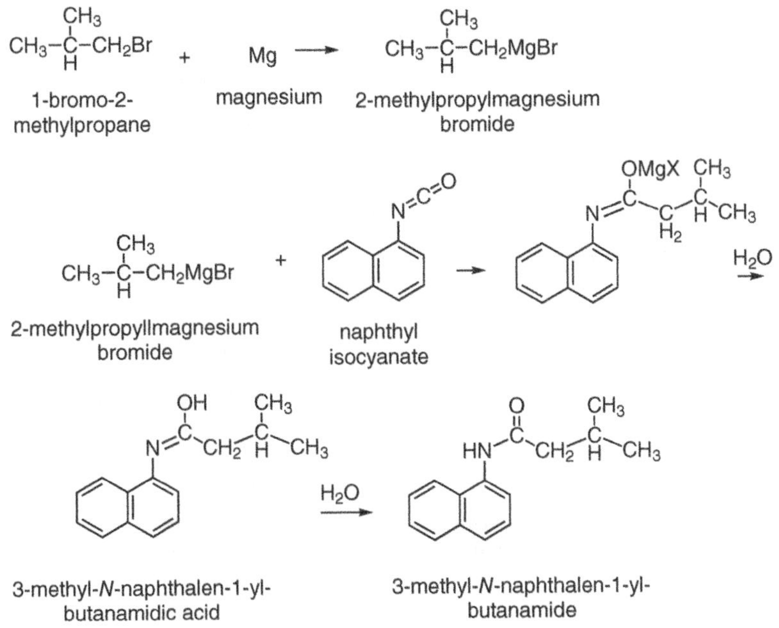

c. The Grignard reagent is prepared from the reaction of 1-bromo-2-methylpropane with magnesium. The reaction of the Grignard reagent with mercuric bromide gives 2-methylpropylmercuric bromide.

CH₃–C(CH₃)(H)–CH₂Br + Mg ⟶ CH₃–C(CH₃)(H)–CH₂MgBr

1-bromo-2-methylpropane   magnesium   2-methylpropylmagnesium bromide

CH₃–C(CH₃)(H)–CH₂MgBr + HgBr₂ ⟶ CH₃–C(CH₃)(H)–CH₂HgBr + MgBr₂

2-methylpropylmagnesium bromide   mercuric bromide   2-methylpropyl-mercuric bromide

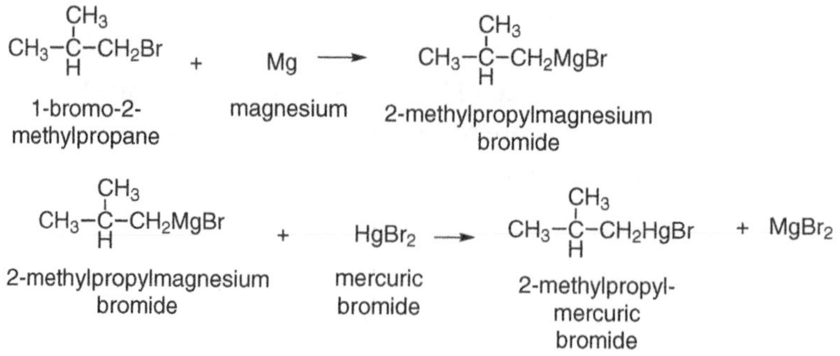

d. 2-Naphthol is treated with sodium hydroxide to form sodium naphthoxide, which then reacts with 1-bromo-2-methylpropane to give the 2-methylpropoxynaphthalene.

2-naphthol   sodium naphthoxide

1-bromo-2-methylpropane   2-methylpropoxynaphthalene

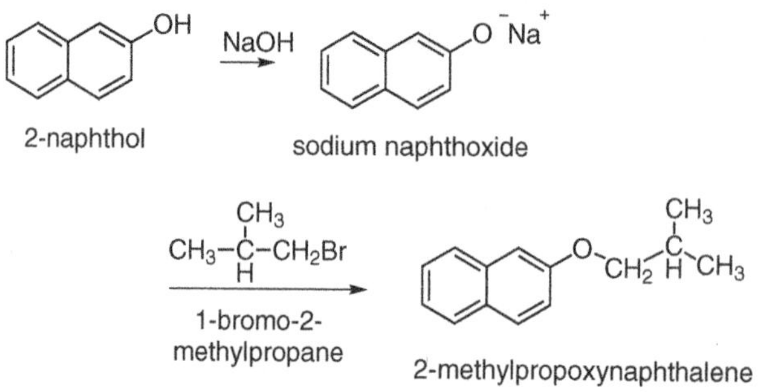

38. a. The Grignard reagent is prepared from the reaction of 2-chloropentane with magnesium. This Grignard reagent reacts with phenyl isocyanate to make the 2-methyl-*N*-phenylpentanamide.

2-chloropentane   magnesium   2-pentylmagnesium chloride

2-pentylmagnesium chloride   phenyl isocyanate

2-methyl-*N*-phenyl-pentanimidic acid   2-methyl-*N*-phenyl-pentanamide

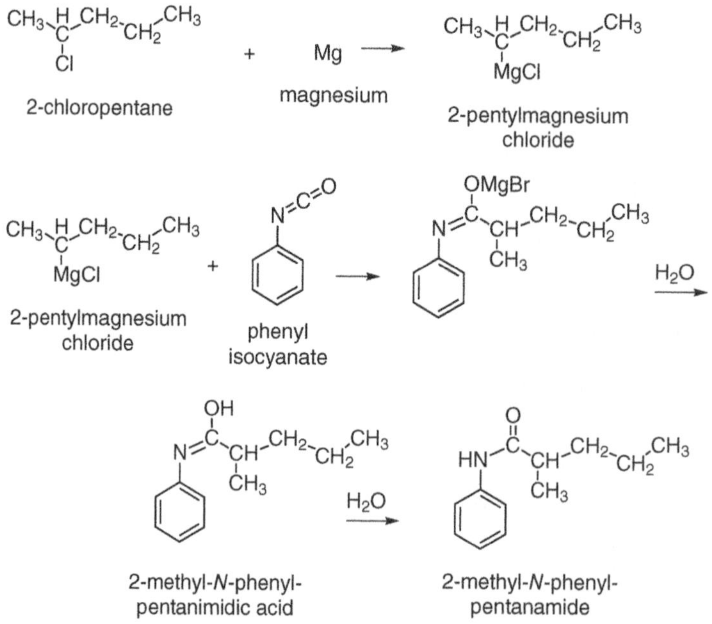

b. 2-Pentylmagnesium chloride is made from the reaction of 2-chloropentane with magnesium. This Grignard reagent undergoes a reaction with naphthyl isocyanate to yield the 2-methyl-*N*-naphthalen-1-ylpentanamide.

2-chloropentane   magnesium   2-pentylmagnesium chloride

2-pentylmagnesium chloride   naphthyl isocyanate

2-methyl-*N*-naphthalen-1-yl-pentanamidic acid   2-methyl-*N*-naphthalen-1-yl-pentanamide

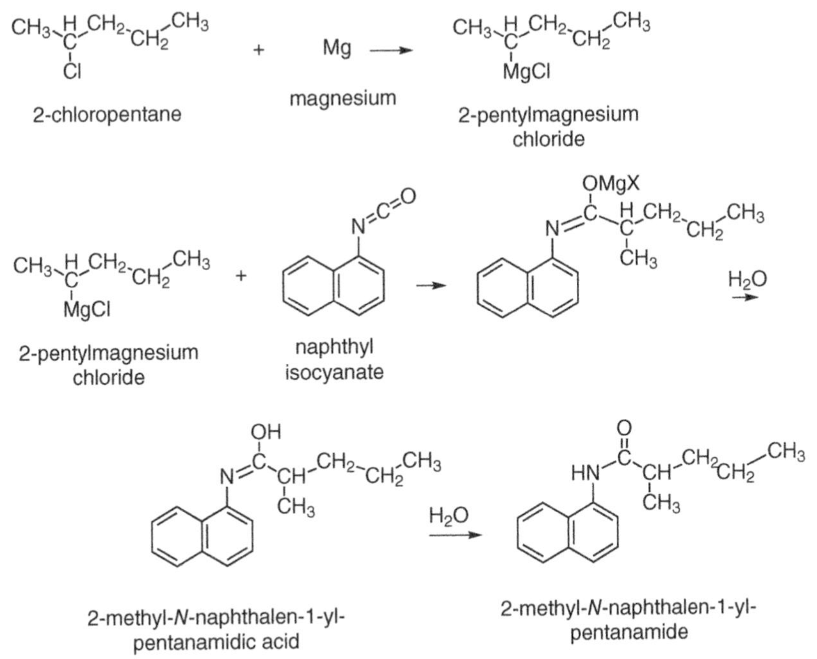

c. The Grignard reagent is prepared from the reaction of 2-chloropentane with magnesium. The reaction of the Grignard reagent with mercuric chloride gives 2-pentylmercuric chloride.

2-chloropentane   magnesium   2-pentylmagnesium chloride

2-pentylmagnesium chloride   mercuric chloride   2-pentylmercuric chloride

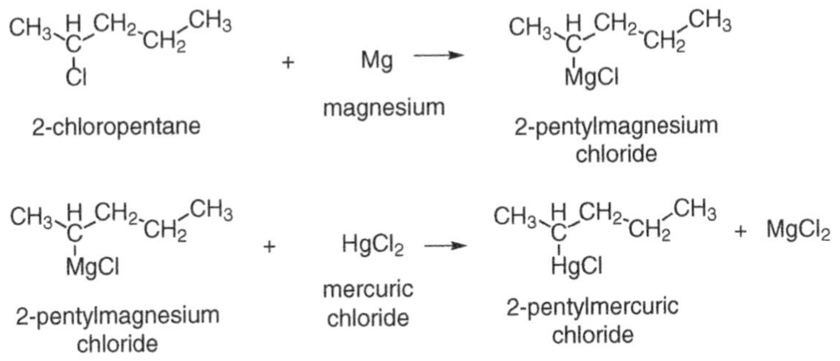

d. 2-Naphthol is treated with sodium hydroxide to form sodium naphthoxide, which then reacts with 2-chloropentane to give the 1-methylbutoxynaphthalene.

2-naphthol   sodium naphthoxide

2-pentylmagnesium chloride   1-methylbutoxynaphthalene

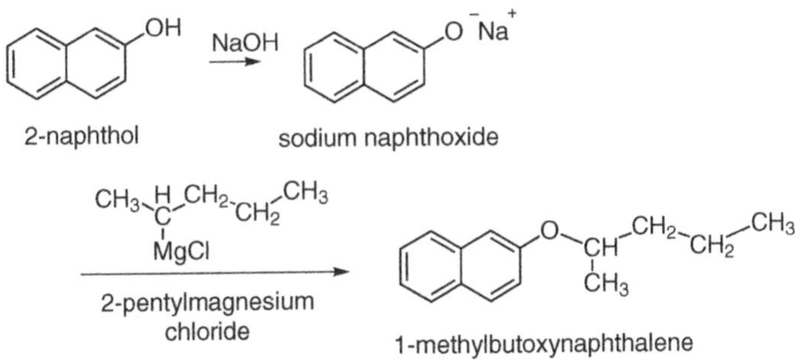

39. a. 3,4-Dichlorotoluene is nitrated to yield 3,4-dichloro-6-nitrotoluene.

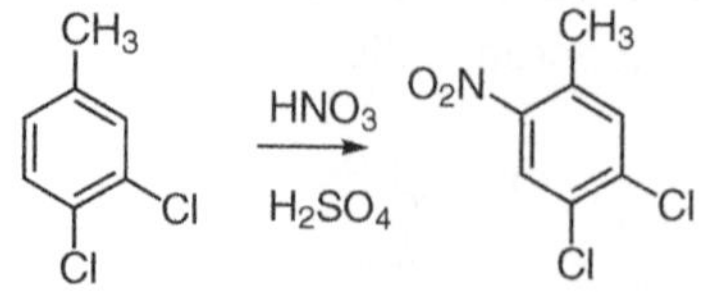

3,4-dichlorotoluene   3,4-dichloro-6-nitrotoluene

b. 3,4-Dichlorobenzene is sulfonated, then treated with ammonium hydroxide to give 3,4-dichloro-2-methylbenzenesulfonamide.

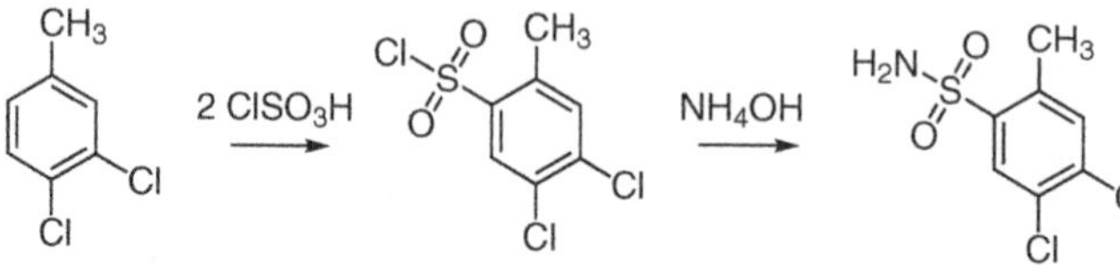

3,4-dichlorotoluene   3,4-dichloro-2-methyl-benzenesulfonyl chloride   3,4-dichloro-2-methyl-benzenesulfonamide

+ HCl + H_2SO_4

c. The oxidation of 3,4-dichlorotoluene produces 3,4-dichlorobenzoic acid.

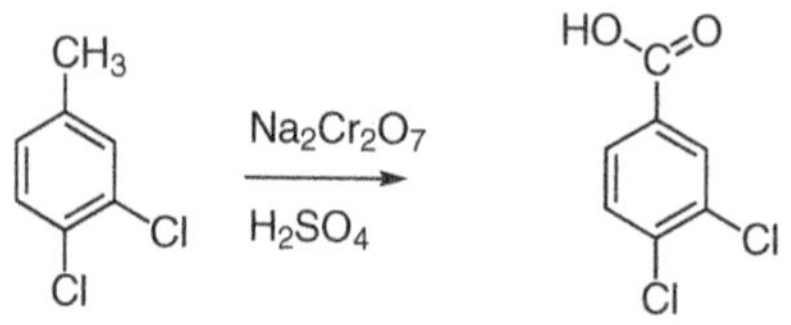

3,4-dichlorotoluene   3,4-dichlorobenzoic acid

40. The value of 148 corresponds to the disubstituted aromatic ring, the two carbonyl groups, and the hydroxy group minus the —H on the aromatic compound.

41. a. The nitration of ethylbenzene gives 1-ethyl-2,4,6-trinitrobenzene.

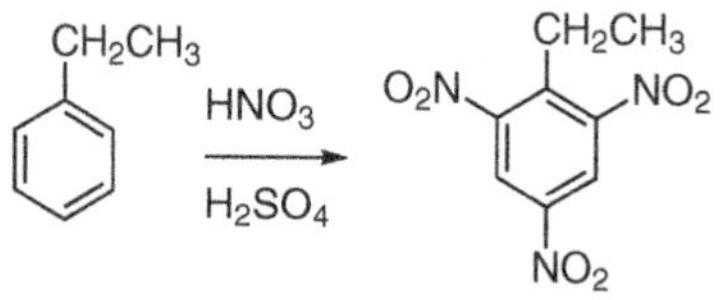

ethylbenzene   1-ethyl-2,4,6-trinitrobenzene

b. Ethyl benzene reacts with phthalic anhydride to give 2-(4-ethylbenzoyl)-benzoic acid.

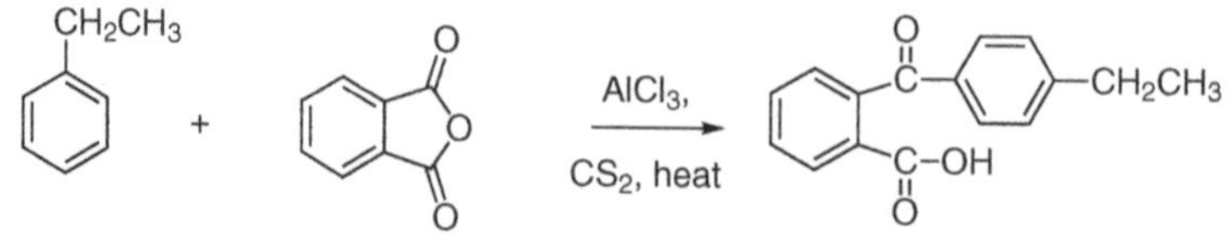

ethylbenzene   phthalic anhydride   2-(4-ethylbenzoyl)benzoic acid

42. a. Benzonitrile can be hydrolyzed in the presence of acid or base to benzoic acid.

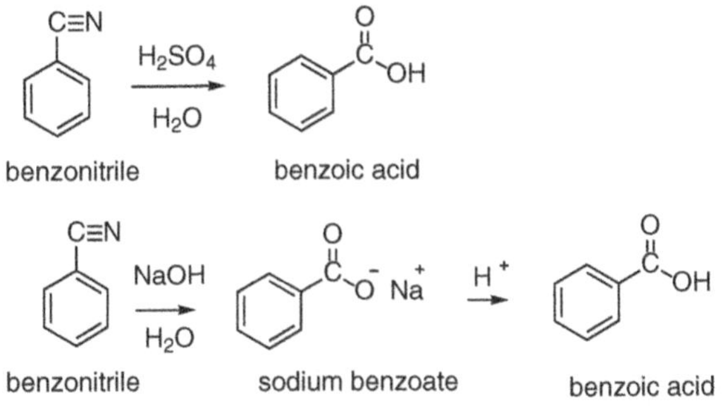

benzonitrile → benzoic acid (H$_2$SO$_4$, H$_2$O)

benzonitrile → sodium benzoate (NaOH, H$_2$O) → benzoic acid (H$^+$)

b. Benzonitrile can be hydrolyzed, treated with thionyl chloride, then ammonia to yield the benzamide. Another method for the formation of the amide is the controlled hydrolysis of the nitrile with a boron trifluoride/acetic acid complex.

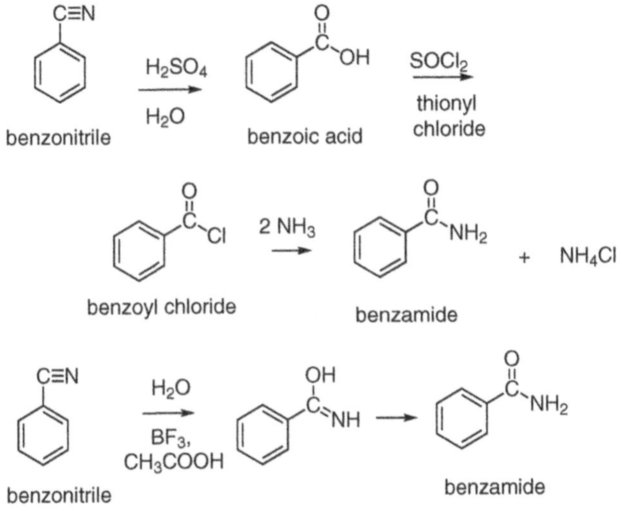

benzonitrile → (H$_2$SO$_4$, H$_2$O) → benzoic acid → (SOCl$_2$, thionyl chloride)

benzoyl chloride → (2 NH$_3$) → benzamide + NH$_4$Cl

benzonitrile → (H$_2$O, BF$_3$, CH$_3$COOH) → → benzamide

c. Benzonitrile is hydrolyzed and treated with thionyl chloride to form benzoyl chloride. Benzoyl chloride reacts with aniline to form benzanilide.

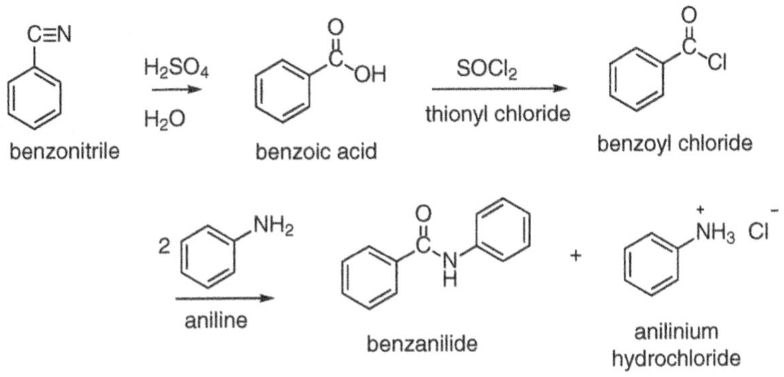

benzonitrile → (H$_2$SO$_4$, H$_2$O) → benzoic acid → (SOCl$_2$, thionyl chloride) → benzoyl chloride

2 aniline (NH$_2$) → benzanilide + anilinium hydrochloride (NH$_3^+$ Cl$^-$)

d. Synthesis of the 4-toluidide derivative is very similar to the procedure above, except 4-toluidine is used instead of aniline.

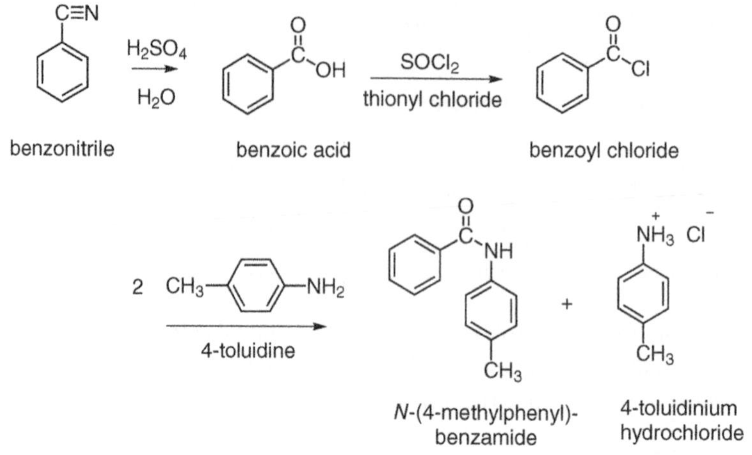

benzonitrile → benzoic acid → benzoyl chloride

2 4-toluidine → *N*-(4-methylphenyl)-benzamide + 4-toluidinium hydrochloride

e. Reduction of benzonitrile with sodium/acetic acid produces benzylamine.

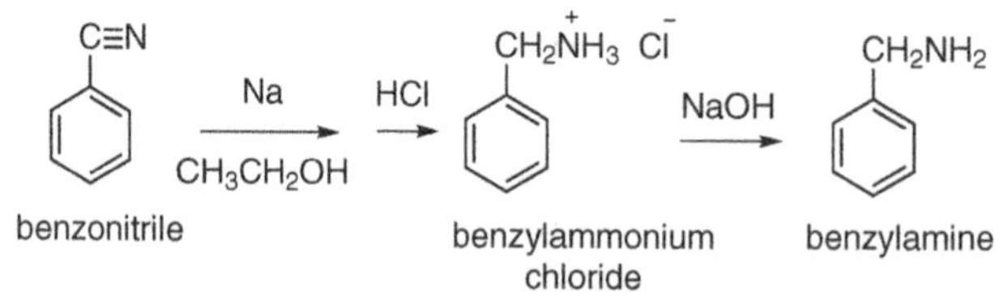

benzonitrile → benzylammonium chloride → benzylamine

f. Benzonitrile is reduced to benzylamine, then reacts with benzoyl chloride to yield the *N*-benzylbenzamide.

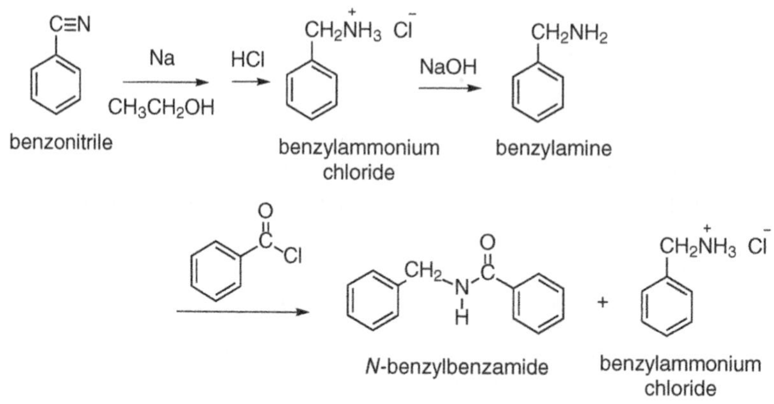

benzonitrile → benzylammonium chloride → benzylamine

N-benzylbenzamide + benzylammonium chloride

g. The benzylamine is formed by a reduction of the benzonitrile. Benzonitrile is treated with benzenesulfonyl chloride to produce *N*-benzylbenzenesulfonamide.

benzonitrile

benzylammonium chloride

benzylamine

benzenesulfonyl chloride

2 NaOH

(soluble)

+ NaCl + 2 H₂O

N-benzylbenzene-sulfonamide

+ NaCl

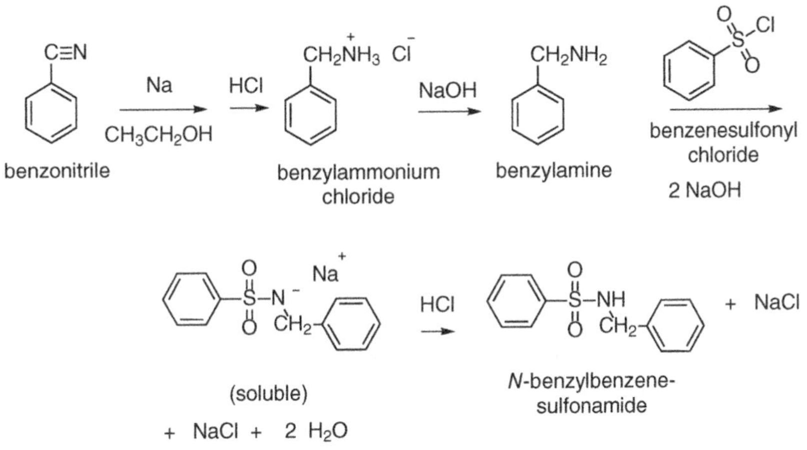

h. Benzonitrile is reduced to benzylamine, then treated with phenyl isocyanate to produce 1-benzyl-3-phenylthiourea.

benzonitrile

benzylammonium chloride

benzylamine

phenyl isocyanate

1-benzyl-3-phenylthiourea

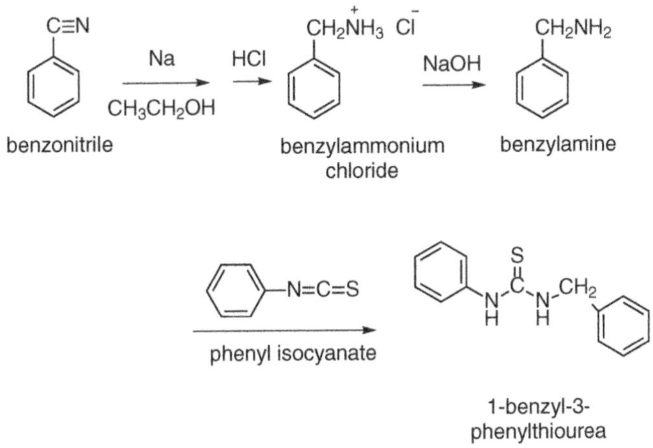

i. Benzonitrile reacts with thioglycolic acid to produce the phenyl α-(imidioylthio) acetic acid hydrochloride.

benzonitrile

thioglycolic acid

phenyl α-(imidioylthio)-acetic acid hydrochloride

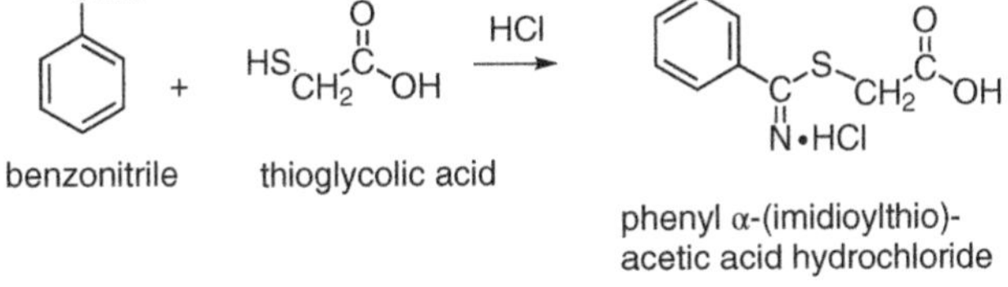

43. a. Pentanenitrile can be hydrolyzed in the presence of acid or base to pentanoic acid.

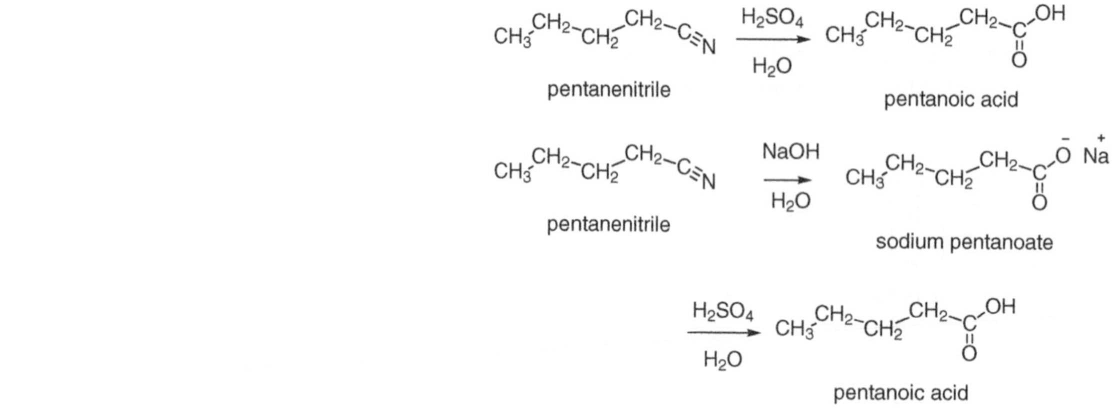

pentanenitrile → pentanoic acid (H₂SO₄ / H₂O)

pentanenitrile → sodium pentanoate (NaOH / H₂O) → pentanoic acid (H₂SO₄ / H₂O)

b. Pentanenitrile can be hydrolyzed, treated with thionyl chloride, then ammonia to yield the pentanamide. Another method for the formation of the amide is the controlled hydrolysis of the nitrile with a boron trifluoride/acetic acid complex.

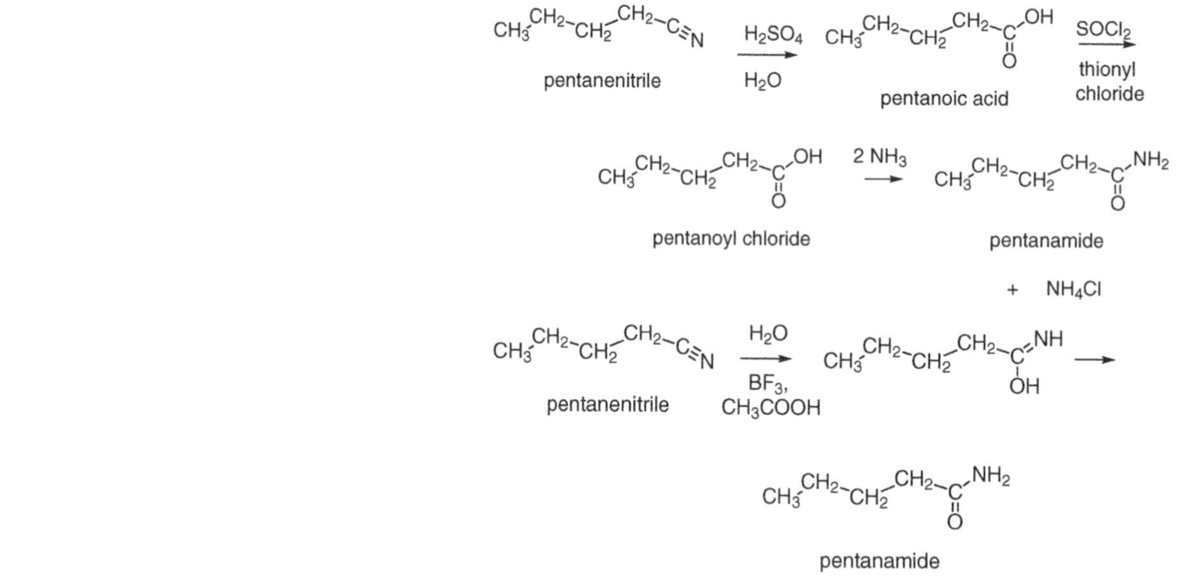

pentanenitrile → pentanoic acid (H₂SO₄ / H₂O) → (SOCl₂, thionyl chloride)

pentanoyl chloride + 2 NH₃ → pentanamide + NH₄Cl

pentanenitrile → (H₂O, BF₃, CH₃COOH) → intermediate → pentanamide

c. Pentanenitrile is hydrolyzed and treated with thionyl chloride to form pentanoyl chloride. Pentanoyl chloride reacts with aniline to form *N*-phenylpentanamide.

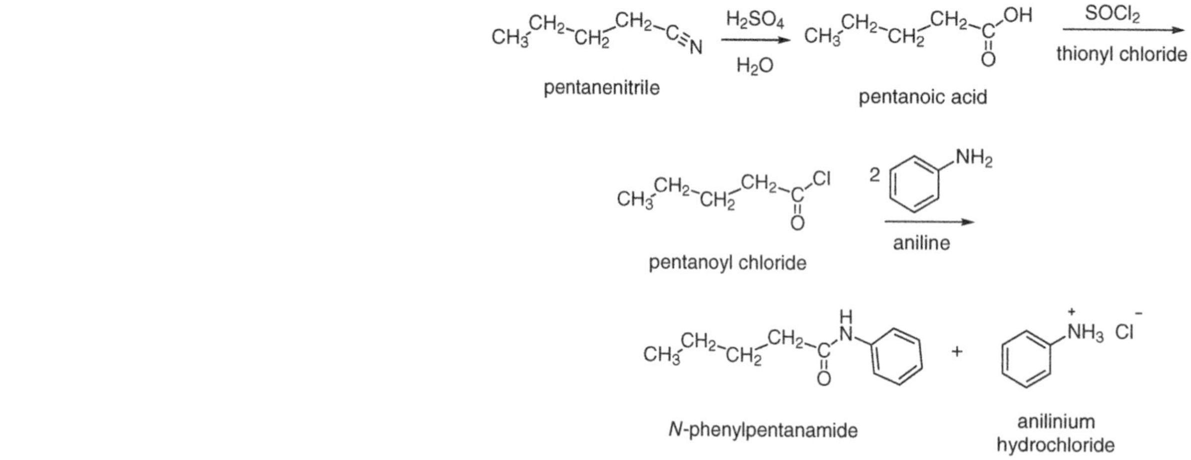

pentanenitrile → pentanoic acid (H₂SO₄ / H₂O) → (SOCl₂, thionyl chloride)

pentanoyl chloride + 2 aniline → *N*-phenylpentanamide + anilinium hydrochloride

d. Synthesis of the 4-toluidide derivative is very similar to the procedure above, except 4-toluidine is used instead of aniline.

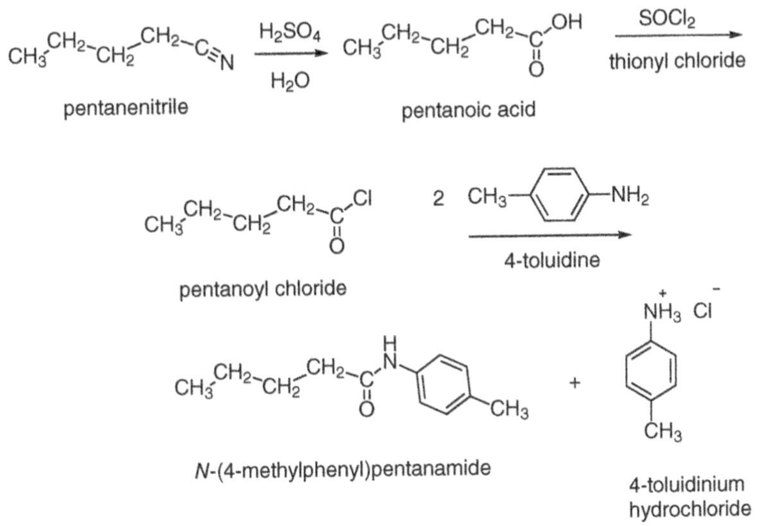

pentanenitrile → pentanoic acid → pentanoyl chloride

2 CH₃—⟨phenyl⟩—NH₂
4-toluidine

N-(4-methylphenyl)pentanamide   +   4-toluidinium hydrochloride

e. Pentanitrile is reduced to pentanamine, then reacts with benzoyl chloride to yield the *N*-pentylbenzamide.

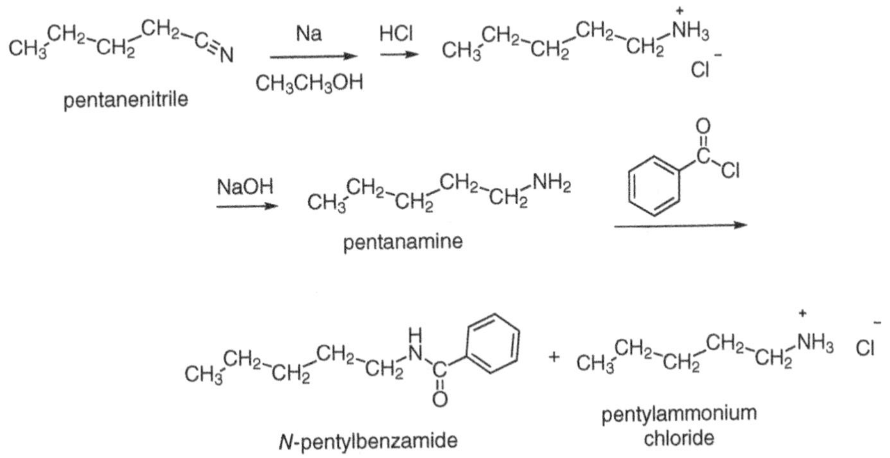

pentanenitrile $\xrightarrow[\text{CH}_3\text{CH}_3\text{OH}]{\text{Na}}$ $\xrightarrow{\text{HCl}}$

$\xrightarrow{\text{NaOH}}$ pentanamine

N-pentylbenzamide   +   pentylammonium chloride

f. The pentanamine is formed by a reduction of the pentanenitrile. Pentanamine is treated with benzenesulfonyl chloride to produce *N*-pentylbenzenesulfonamide.

pentanenitrile $\xrightarrow[\text{CH}_3\text{CH}_3\text{OH}]{\text{Na}}$ $\xrightarrow{\text{HCl}}$

$\xrightarrow{\text{NaOH}}$ pentanamine

benzenesulfonyl chloride
2 NaOH

(soluble) $\xrightarrow{\text{HCl}}$ *N*-pentylbenzenesulfonamide

+ NaCl + 2 H₂O   + NaCl

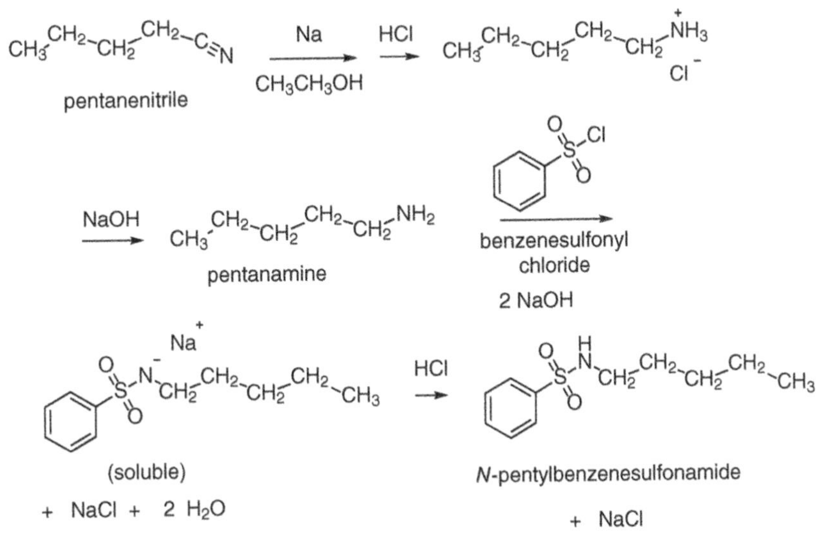

g. Pentanenitrile is reduced to pentanamine, then treated with phenyl isocyanate to produce 1-pentyl-3-phenylthiourea.

pentanenitrile

pentanamine

phenyl isocyanate

1-pentyl-3-phenylthiourea

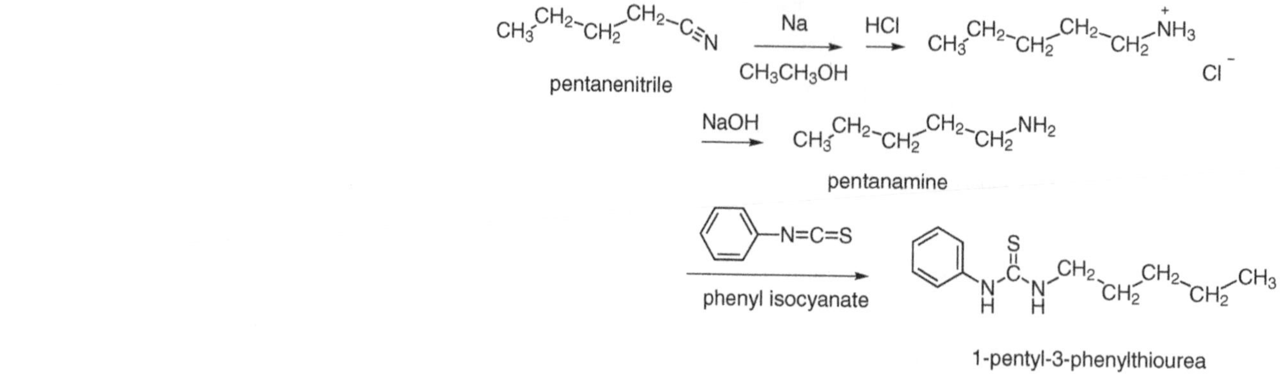

h. Pentanenitrile reacts with thioglycolic acid to produce the pentyl α-(imidioylthio) acetic acid hydrochloride.

pentanenitrile

thioglycolic acid

pentyl α-(imidioylthio)-
acetic acid hydrochloride

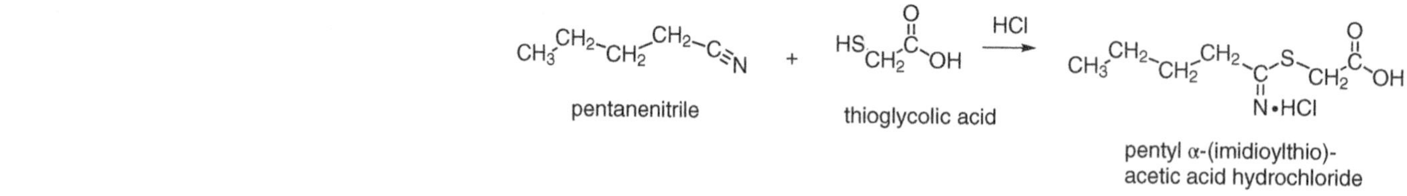

44. a. The nitro group in 2,4-dichloro-1-nitrobenzene is reduced to an amine, then reacts with acetic anhydride to form *N*-(2,4-dichlorophenyl)acetamide.

2,4-dichloro-1-
nitrobenzene

2,4-dichloroanilium
chloride

2,4-dichloroaniline

acetic anhydride

N-(2,4-dichlorophenyl)-
acetamide

acetic acid

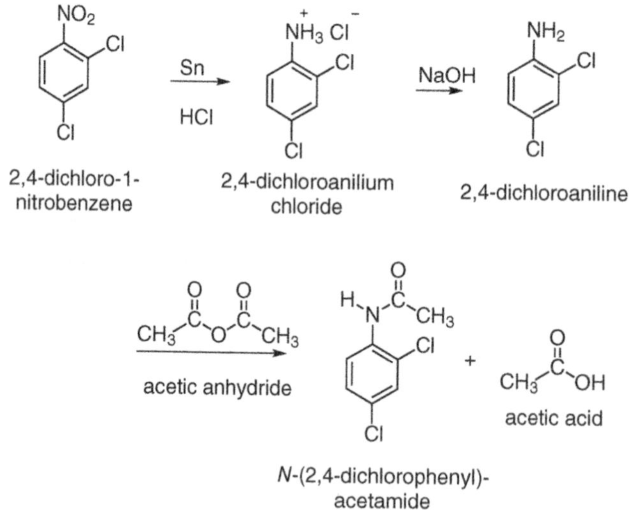

b. The nitro group in 2,4-dichloro-1-nitrobenzene is reduced to an amine, then reacts with benzoyl chloride to form *N*-(2,4-dichlorophenyl)benzamide.

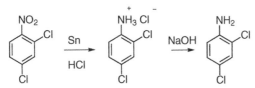

c. The nitro group in 2,4-dichloro-1-nitrobenzene is reduced to an amine, then reacts with benzenesulfonyl chloride to form *N*-(2,4-dichlorophenyl)benzenesulfonamide.

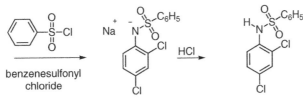

45. a. 4-Nitrophenol reacts with phenyl isocyanate to form phenylcarbamic acid 4-nitrophenyl ester.

b. Naphthalen-2-ylcarbamic acid 4-nitrophenyl ester is formed from the reaction of 4-nitrophenol with 1-naphthyl isocyanate.

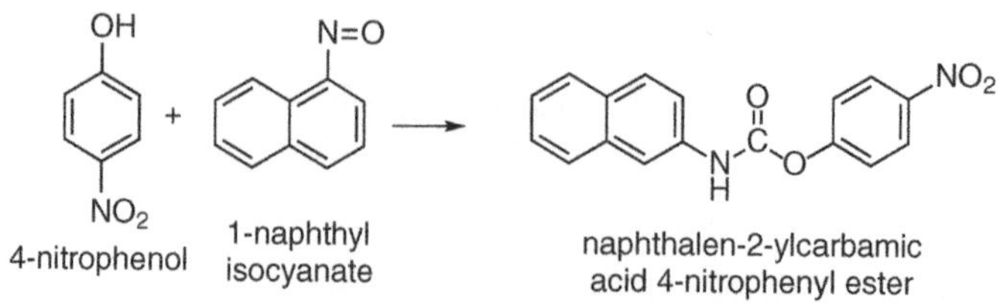

4-nitrophenol   1-naphthyl isocyanate   naphthalen-2-ylcarbamic acid 4-nitrophenyl ester

c. 4-Nitrophenol reacts with benzoyl chloride to produce 4-nitrophenyl benzoate.

4-nitrophenol   benzoyl chloride   4-nitrophenyl benzoate

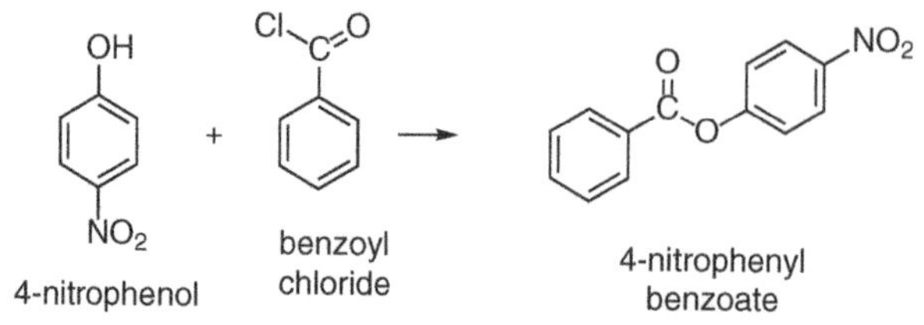

d. The reaction of 4-nitrophenol with 4-nitrobenzoyl chloride produces 4-nitrophenyl 4-nitrobenzoate.

4-nitrophenol   4-nitrobenzoyl chloride   4-nitrophenyl 4-nitrobenzoate

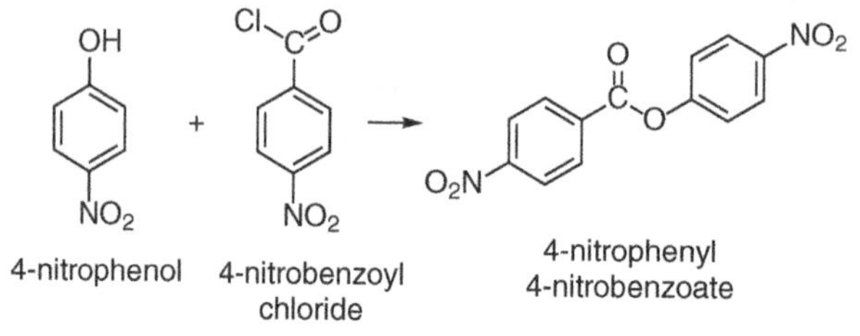

e. 3,5-Dinitrobenzoyl chloride reacts with 4-nitrophenol to yield 4-nitrophenyl 3,5-dinitrobenzoate.

4-nitrophenol   3,5-ninitrobenzoyl chloride   4-nitrophenyl 3,5-dinitrobenzoate

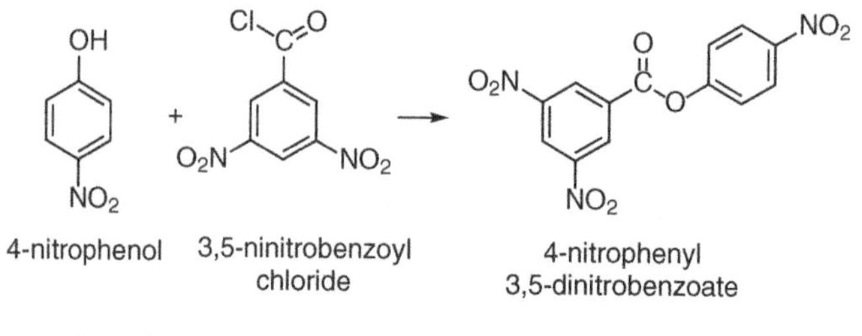

f. 4-Nitrophenol is treated with acetic anhydride to form the 4-nitrophenyl acetate.

4-nitrophenol   acetic anhydride   4-nitrophenyl acetate   acetic acid

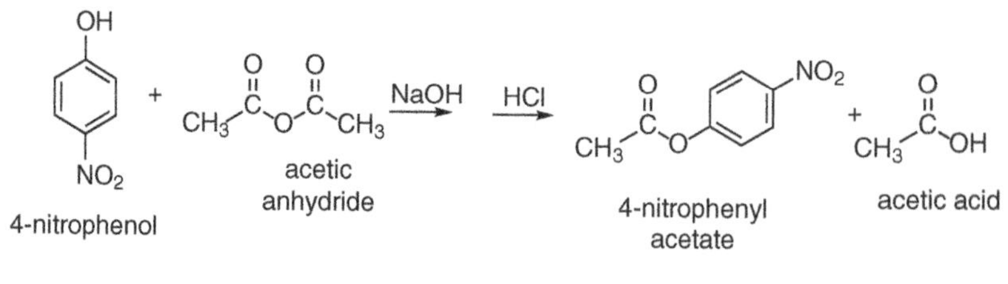

g. The base hydrolysis of 4-nitrophenol gives sodium 4-nitrophenolate. This salt reacts with chloroacetic acid to produce (4-nitrophenoxy)acetic acid.

4-nitrophenol

sodium
4-nitrophenolate

chloroacetic acid

(4-nitrophenoxy)acetic acid

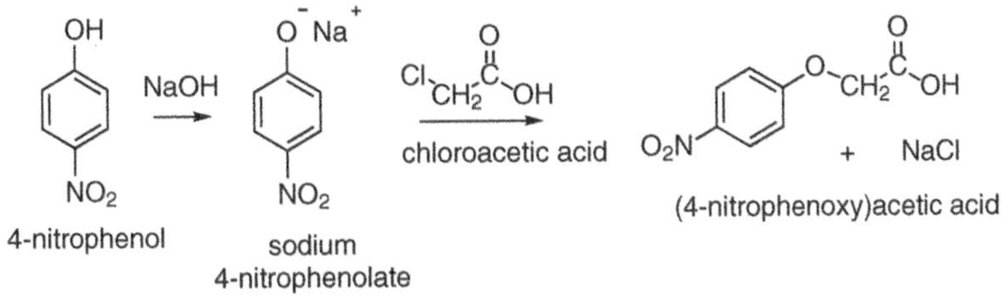

h. 4-Nitrophenol is brominated to yield 2,6-dibromo-4-nitrophenol.

4-nitrophenol

2,6-dibromo-
4-nitrophenol

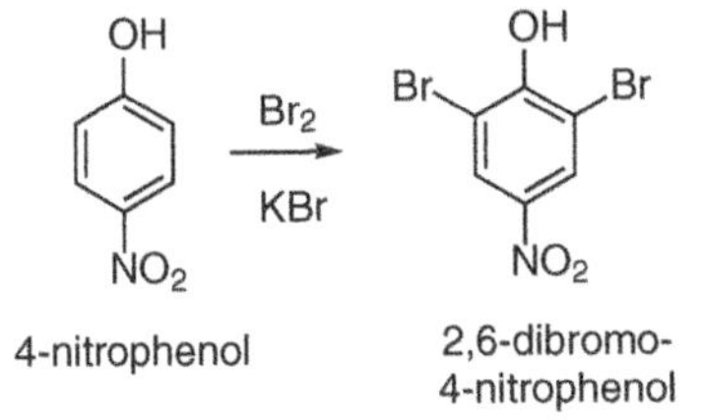

46. a. 4-Methylbenzenesulfonamide is hydrolyzed to 4-methylbenzenesulfonic acid.

4-methylbenzene-
sulfonamide

4-methylbenzene-
sulfonic acid

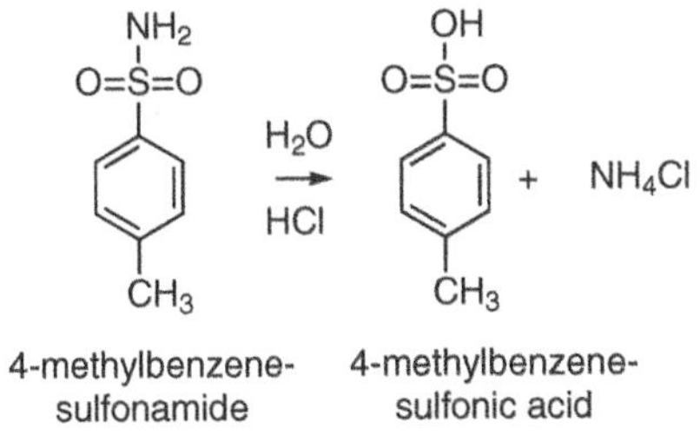

b. To form the 4-methylbenzenesulfonyl chloride, 4-methylbenzene-sulfonamide is hydrolyzed, then treated with phosphorus pentachloride.

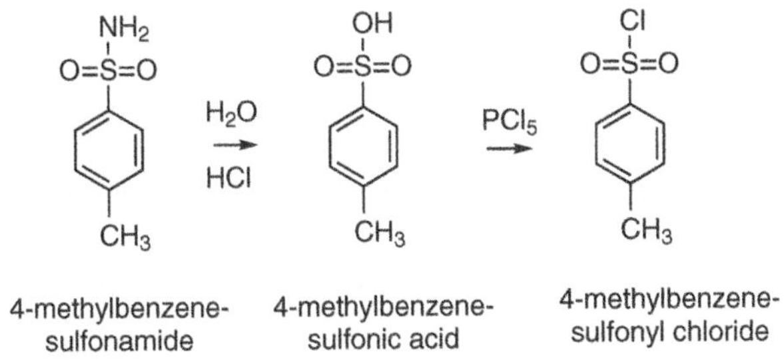

4-methylbenzene-
sulfonamide

4-methylbenzene-
sulfonic acid

4-methylbenzene-
sulfonyl chloride

c. 4-Methylbenzensulfonamide is hydrolyzed then treated with phosphorus pentachloride to give the 4-methylbenzenesulfonyl chloride. The sulfonyl chloride is reacted with aniline to give the 4-methyl-N-phenyl-benzenesulfonamide.

4-methylbenzene-
sulfonamide

4-methylbenzene-
sulfonic acid

4-methylbenzene-
sulfonyl chloride

4-methyl-N-phenyl-
benzenesulfonamide

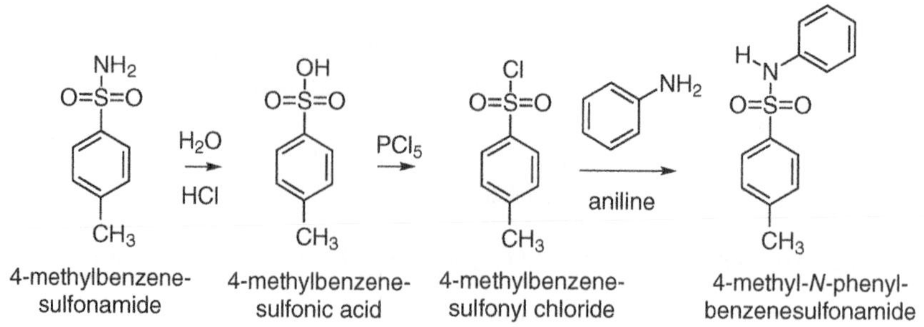

d. Treatment of 4-methylbenzenesulfonamide with xanthydrol gives N-xanthyl-4-methylbenzenesulfonamide.

4-methylbenzene-
sulfonamide

xanthydrol

N-xanthyl-4-methyl-
benzenesulfonamide

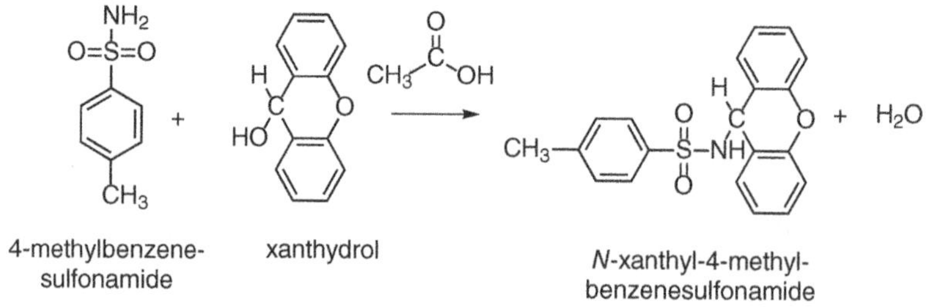

47. a. 4-Bromobenzenesulfonic acid is treated with phosphorus pentachloride to give 4-bromobenzenesulfonyl chloride.

4-bromobenzene-
sulfonic acid

4-bromobenzene-
sulfonyl chloride

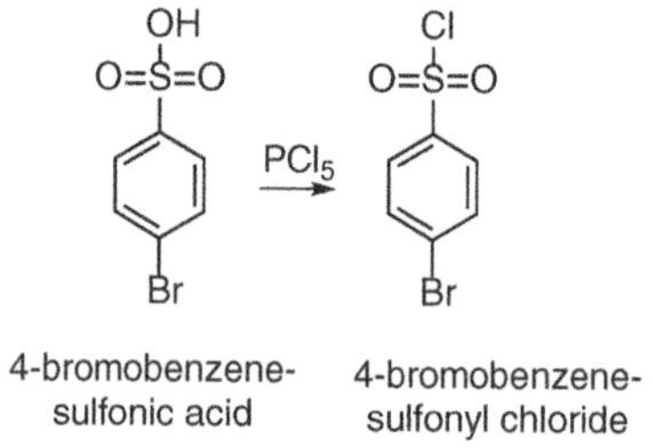

b. 4-Bromobenzenesulfonyl chloride is formed by the treatment of 4-bromo-benzenesulfonic acid with phosphorus pentachloride. The sulfonyl chloride is reacted with ammonium carbonate to form 4-bromobenzenesulfonamide.

4-bromobenzene-
sulfonic acid

4-bromobenzene-
sulfonyl chloride

4-bromobenzene-
sulfonamide

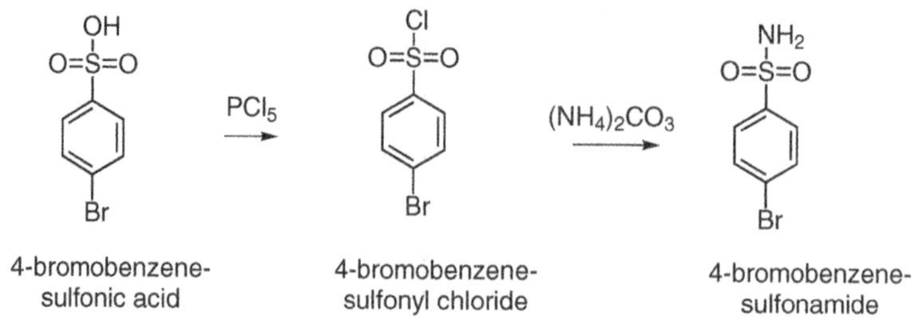

c. To make the sulfonamide, 4-bromobenzenesulfonic acid is treated with phosphorus pentachloride and the sulfonyl chloride is treated with aniline to make the 4-bromo-*N*-phenylbenzenesulfonamide.

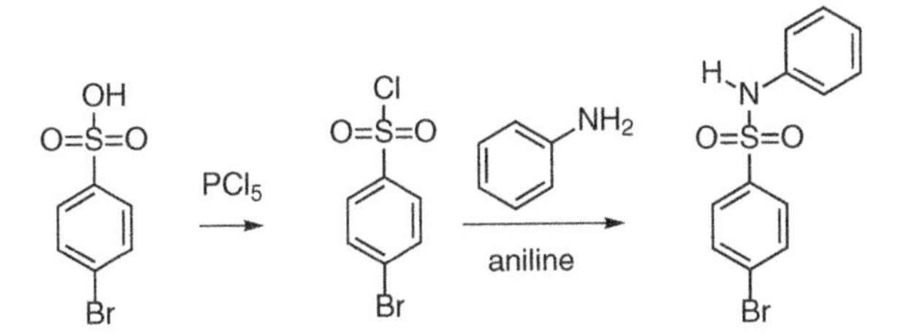

4-bromobenzene-
sulfonic acid

4-bromobenzene-
sulfonyl chloride

4-bromo-*N*-phenyl-
benzenesulfonamide

d. 4-Bromobenzenesulfonic acid is treated with sodium hydroxide to make the sodium salt. The salt reacts with *S*-benzylthiuronium chloride to form the *S*-benzylthiuronium 4-bromobenzenesulfonamide.

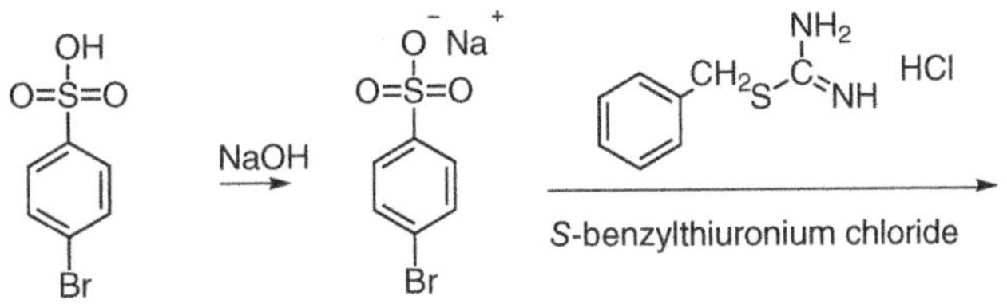

4-bromobenzene-
sulfonic acid

sodium salt of
4-bromobenzene-
sulfonic acid

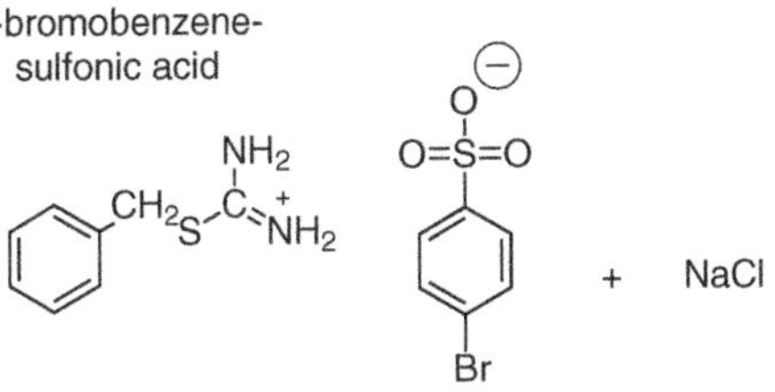

S-benzylthiuronium chloride

S-benzylthiuronium 4-bromobenzenesulfonate

e. 4-Bromobenzenesulfonic acid reacts with 4-toludine to produce the 4-toluidine salt.

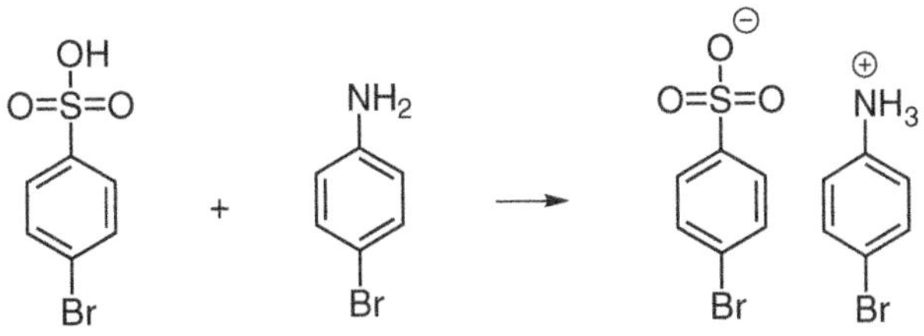

4-bromobenzene-
sulfonic acid

4-toluidine

4-toluidine salt of
4-bromobenzenesulfonic acid

48. a. The hydrolysis of 4-ethylbenzenesulfonyl chloride produces 4-ethyl-benzene-sulfonic acid.

4-ethylbenzene-
sulfonyl chloride

4-ethylbenzene-
sulfonic acid

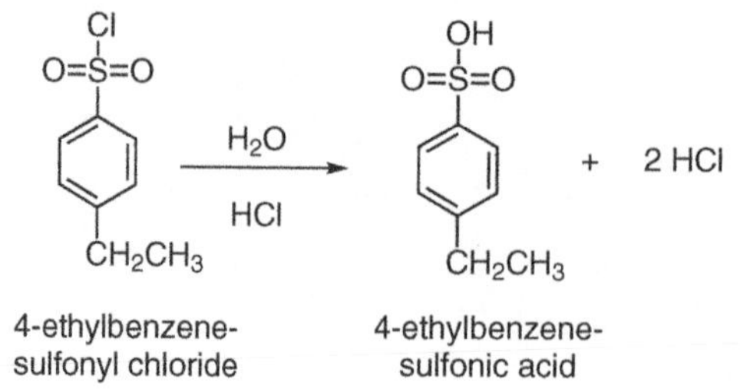

b. 4-Ethylbenzenesulfonyl chloride is treated with ammonium carbonate to produce 4-ethylbenzenesulfonamide.

4-ethylbenzene-
sulfonyl chloride

4-ethylbenzene-
sulfonamide

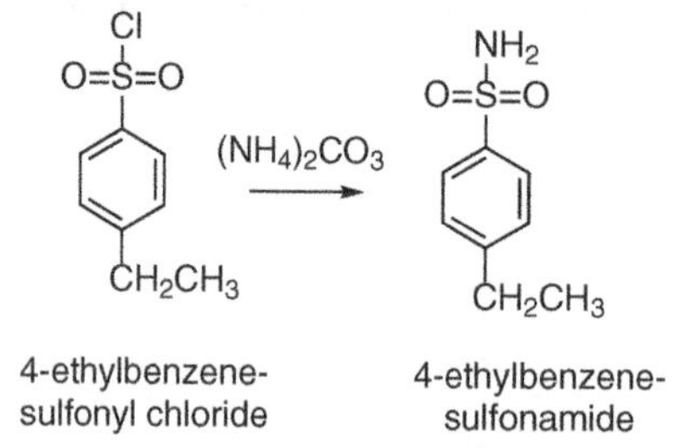

c. The treatment of 4-ethylbenzenesulfonyl chloride with aniline gives 4-ethyl-*N*-phenylbenzenesulfonamide.

4-ethylbenzene-
sulfonyl chloride

4-ethyl-*N*-phenyl-
benzenesulfonamide

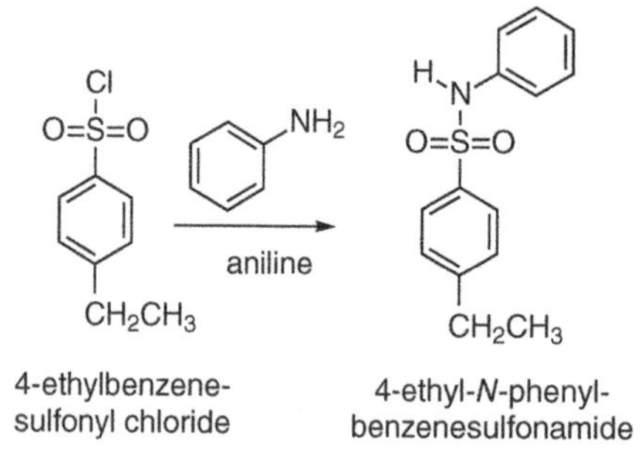

Chapter 13[*]

Structural Problems

PROBLEM SET 1

1. The structures for compounds A, B, and C are below.

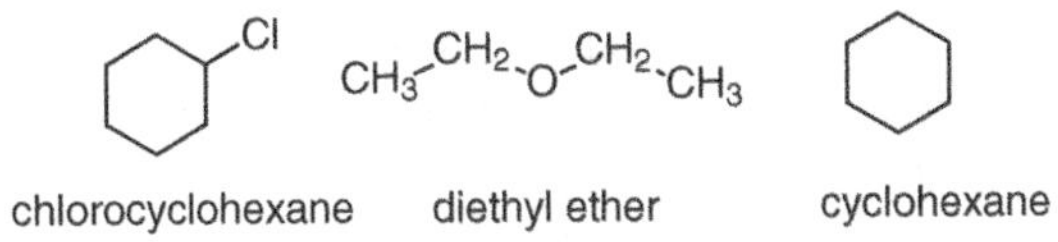

chlorocyclohexane   diethyl ether   cyclohexane

Compound A is in solubility class I, which consists of saturated hydrocarbons, haloalkanes, aryl halides, other deactivated compounds, and diaryl ethers. Since a white precipitate was formed from the addition of silver nitrate to a sodium fusion solution of the unknown, a chlorine is indicated. Thus, compound A is chlorocyclohexane.

Compound B is in solubility class I, which consists of saturated hydrocarbons, haloalkanes, aryl halides, other deactivated compounds, and diaryl ethers. A negative silver nitrate test indicates the absence of a halogen. Therefore, compound B is cyclohexane.

Compound C is in solubility class N, which contains alcohols, aldehydes, ketones, esters with one functional group, and more than five carbons but fewer than nine carbons, ethers, epoxides, alkenes, alkynes, and some aromatic compounds. A negative silver nitrate test indicates the absence of a halogen. Thus, compound C is diethyl ether.

---

*Companion website: www.wiley.com/go/hermann/identorganiccomp9e

Student Solutions Manual to Accompany The Systematic Identification of Organic Compounds, Ninth Edition.
Christine K. F. Hermann, Terence C. Morrill, Ralph L. Shriner, and Reynold C. Fuson.
© 2023 John Wiley & Sons, Inc. Published 2023 by John Wiley & Sons, Inc.

2. The possible structures are below.

CH3–C(=O)–CH3   CH3–C(CH3)(OH)–CH3   (cyclohexane structure)

acetone   2-methyl-2-propanol   cyclohexane

(cyclohexene structure)   CH3–CH2–OH

cyclohexene   ethanol

Compound A is in solubility class A_1, A_2, B, MN, N, or I. It does not have a multiple bond (negative bromine test), or an active hydrogen (negative acetyl chloride test). It is not an aldehyde or a ketone (negative 2,4-dinitrophenylhydrazine test). Compound A is not a methyl ketone and it does not have a methyl next to a –CHOH (negative iodoform test). Compound A must be cyclohexane.

Compound B is in solubility class A_1, A_2, B, MN, N, or I. It contains a multiple bond (positive bromine test) but does not have an active hydrogen (negative acetyl chloride test). It is not an aldehyde or a ketone (negative 2,4-dinitrophenylhydrazine test). Compound B is not a methyl ketone and it does not have a methyl next to a –CHOH (negative iodoform test). Compound B is cyclohexene.

Compound C is in solubility class S_2, S_A, S_B, or S_1. It contains an active hydrogen (positive acetyl chloride test) but is not an aldehyde or a ketone (negative 2,4-dinitrophenylhydrazine test). It is either a methyl ketone or a methyl next to a –CHOH (positive iodoform test). Since compound C cannot be a ketone, compound C must be an alcohol that gives a positive iodoform test. Therefore, compound C is ethanol.

Compound D is in solubility class S_2, S_A, S_B, or S_1. It does not contain an active hydrogen (negative acetyl chloride test) but is an aldehyde or ketone (positive 2,4-dinitrophenylhydrazine test). It is either a methyl ketone or has a methyl next to a –CHOH (positive iodoform test). From the data, compound D must be acetone (propanone).

Compound E is in solubility class S_2, S_A, S_B, or S_1. It contains an active hydrogen (positive acetyl chloride test), but is not an aldehyde or a ketone (negative 2,4-dinitrophenylhydrazine test). Compound E is not a methyl ketone and it does not have a methyl next to a –CHOH (negative iodoform test). The only possibility for compound E is 2-methyl-2-propanol (*t*-butyl alcohol).

3. Compounds A, B, and C are in solubility class S_1 which consists of monofunctional alcohols, aldehydes, ketones, esters, nitriles, and amides with five carbons or fewer. Compounds A, B, and C do not contain aldehydes or ketones (negative 2,4-dinitrophenylhydrazine test) but contain an alcohol, phenol, primary amine, or secondary amine (positive acetyl chloride test). Compound A is a primary alcohol (positive chromium trioxide test, negative hydrochloric acid, zinc chloride test). Compound B is a secondary alcohol (positive chromium trioxide test, positive hydrochloric acid, zinc chloride test). Compound C is a tertiary alcohol (negative chromium trioxide test, positive hydrochloric acid, zinc chloride test).

The compounds that match the boiling points are listed below.

CH3–CH2–OH   CH3–CH(OH)–CH3   CH3–C(CH3)(OH)–CH3

Compound A: Compound B: Compound C:
ethanol   2-propanol   2-methyl-2-propanol

4. Compounds A, B, and C are in solubility class I, which is composed of saturated hydrocarbons, haloalkanes, aryl halides, other deactivated compounds, and diaryl ethers. Compounds A, B, and C are not aromatic (negative chloroform and aluminum chloride test). Compound A contains chlorine (a white solid with silver nitrate). Compound B contains bromine (a pale-yellow solid with silver nitrate). Compound C contains iodine (a yellow solid with silver nitrate).

The alkyl halides with the closest boiling points are listed below.

$$CH_3-CH_2-CH_2-CH_2-CH_2-CH_2-CH_2-Cl \qquad CH_3-CH_2-CH_2-CH_2-CH_2-CH_2-Br$$

Compound A:
1-chloroheptane

Compound B:
1-bromohexane

$$CH_3-CH_2-CH_2-CH_2-CH_2-I$$

Compound C:
1-iodopentane

5. Compounds A, B, and C are in solubility class S_1, which consists of monofunctional alcohols, aldehydes, ketones, esters, nitriles, and amides with five carbons or fewer. Compounds A, B, and C are not an alcohol, phenol, primary amine, or secondary amine (negative acetyl chloride test), or an aldehyde or ketone (negative 2,4-dinitrophenylhydrazine test and no IR peak between 1600 and 1800 cm^{-1}), but are nitriles (IR peak around 2250 cm^{-1}). Compound A contains a double bond or a triple bond (positive bromine test), but does not contain halogen. Compound B does not contain a double bond or a triple bond (negative bromine test), but contains halogen. Compound C does not contain a double bond or a triple bond (negative bromine test), and does not contain halogen.

$$CH_2{=}CH{-}C{\equiv}N \qquad FCH_2{-}C{\equiv}N \qquad CH_3{-}C{\equiv}N$$

Compound A:
propenenitrile

Compound B:
fluoroacetonitrile

Compound C:
acetonitrile

PROBLEM SET 2

1. The unknown is in solubility class A_1, which consists of strong organic carboxylic acids with more than six carbons, phenols with electron-withdrawing groups in the ortho/para positions, or β-diketones. The compound does not have an active hydrogen (negative acetyl chloride test), contains a carboxylic acid (positive sodium bicarbonate test), contains a double or triple bond (positive bromine test), but is not an aldehyde or ketone (negative 2,4-dinitrophenylhydrazine test). A carboxylic acid with a melting point of 170 °C is 4-methoxycinnamic acid.

IR spectrum:

Frequency	Bond	Compound type
823	C—H out-of-plane bend	*p*-Disubstituted aromatic
976	C=C stretch	RCH=CHR (*trans*)
1029	C—O stretch	Aryl alkyl ether
1218	C—O stretch	Aryl alkyl ether
1315	C—O stretch	Carboxylic acid
1444	C=C stretch	Aromatic
1444	C—H bend	—CH$_3$
1444	O—H bend	Carboxylic acid
1598	C=C stretch	Aromatic
1678	C=C stretch	RCH=CHR (*trans*)
1678	C=O stretch	α,β-Unsaturated carboxylic acid
2844	C—H stretch	Alkyl
2844	O—H stretch	Carboxylic acid
3070	C—H stretch	Alkene
3070	C—H stretch	Aromatic

¹H NMR spectrum:

	Chemical shift (Actual)	(Calculated)	Splitting	Integration	Interpretation
a	3.85	3.7	s	3H	CH$_3$ isolated
b	6.32	6.34	d	1H	Aromatic
c	6.92	7.00	d	2H	Aromatic
d	7.51	7.19	d	2H	Aromatic
e	7.75	7.60	d	1H	Aromatic
f	11.49	10–13	bs	1H	OH isolated

Calculations:

4-methoxycinnamic acid

1	2	3	4	5	6
0.9	7.27	7.27	5.28	5.28	
2.8	−0.27	−0.08	1.35	0.37	
	0.00	0.00	0.97	0.69	
3.7	7.00	7.19	7.60	6.34	10–13

Labeled structure:

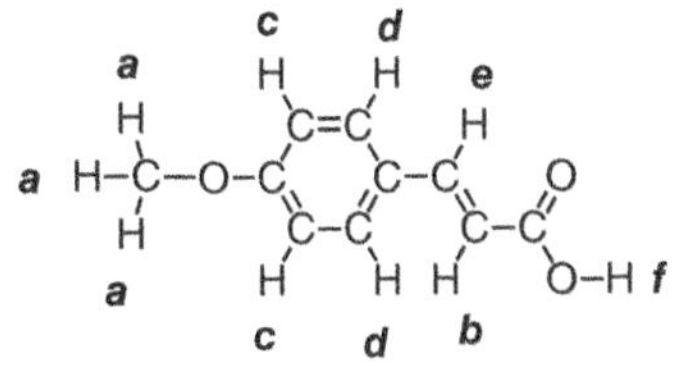

4-methoxycinnamic acid

^{13}C NMR and DEPT spectra:

	Chemical shift		DEPT	Interpretation
	(Actual)	(Calculated)		
A	55.42	55.9	CH$_3$	Ether
B	114.42	114.5	2 CH	Aromatic
C	114.59	117.3	CH	Alkene
D	126.83	130.5	C	Aromatic
E	130.12	127.7	2 CH	Aromatic
F	146.73	145.6	CH	Alkene
G	161.77	159.6	C	Aromatic
H	172.16	160–185	C	Carboxylic acid

Calculations:

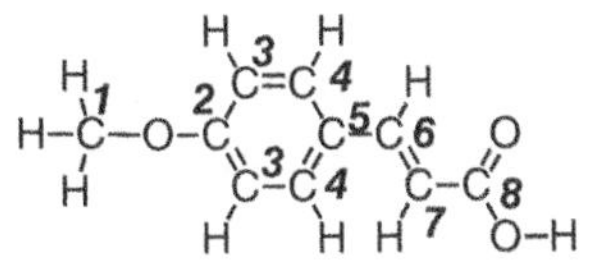

4-methoxycinnamic acid

1	2	3	4	5	6	7	8
−2.1	128.7	128.7	128.7	128.7	123.3	123.3	
58	31.4	−14.4	1.0	−7.7	9.8	5.0	
	−0.5	0.2	−2.0	9.5	12.5	−11.0	
55.9	159.6	114.5	127.7	130.5	145.6	117.3	160–185

Labeled structure:

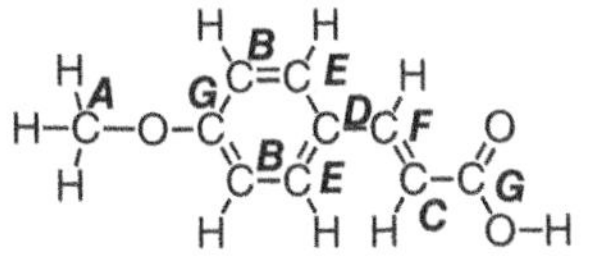

4-methoxycinnamic acid

COSY:

H_a is adjacent to nothing.

H_b is adjacent to H_e.

H_c is adjacent to H_d.

H_d is adjacent to H_c.

H_e is adjacent to H_b.

H_f is adjacent to nothing.

4-methoxycinnamic acid

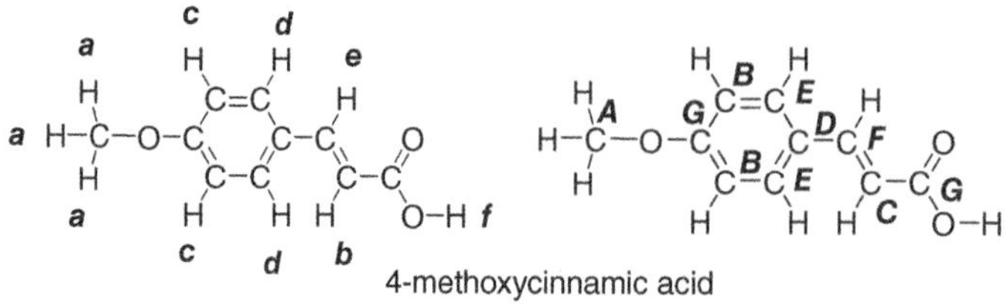

HSQC:

H_a is attached to C_A.

H_b is attached to C_C.

H_c is attached to C_B.

H_d is attached to C_E.

H_e is attached to C_F.

H_f is adjacent to nothing.

2. The unknown compound is in solubility class B, which consists of aliphatic amines with eight or more carbons, anilines, and some ethers. The compound has an active hydrogen (positive acetyl chloride test) and was not an aldehyde or ketone (negative 2,4-dinitrophenylhydrazine test). The Hinsberg test indicates a primary amine. A primary amine with a benzamide derivative that melts at 147 °C matches 2-ethylaniline.

IR spectrum:

Frequency	Bond	Compound type
746	C—H out-of-plane bend	*p*-Disubstituted aromatic
1279	C—N stretch	Primary aromatic amine
1376	C—H bend	—CH$_3$
1456	C=C stretch	Aromatic
1456	C—H bend	—CH$_3$
1586	C=C stretch	Aromatic
1586	N—H bend	Primary amine
2969	C—H stretch	Alkyl
3026	C—H stretch	Aromatic
3381	N—H stretch	Primary amine
3469	N—H stretch	Primary amine

¹H NMR spectrum:

	Chemical shift (Actual)	(Calculated)	Splitting	Integration	Interpretation
a	1.30	1.3	t	3H	CH_3 adjacent to CH_2
b	2.56	2.6	q	2H	CH_2 adjacent to CH_3
c	3.64	1.0–5.0	bs	2H	NH_2 isolated
d	6.71–6.72	6.46	m	1H	Aromatic
e	6.79–6.82	6.58	m	1H	Aromatic
f	7.07–7.13	6.85, 6.88	m	2H	Aromatic

Calculations:

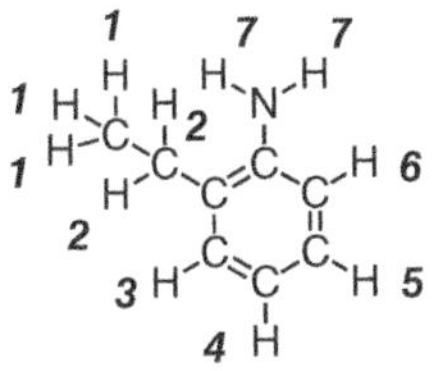

2-ethylaniline

1	2	3	4	5	6	7
0.9	1.2	7.27	7.27	7.27	7.27	
0.4	1.4	−0.15	−0.06	−0.18	−0.06	
		−0.24	−0.63	−0.24	−0.75	
1.3	2.6	6.88	6.58	6.85	6.46	1.0–5.0

Labeled structure:

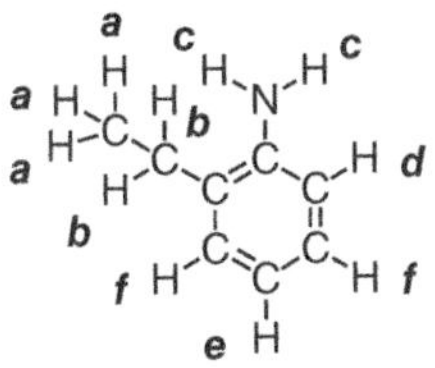

2-ethylaniline

¹³C NMR and DEPT spectra:

	Chemical shift (Actual)	(Calculated)	DEPT	Interpretation
A	13.07	14.9	CH_3	Alkyl
B	24.07	28.9	CH_2	Alkyl
C	115.43	115.3	CH	Aromatic
D	118.86	118.8	CH	Aromatic
E	126.88	126.8	CH	Aromatic
F	128.12	126.9	C	Aromatic
G	128.42	129.0	CH	Aromatic
H	144.09	147.4	C	Aromatic

Calculations:

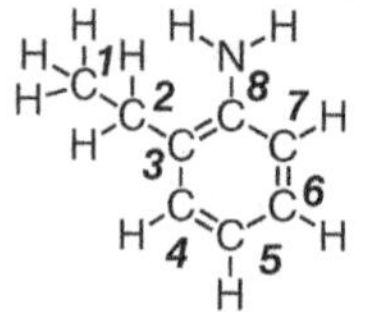

2-ethylaniline

1	2	3	4	5	6	7	8
5.9	5.9	128.7	128.7	128.7	128.7	128.7	128.7
9	23	11.5	−0.6	−0.1	−2.8	−0.1	−0.6
		−13.3	0.9	−9.8	0.9	−13.3	18
14.9	28.9	126.9	129.0	118.8	126.8	115.3	146.1

Labeled structure:

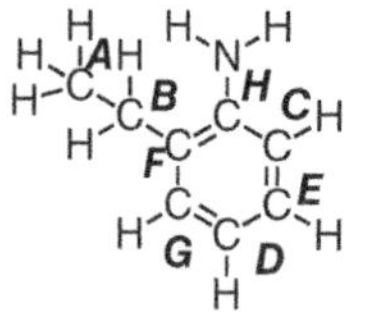

2-ethylaniline

COSY:

H_a is adjacent to H_b.

H_b is adjacent to H_a.

H_c is adjacent to nothing.

H_d is adjacent to H_f.

H_e is adjacent to H_f.

H_f is adjacent to H_d, H_e.

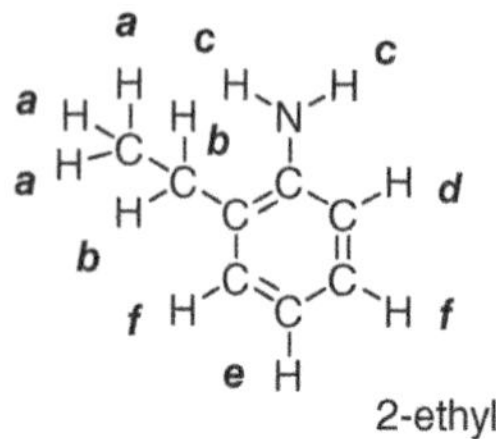

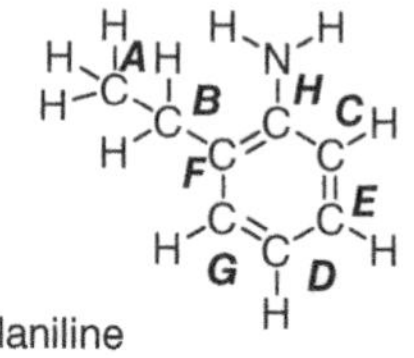

2-ethylaniline

HSQC:

H_a is attached to C_A.

H_b is attached to C_B.

H_c is attached to nothing.

H_d is attached to C_C.

H_e is attached to C_D.

H_f is attached to C_E, C_G.

2-Ethylaniline react with benzoyl chloride to yield *N*-benzoyl-2-ethylaniline.

2-ethylaniline   benzoyl chloride   *N*-benzoyl-2-ethylaniline   2-ethylanilinium chloride

3. The unknown is in solubility class A$_1$, which includes strong organic acids: carboxylic acids with more than six carbons; phenols with electron-withdrawing groups in the ortho and/or para positions; and diketones. The compound is a carboxylic acid (positive sodium test) and an alcohol or aldehyde (positive chromic acid test). 2-Hydroxybenzaldehyde gives both the oxime and the phenylurethane derivatives.

IR spectrum:

Frequency	Bond	Compound type
666	O—H bend	Phenol
759	C—H out-of-plane bend	*o*-Disubstituted aromatic
1202	C—O stretch	Phenol
1384	C—O stretch	Phenol
1460	C=C stretch	Aromatic
1582	C=C stretch	Aromatic
1662	C=O stretch	Aryl aldehyde
2747	C—H stretch	Aldehyde
2848	C—H stretch	Aldehyde
3070	C—H stretch	Aromatic
3191	O—H stretch	Phenol

^{1}H NMR spectrum:

	Chemical shift (Actual)	Chemical shift (Calculated)	Splitting	Integration	Interpretation
a	6.97–7.03	6.98, 7.08	m	2H	Aromatic
b	7.50–7.56	7.40, 7.71	m	2H	Aromatic
c	9.88	4.5–7.7	s	1H	CH isolated
d	11.02	9.0–10.0	s	1H	CH isolated

Calculations:

2-hydroxybenzaldehyde

1	2	3	4	5	6
7.27	7.27	7.27	7.27		
0.58	0.21	0.27	0.21		
−0.14	−0.40	−0.14	−0.50		
7.71	7.08	7.40	6.98	4.5−7.7	9.0−10.0

Labeled structure:

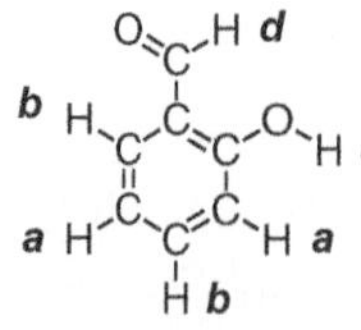

2-hydroxybenzaldehyde

^{13}C NMR and DEPT spectra:

	Chemical shift (Actual)	(Calculated)	DEPT	Interpretation
A	117.61	116.6	CH	Aromatic
B	119.86	122.0	CH	Aromatic
C	120.68	124.6	C	Aromatic
D	133.75	131.4	CH	Aromatic
E	137.00	135.6	CH	Aromatic
F	161.63	156.9	C	Aromatic
G	196.63	175–220	CH	Aldehyde

Calculations:

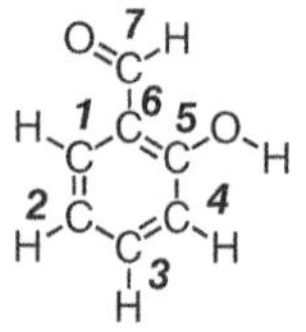

2-hydroxybenzaldehyde

1	2	3	4	5	6	7
128.7	128.7	128.7	128.7	128.7	128.7	
1.3	0.6	5.5	0.6	1.3	8.6	
1.4	−7.3	1.4	−12.7	26.9	−12.7	
131.4	122.0	135.6	116.6	156.9	124.6	175–220

Labeled structure:

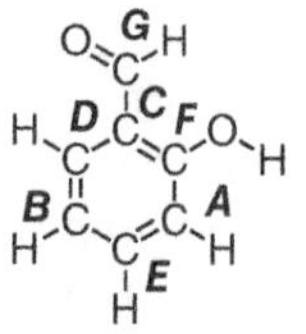

2-hydroxybenzaldehyde

COSY:

H_a is adjacent to H_b.

H_b is adjacent to H_a.

H_c is adjacent to nothing.

H_d is adjacent to nothing.

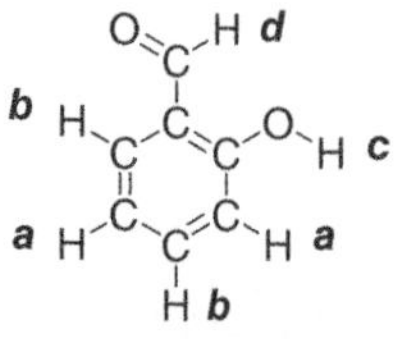

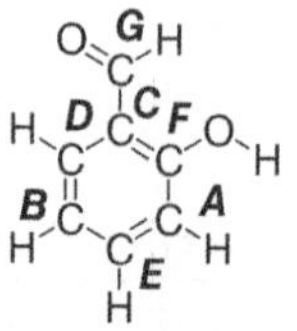

2-hydroxybenzaldehyde

HSQC:

H_a is attached to C_A, C_B.

H_b is attached to C_D, C_E.

H_c is attached to nothing.

H_d is attached to C_G.

2-Hydroxybenzaldehyde reacts with hydroxylamine to form the oxime of 2-hydroxybenzaldehyde.

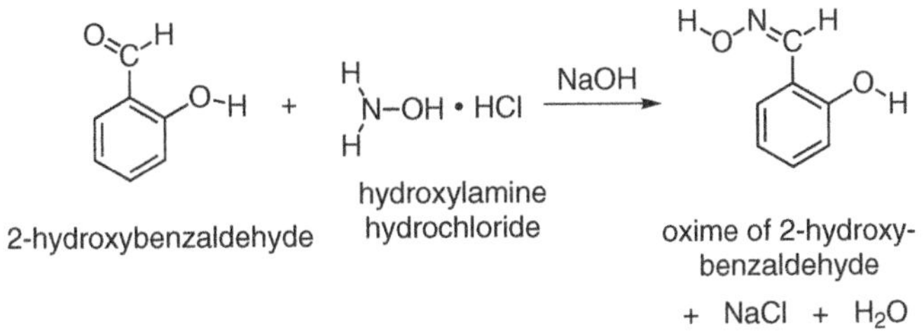

2-hydroxybenzaldehyde   hydroxylamine hydrochloride   oxime of 2-hydroxy-benzaldehyde

+ NaCl + H₂O

2-Hydroxybenzaldehyde reacts with phenyl isocyanate to produce 2-formylphenyl phenyl carbamate.

2-hydroxybenzaldehyde   phenyl isocyanate   2-formylphenyl phenyl carbamate

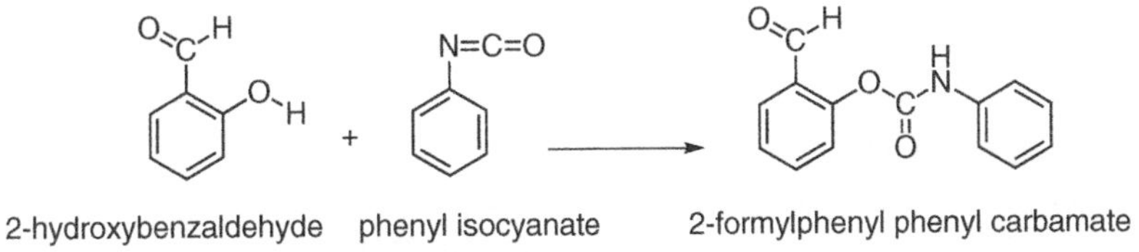

4. The unknown is in solubility class S_B, which includes monofunctional amines with six carbons or fewer. Only tertiary amines give soluble products with the Hinsberg test. Triethylamine fits the boiling point of the unknown and the melting point of the methyl iodide derivative.

IR spectrum:

Frequency	Bond	Compound type
1070	C—N stretch	3° aliphatic amine
1202	C—H bend	—CH₂
1382	C—H bend	—CH₃
1446	C—H bend	—CH₃
1470	C—H bend	—CH₂
2970	C—H stretch	Alkyl

¹H NMR spectrum:

	Chemical shift (Actual)	(Calculated)	Splitting	Integration	Interpretation
a	0.92	1.1	t	9H	3 (CH₃ adjacent to CH₂)
b	2.42	2.7	q	6H	3 (CH₂ adjacent to CH₃)

Calculations:

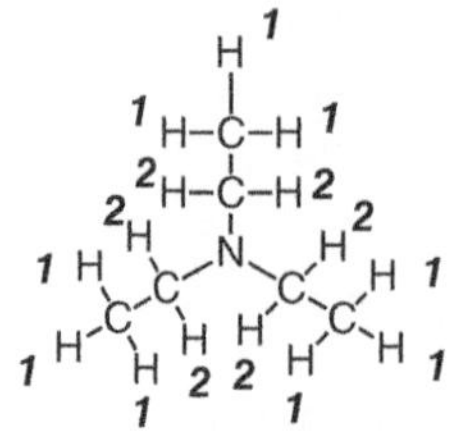

triethylamine

1	2
0.9	1.2
0.2	1.5
1.1	2.7

Labeled structure:

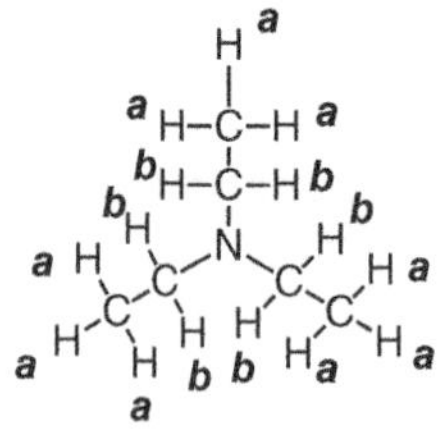

triethylamine

^{13}C NMR and DEPT spectra:

	Chemical shift		DEPT	Interpretation
	(Actual)	(Calculated)		
A	11.57	11.9	3 CH_3	Alkyl
B	46.21	47.9	3 CH_2	Amine

Calculations:

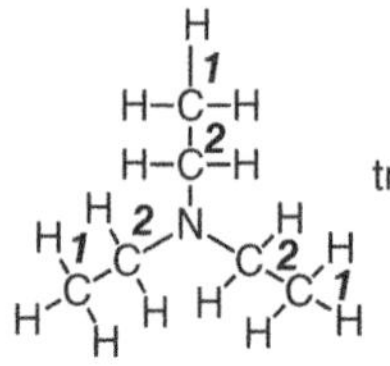

triethylamine

1	2
5.9	5.9
6	42
11.9	47.9

Labeled structure:

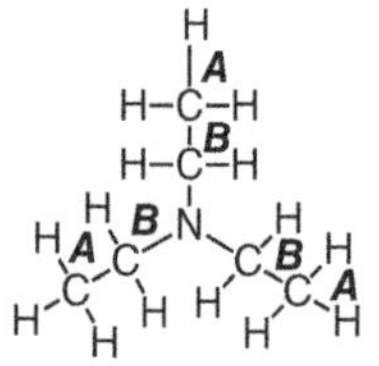

triethylamine

H_a is adjacent to H_b.

H_b is adjacent to H_a.

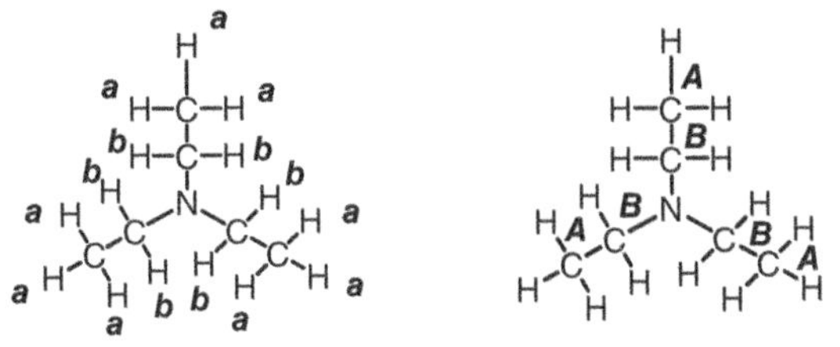

triethylamine

HSQC:

H_a is attached to C_A.

H_b is attached to C_B.

Triethylamine reacts with methyl iodide to form the triethylmethylammonium chloride.

triethylamine   methyl iodide   triethylmethylammonium chloride

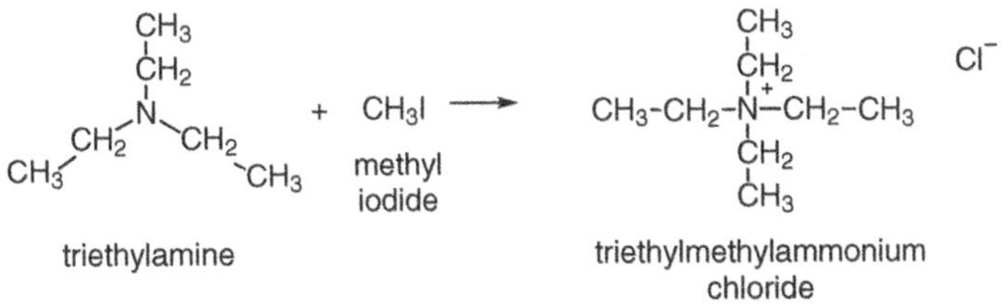

5. The unknown is in solubility class A_1, which includes strong carboxylic acids with more than six carbons; phenols with electron-withdrawing groups in the ortho and para positions; and diketones. A carboxylic acid is present (positive sodium bicarbonate test). No aldehyde or ketone is present (negative 2,4-dinitrophenylhydrazine test). A double or triple bond is present (positive potassium permanganate test). The melting point of the unknown and the derivative indicates *trans*-cinnamic acid.

IR spectrum:

Frequency	Bond	Compound type
707	C—H out-of-plane bend	Monosubstituted aromatic
768	C—H out-of-plane bend	Monosubstituted aromatic
978	C—H out-of-plane bend	RCH=CHR (*trans*)
1221	C—O stretch	Carboxylic acid
1419	C=C stretch	Aromatic
1419	O—H bend	Carboxylic acid
1577	C=C stretch	Aromatic
1670	C=C stretch	RCH=CHR (*trans*)
1670	C=O stretch	Aryl carboxylic acid
2824	O—H stretch	Carboxylic acid
3022	C—H stretch	Alkene
3022	C—H stretch	Aromatic

¹H NMR spectrum:

	Chemical shift (Actual)	(Calculated)	Splitting	Integration	Interpretation
a	6.47	6.34	d	1H	CH adjacent to CH
b	7.42–7.56	7.27	m	5H	Aromatic
c	7.81	7.60	d	1H	CH adjacent to CH
d	11.91	10–13	s	1H	OH isolated

Calculations:

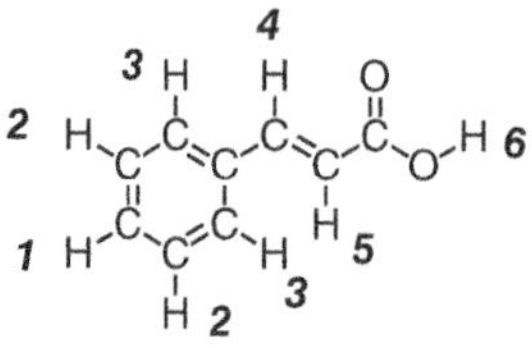

trans-cinnamic acid

1	2	3	4	5	6
7.27	7.27	7.27	5.28	5.28	
0.00	0.00	0.00	1.35	0.37	
			0.97	0.69	
7.27	7.27	7.27	7.60	6.34	10–13

Labeled structure:

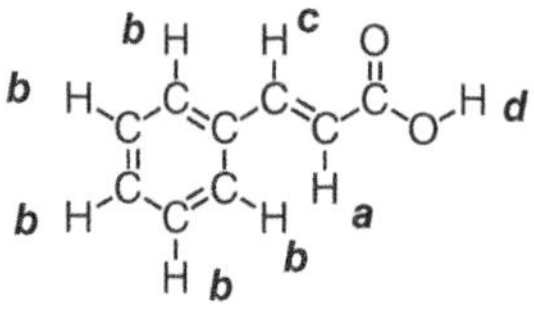

trans-cinnamic acid

¹³C NMR and DEPT spectra:

	Chemical shift (Actual)	(Calculated)	DEPT	Interpretation
A	117.34	117.3	CH	Alkene
B	128.40	126.7	2 CH	Aromatic
C	128.99	128.9	2 CH	Aromatic
D	130.79	128.2	CH	Aromatic
E	134.06	138.2	C	Aromatic
F	147.15	145.6	CH	Alkene
G	172.60	160–185	C	Carboxylic acid

Calculations:

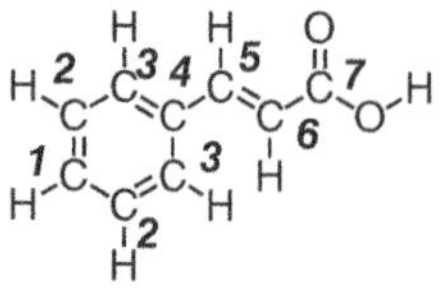

trans-cinnamic acid

1	2	3	4	5	6	7
128.7	128.7	128.7	128.7	123.3	123.3	
−0.5	0.2	−2.0	9.5	12.5	−11.0	
				9.8	5.0	
128.2	128.9	126.7	138.2	145.6	117.3	160–185

Labeled structure:

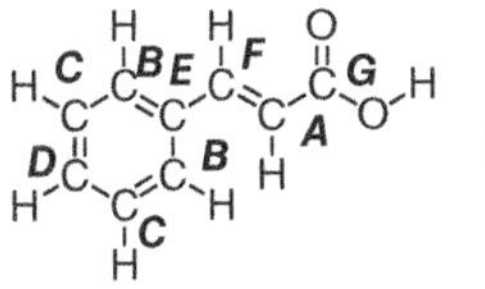

trans-cinnamic acid

COSY:

H_a is adjacent to H_c.

H_b is adjacent to nothing.

H_c is adjacent to H_a.

H_d is adjacent to nothing.

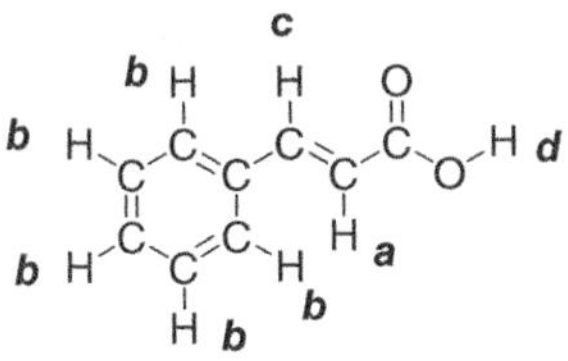

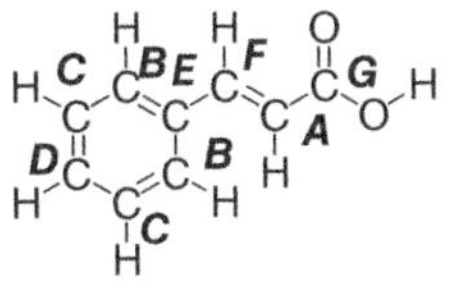

trans-cinnamic acid

HSQC:

H_a is attached to C_A.

H_b is attached to C_B, C_C, **and** C_D.

H_c is attached to C_F.

H_d is attached to nothing.

Trans-cinnamic acid reacts with thionyl chloride to yield trans-cinnamoyl chloride, which reacts with 4-toluidide to form *N*-tolyl-*trans*-cinnamamide.

trans-cinnamic acid   SOCl₂ thionyl chloride   *trans*-cinnamoyl chloride   4-toluidine   *N*-tolyl-*trans*-cinnamamide

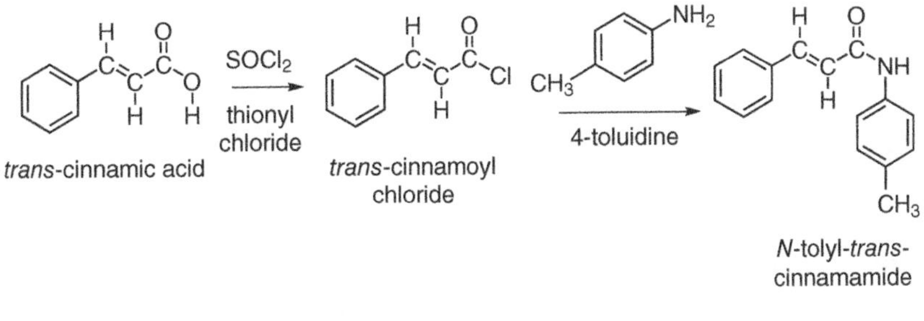

6. The unknown is in solubility class S_1, which indicates a monofunctional alcohol, aldehyde, ketone, ester, nitrile, or amide with five carbons or fewer. No active hydrogen is present (negative acetyl chloride test). The unknown is an aldehyde or ketone (positive hydroxylamine hydrochloride test) but limits it to a ketone (negative Tollens test). Cyclopentanone matches the boiling point and the melting point of the derivative.

IR spectrum:

Frequency	Bond	Compound type
1149	C—H bend	—CH$_2$
1456	C—H bend	Cyclopentane
1470	C—H bend	—CH$_2$
1743	C=O stretch	Cyclopentanone
2887	C—H stretch	Alkyl

¹H NMR spectrum:

	Chemical shift		Splitting	Integration	Interpretation
	(Actual)	(Calculated)			
a	1.82	1.5	d of d	4H	2 (CH$_2$ adjacent to 2CH$_2$)
b	2.00	2.4	t	4H	2 (CH$_2$ adjacent to CH$_2$)

Calculations:

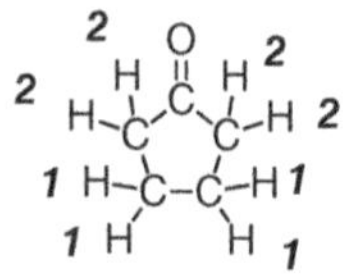

cyclopentanone

1	*2*
1.2	1.2
0.3	1.2
1.5	2.4

Labeled structure:

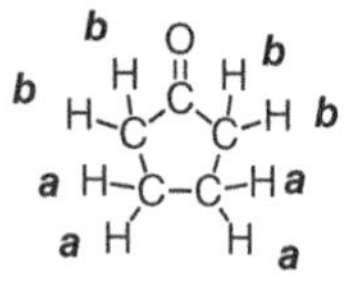

cyclopentanone

¹³C NMR and DEPT spectra:

	Chemical shift		DEPT	Interpretation
	(Actual)	(Calculated)		
A	23.10	6.9	2 CH$_2$	Alkyl
B	38.17	30.0	2 CH$_2$	Alkyl
C	219.34	180–220	C	Ketone

Calculations:

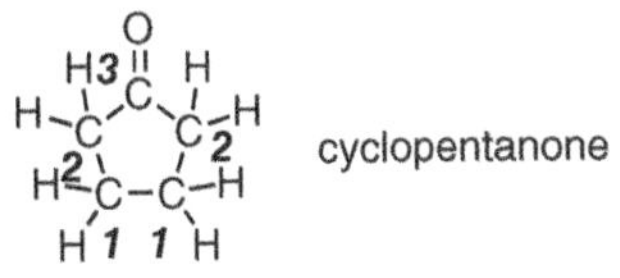

cyclopentanone

1	2	3
5.9	5.9	
1.0	30.0	
6.9	35.9	180–220

Labeled structure:

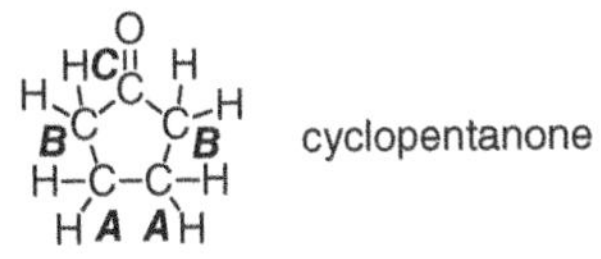

cyclopentanone

COSY:

H_a is adjacent to H_b.

H_b is adjacent to H_a.

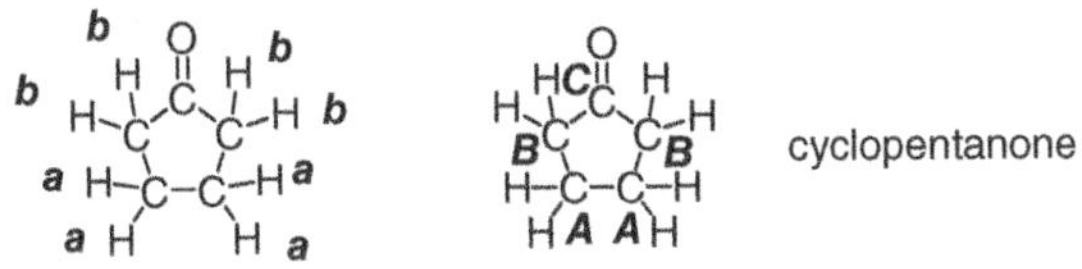

cyclopentanone

HSQC:

H_a is attached to C_A.

H_b is attached to C_B.

Cyclopentanone reacts with semicarbazide hydrochloride to yield the semicarbazone of cyclopentanone.

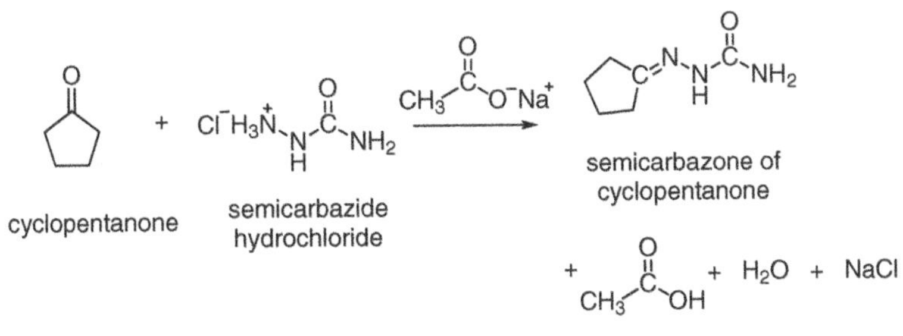

7. The unknown is in solubility class A₁, which includes strong organic acids with more than six carbons; phenols with electron-withdrawing groups; and β-diketones. The compound is a carboxylic acid (positive sodium bicarbonate test), not a phenol (negative ferric chloride test), and not an aldehyde or ketone (negative 2,4-dinitrophenylhydrazine test). With the melting point of 77 °C and a 4-toluidide derivative melting point of 135 °C, the unknown can be either phenylacetic acid or 2-hydroxy-2-methylpropanoic acid. The IR and NMR spectra eliminate the 2-hydroxy-2-methylpropanoic acid as a possibility since aromatic peaks are present.

IR spectrum:

Frequency	Bond	Compound type
698	C–H out-of-plane bend	Monosubstituted aromatic
751	C–H out-of-plane bend	Monosubstituted aromatic
1186	C–H bend	$-CH_2$
1227	C–O stretch	Carboxylic acid
1408	O–H bend	Carboxylic acid
1456	C=C stretch	Aromatic
1602	C=C stretch	Aromatic
1690	C=O stretch	Aliphatic carboxylic acid
2921	C–H stretch	Alkyl
3030	C–H stretch	Aromatic
3030	O–H stretch	Carboxylic acid

^{1}H NMR spectrum:

	Chemical shift		Splitting	Integration	Interpretation
	(Actual)	(Calculated)			
a	3.67	3.7	s	2H	CH_2 isolated
b	7.29–7.35	7.09, 7.10, 7.27	m	5H	Aromatic
c	11.44	10–13	s	1H	OH isolated

Calculations:

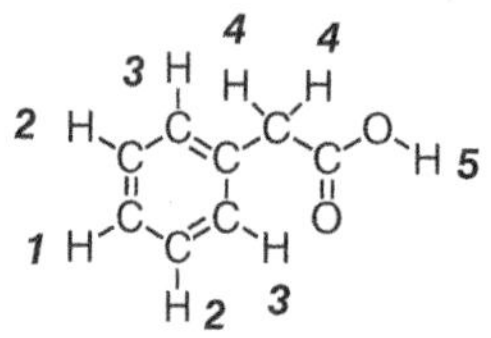

phenylacetic acid

1	2	3	4	5
7.27	7.27	7.27	1.2	
−0.18	0.00	−0.17	1.4	
			1.1	
7.09	7.27	7.10	3.7	10–13

Labeled structure:

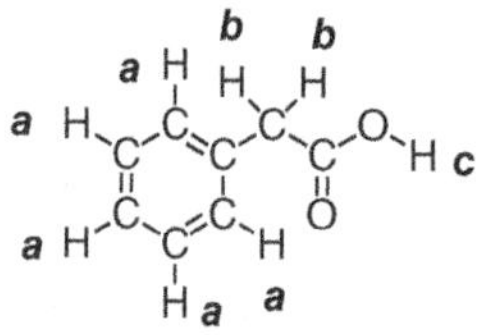

phenylacetic acid

^{13}C NMR and DEPT spectra:

	Chemical shift		DEPT	Interpretation
	(Actual)	(Calculated)		
A	41.11	40.9	CH_2	Alkyl
B	127.40	125.8	CH	Aromatic
C	128.69	128.6	2 CH	Aromatic
D	129.42	129.4	2 CH	Aromatic
E	133.26	137.6	C	Aromatic
F	178.24	160–185	C	Carboxylic acid

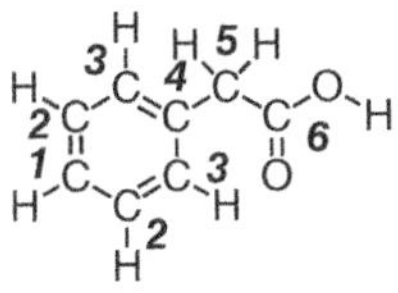

phenylacetic acid

1	2	3	4	5	6
128.7	128.7	128.7	128.7	−2.1	
−2.9	−0.1	0.7	8.9	23	
				21	
125.8	128.6	129.4	137.6	40.9	160–185

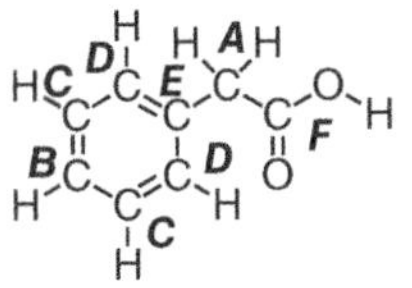

phenylacetic acid

COSY:

H_a is adjacent to nothing.

H_b is adjacent to nothing.

H_c is adjacent to nothing.

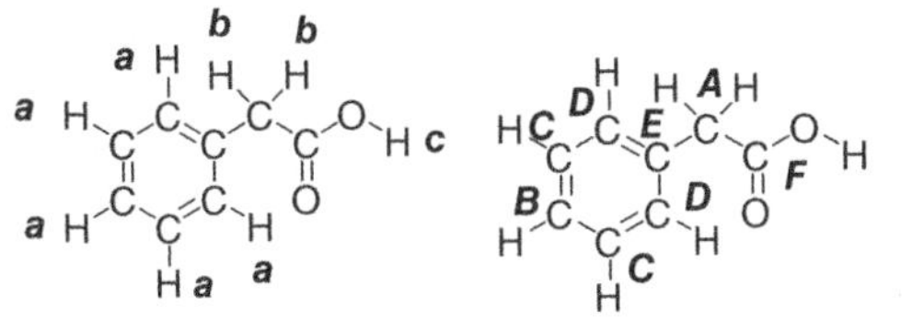

phenylacetic acid

HSQC:

H_a is attached to C_A.

H_b is attached to C_B, C_C, and C_D.

H_c is attached to nothing.

Phenylacetic acid reacts with 4-toluidine to produce N-(4-tolyl)benzamide.

phenylacetic acid   phenylacetyl chloride   N-(4-tolyl)benzamide

8. The unknown is in solubility class A_1, which includes strong organic acids with more than six carbons; phenols with electron-withdrawing groups; and β-diketones. The compound is a carboxylic acid (positive sodium bicarbonate test). The compound tested positive for anhydride, acyl halide, ester, amide, nitrile, sulfonic acid, or sulfonyl chloride (positive ferric chloride test). Acetylsalicylic acid matches the melting point and the melting point of the derivative.

IR spectrum:

Frequency	Bond	Compound type
755	C—H out-of-plane bend	o-Disubstituted aromatic
1182	C—O stretch	Phenyl acetate
1307	C—O stretch	Carboxylic acid
1372	C—H bend	—CH$_3$
1420	O—H bend	Carboxylic acid
1456	C=C stretch	Aromatic
1456	C—H bend	—CH$_3$
1606	C=C stretch	Aromatic
1682	C=O stretch	Aryl carboxylic acid
1751	C=O stretch	Phenyl acetate
2828	C—H stretch	Alkyl
2828	O—H stretch	Carboxylic acid
3006	C—H stretch	Aromatic

¹H NMR spectrum:

| | Chemical shift | | Splitting | Integration | Interpretation |
	(Actual)	(Calculated)			
a	2.35	2.00	s	3H	CH$_3$ isolated
b	7.13–7.15	7.19	m	1H	Aromatic
c	7.34–7.38	7.41	m	1H	Aromatic
d	7.61–7.65	7.47	m	1H	Aromatic
e	8.11–8.14	8.07	m	1H	Aromatic
f	9.66	10–13	bs	1H	CH isolated

Calculations:

acetylsalicylic acid

1	2	3	4	5	6
0.9	7.27	7.27	7.27	7.27	
1.1	−0.22	0.00	0.00	0.00	
	0.14	0.20	0.14	0.80	
2.0	7.19	7.47	7.41	8.07	10–13

Labeled structure:

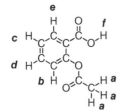

acetylsalicylic acid

^{13}C NMR and DEPT spectra:

	Chemical shift		DEPT	Interpretation
	(Actual)	(Calculated)		
A	21.03	17.9	CH_3	Alkyl
B	122.24	124.4	C	Aromatic
C	124.03	122.3	CH	Aromatic
D	126.19	126.4	CH	Aromatic
E	132.53	131.5	CH	Aromatic
F	134.91	135.1	CH	Aromatic
G	151.27	153.2	C	Aromatic
H	169.76	150–185	C	Ester
I	169.95	160–185	C	Carboxylic acid

Calculations:

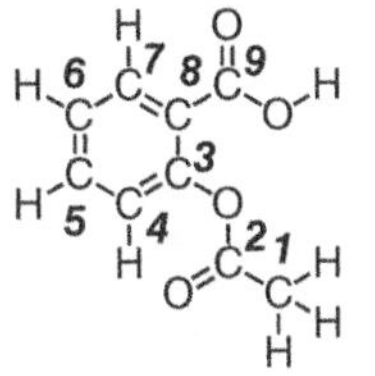

acetylsalicylic acid

1	2	3	4	5	6	7	8	9
−2.1		128.7	128.7	128.7	128.7	128.7	128.7	
20.0		1.5	0.0	5.1	0.0	1.5	2.1	
		23	−6.4	1.3	−2.3	1.3	−6.4	
17.9	150–185	153.2	122.3	135.1	126.4	131.5	124.4	160–185

Labeled structure:

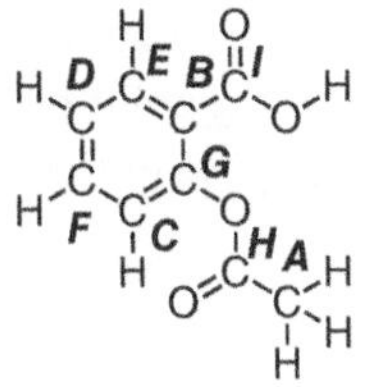

acetylsalicylic acid

COSY:

H_a is adjacent to nothing.

H_b is adjacent to H_d.

H_c is adjacent to H_d, H_e.

H_d is adjacent to H_b, H_c.

H_e is adjacent to H_c.

H_f is adjacent to nothing.

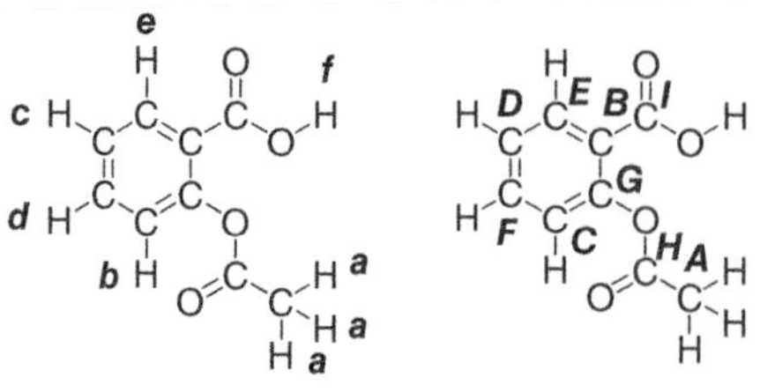

acetylsalicylic acid

HSQC:

H_a is attached to C_A.

H_b is attached to C_C.

H_c is attached to C_D.

H_d is attached to C_F.

H_e is attached to C_E.

H_f is attached to nothing.

Sodium hydroxide is slowly added to acetylsalicylic acid to produce the sodium acetylsalicylate, which reacts with 4-nitrobenzyl chloride to yield 4-nitrobenzyl acetylsalicylate.

acetylsalicylic acid sodium acetylsalicylate 4-nitrobenzyl chloride 4-nitrobenzyl acetylsalicylate

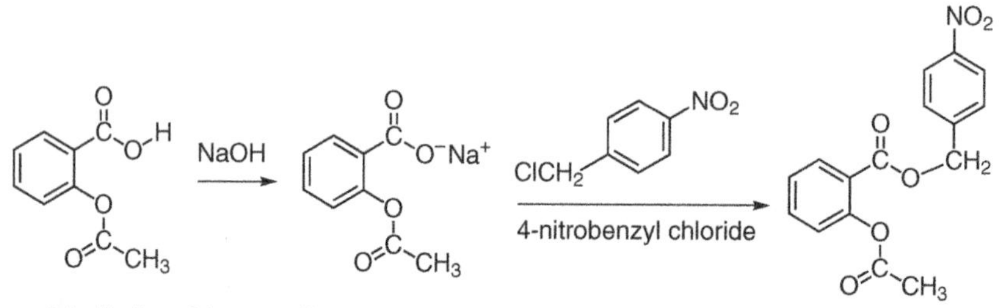

9. The unknown sample is in solubility class N, which includes alcohols, aldehydes, ketones, esters with one functional group and more than five but fewer than nine carbons, ethers, epoxides, alkenes, alkynes, and some aromatic compounds. The compound does not contain an active hydrogen (negative acetyl chloride test), an aldehyde or ketone (negative 2,4-dinitrophenylhydrazine test), or a double bond or triple bond (negative potassium permanganate test). With the boiling point of the sample, melting point of the derivative, and the classification tests, the only possibility is methyl benzoate.

IR spectrum:

Frequency	Bond	Compound type
706	C—H out-of-plane bend	Monosubstituted aromatic
730	C—H out-of-plane bend	Monosubstituted aromatic
1275	C—O stretch	Ester of aromatic acid
1452	C=C stretch	Aromatic
1452	C—H bend	—CH_3
1602	C=C stretch	Aromatic
1719	C=O stretch	Aryl ester
2953	C—H stretch	Alkyl
3079	C—H stretch	Aromatic

^1H NMR spectrum:

	Chemical shift		Splitting	Integration	Interpretation
	(Actual)	(Calculated)			
a	3.92	4.0	s	3H	CH_3 isolated
b	7.42–7.46	7.34	m	2H	Aromatic
c	7.54–7.57	7.47	m	1H	Aromatic
d	8.03–8.05	8.01	m	2H	Aromatic

Calculations:

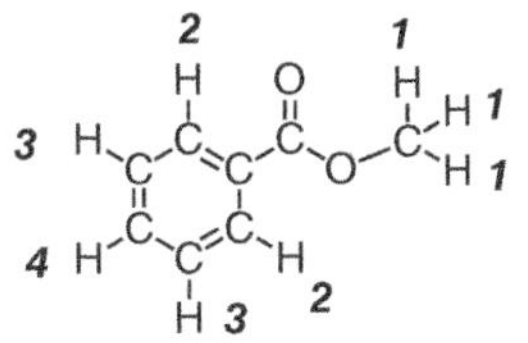

methyl benzoate

1	2	3	4
0.9	7.27	7.27	7.27
3.1	0.74	0.07	0.20
4.0	8.01	7.34	7.47

Labeled structure:

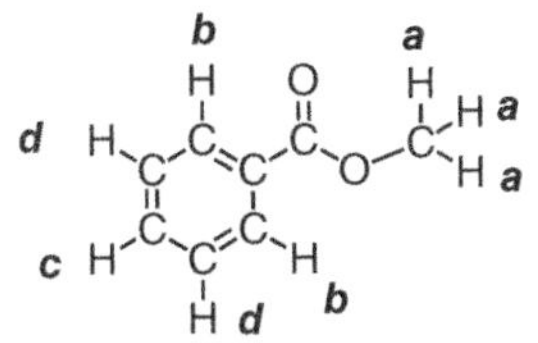

methyl benzoate

^{13}C NMR and DEPT spectra:

	Chemical shift		DEPT	Interpretation
	(Actual)	(Calculated)		
A	52.10	48.9	CH_3	Ester
B	128.36	128.6	2 CH	Aromatic
C	129.58	129.9	2 CH	Aromatic
D	130.18	130.7	C	Aromatic
E	132.91	133.5	CH	Aromatic
F	167.13	150–185	C	Ester

Calculations:

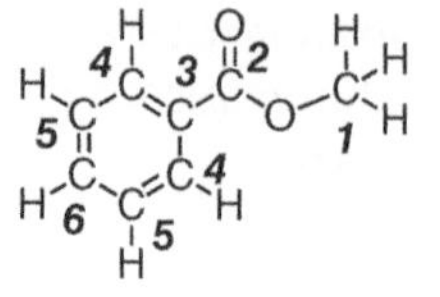

methyl benzoate

1	2	3	4	5	6
−2.1		128.7	128.7	128.7	128.7
51		2.0	1.2	−0.1	4.8
48.9	150–185	130.7	129.9	128.6	133.5

Labeled structure:

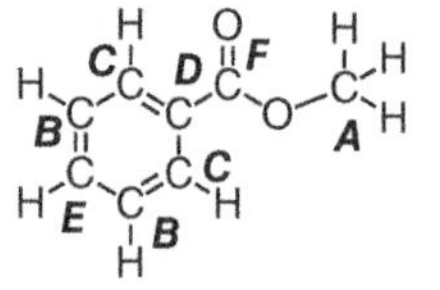

methyl benzoate

COSY:

H_a is adjacent to nothing.

H_b is adjacent to H_c, H_d.

H_c is adjacent to H_b.

H_d is adjacent to H_b.

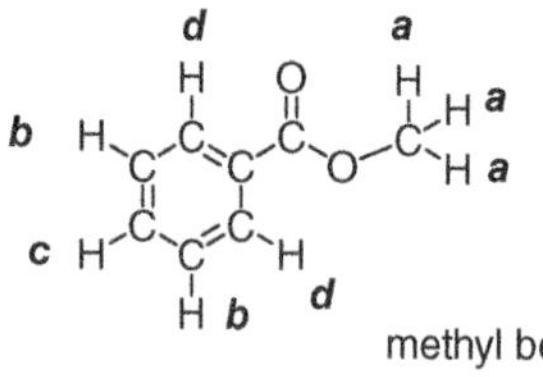

methyl benzoate

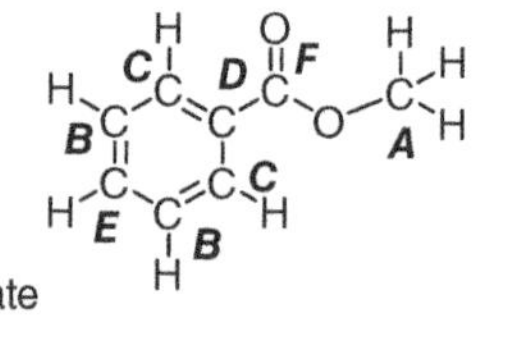

HSQC:

H_a is attached to C_A.

H_b is attached to C_B.

H_c is attached to C_E.

H_d is attached to C_C.

Methyl benzoate reacts with benzylamine to give *N*-benzylbenzamide.

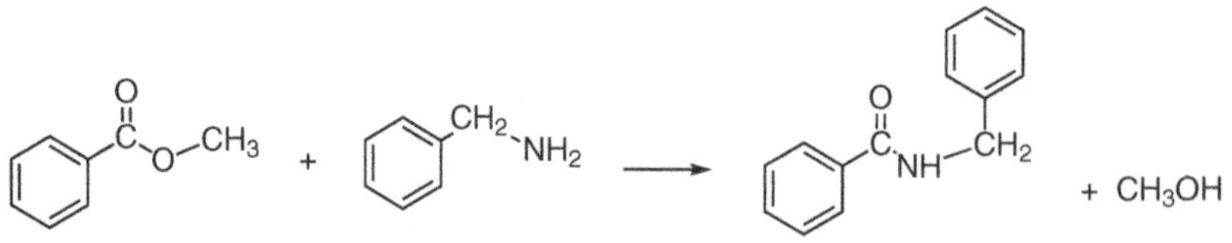

| methyl benzoate | benzylamine | *N*-benzylbenzamide | methanol |

10. The unknown compound is in solubility class S_2, which includes salts of organic acids, amino hydrochlorides, amino acids, and polyfunctional compounds with hydrophilic functional groups. The compound is not an aldehyde or ketone (negative 2,4-dinitrophenylhydrazine test) but has active hydrogen (positive acetyl chloride test). The compound is a carboxylic acid (positive sodium bicarbonate test). The nitrous acid test indicates a primary amine. 3-Aminobenzoic acid matches the melting point of both the sample and the derivative.

IR spectrum:

Frequency	Bond	Compound type
671	C—H out-of-plane bend	*m*-Disubstituted aromatic
756	C—H out-of-plane bend	*m*-Disubstituted aromatic
1222	C—O stretch	Carboxylic acid
1287	C—N stretch	Primary aromatic amine
1449	C=C stretch	Aromatic
1449	O—H bend	Carboxylic acid
1599	C=C stretch	Aromatic
1599	N—H bend	Primary amine
1619	C=O stretch	Aryl carboxylic acid
2960	O—H stretch	Carboxylic acid
3398	N—H stretch	Primary amine
3487	N—H stretch	Primary amine

¹H NMR spectrum:

	Chemical shift (Actual)	Chemical shift (Calculated)	Splitting	Integration	Interpretation
a	5.31	1.0–5.0	bs	2H	NH₂ isolated
b	6.76–6.78	6.72	m	1H	Aromatic
c	7.09–7.18	7.17, 7.22, 7.44	m	3H	Aromatic
d	12.56	10–13	bs	1H	OH isolated

Calculations:

3-aminobenzoic acid

1	**2**	**3**	**4**	**5**	**6**
	7.27	7.27	7.27		7.27
	0.80	0.14	0.20		0.80
	−0.63	−0.24	−0.75		−0.75
10–13	7.44	7.17	6.72	1.0–5.0	7.22

Labeled structure:

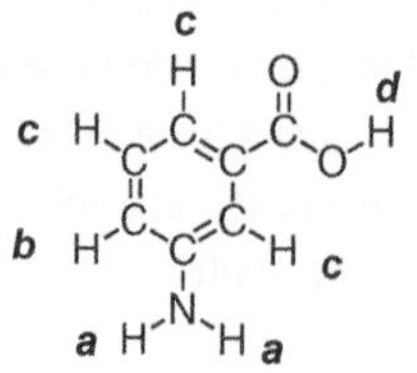

3-aminobenzoic acid

¹³C NMR and DEPT spectra:

| | Chemical shift | | DEPT | Interpretation |
	(Actual)	(Calculated)		
A	114.89	116.9	CH	Aromatic
B	117.10	120.4	CH	Aromatic
C	118.44	120.5	CH	Aromatic
D	129.33	129.6	CH	Aromatic
E	131.78	131.7	C	Aromatic
F	149.29	146.7	C	Aromatic
G	168.35	160–185	C	Carboxylic acid

Calculations:

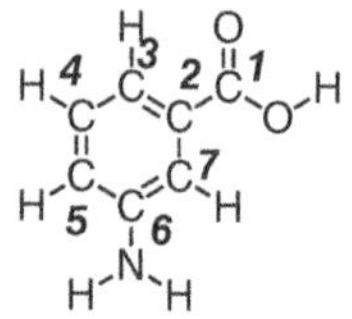

3-aminobenzoic acid

1	2	3	4	5	6	7
	128.7	128.7	128.7	128.7	128.7	128.7
	0.9	−9.8	0.9	−13.3	18.0	−13.3
	2.1	1.5	0.0	5.1	0.0	1.5
160–185	131.7	120.4	129.6	120.5	146.7	116.9

Labeled structure:

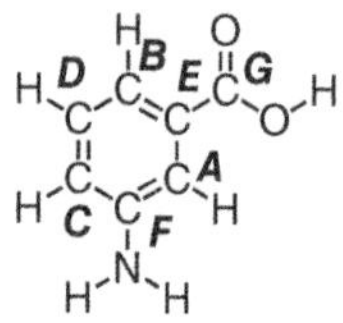

3-aminobenzoic acid

COSY:

H_a is adjacent to nothing.

H_b is adjacent to H_c.

H_c is adjacent to H_b.

H_d is adjacent to nothing.

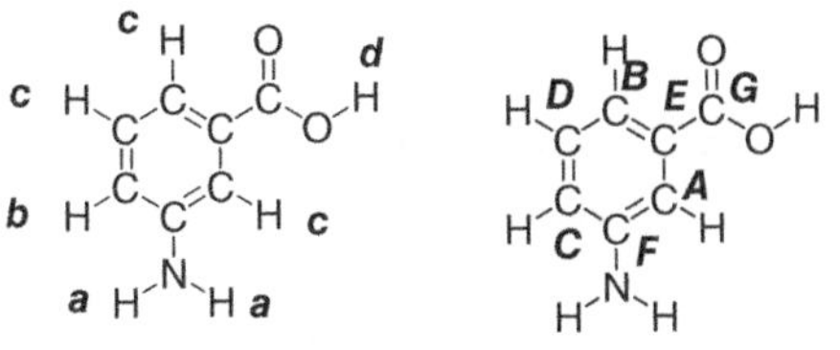

3-aminobenzoic acid

HSQC:

H_a is attached to nothing.

H_b is attached to C_C.

H_c is attached to C_A, C_B, C_D

H_d is attached to nothing.

3-Aminobenzoic acid is reacted with thionyl chloride to give the 3-aminobenzoyl chloride, which reacts with aniline to give the 3-amino-*N*-phenylbenzamide.

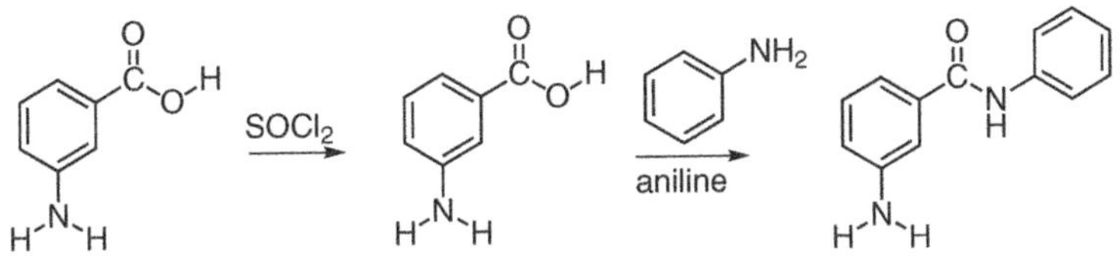

3-aminobenzoic acid   3-aminobenzoyl chloride   3-amino-*N*-phenyl-
                                                                    benzamide

PROBLEM SET 3

1. The unknown sample is in the solubility class S_B, which includes monofunctional amines with six carbons or fewer. The compound has an active hydrogen (positive sodium test and positive sodium test). The compound is not an alcohol or aldehyde (negative Jones test). Propylamine matches the boiling point of the unknown and the melting point of the benzamide derivative.

IR spectrum:

Frequency	Bond	Compound type
1071	C—N stretch	Primary aliphatic amine
1300	C—H bend	—CH_2
1383	C—H bend	—CH_3
1456	C—H bend	—CH_2
1597	N—H bend	Primary amine
2873	C—H stretch	Alkyl
3291	N—H stretch	Primary amine
3369	N—H stretch	Primary amine

^{1}H NMR spectrum:

	Chemical shift (Actual)	Chemical shift (Calculated)	Splitting	Integration	Interpretation
a	0.86	1.0	t	3H	CH_3 adjacent to CH_2
b	1.10	1.0–5.0	s	2H	NH_2 isolated
c	1.41	1.4	sext	2H	CH_2 adjacent to 5H
d	2.61	2.7	t	2H	CH_2 adjacent to CH_2

Calculations:

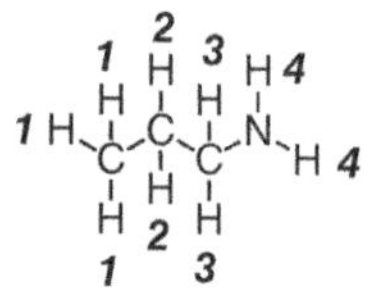

propylamine

1	2	3	4
0.9	1.2	1.2	
0.1	0.2	1.5	
1.0	1.4	2.7	1.0–5.0

Labeled structure:

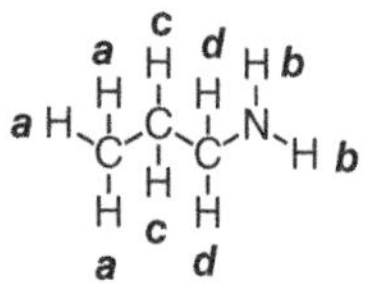

propylamine

^{13}C NMR and DEPT spectra:

	Chemical shift		DEPT	Interpretation
	(Actual)	(Calculated)		
A	11.25	10.6	CH_3	Alkyl
B	26.87	27.1	CH_2	Alkyl
C	44.13	44.6	CH_2	Amine

Calculations:

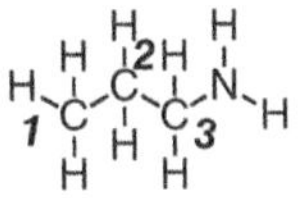

propylamine

1	2	3
15.6	16.1	15.6
−5	11	29
10.6	27.1	44.6

Labeled structure:

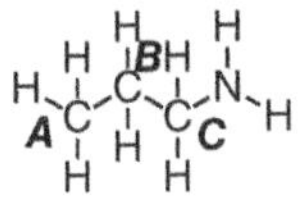

propylamine

COSY:

H_a is adjacent to H_c.

H_b is adjacent to nothing.

H_c is adjacent to H_a, H_d.

H_d is adjacent to H_c.

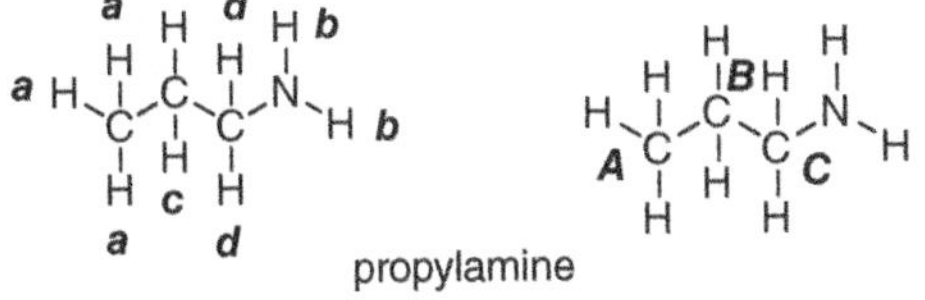

propylamine

HSQC:

H_a is attached to C_A.

H_b is attached to nothing.

H_c is attached to C_B.

H_d is attached to C_C.

Propylamine reacts with benzoyl chloride to form *N*-propylbenzamide.

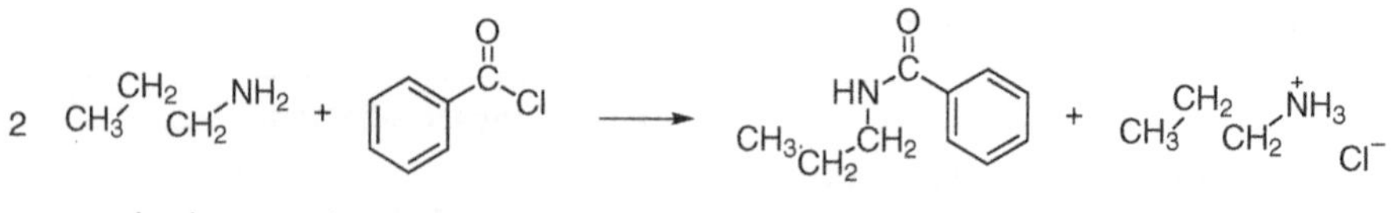

2 CH₃CH₂CH₂NH₂ + (benzoyl chloride) → (benzamide) + CH₃CH₂CH₂N⁺H₃ Cl⁻

propylamine   benzoyl chloride   benzamide   propylammonium
                                                           chloride

2. The unknown is in solubility class S_A, which includes monofunctional carboxylic acids with five carbons or fewer, and arylsulfonic acids. The sample tested contains a carboxylic acid (positive sodium bicarbonate test). 3-Methylbutanoic acid matches the boiling point and the melting point of the derivative.

IR spectrum:

Frequency	Bond	Compound type
930	C—H bend	—CH(CH₃)₂
1217	C—H bend	—CH₂
1305	C—O stretch	Carboxylic acid
1373	C—H bend	—CH(CH₃)₂
1387	C—H bend	—CH(CH₃)₂
1412	O—H bend	Carboxylic acid
1470	C—H bend	—CH₂
1704	C=O stretch	Aliphatic carboxylic acid
2965	O—H stretch	Carboxylic acid

¹H NMR spectrum:

	Chemical shift (Actual)	Chemical shift (Calculated)	Splitting	Integration	Interpretation
a	0.97	1.0	d	6H	2CH₃ adjacent to CH
b	2.10	1.8	nonet	1H	CH adjacent to 8H
c	2.22	2.3	d	2H	CH₂ adjacent to CH
d	11.67	10–13	s	1H	OH isolated

Calculations:

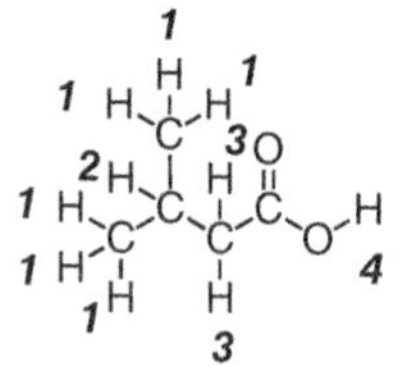

3-methylbutanoic acid

1	2	3	4
0.9	1.5	1.2	
0.1	0.3	1.1	
1.0	1.8	2.3	10–13

Labeled structure:

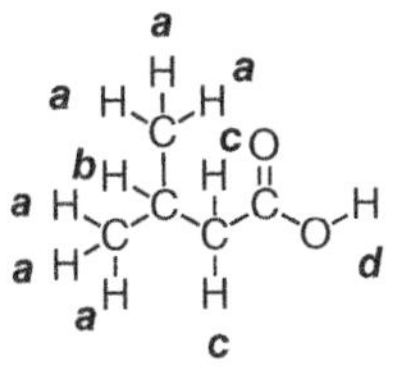

3-methylbutanoic acid

^{13}C NMR and DEPT spectra:

	Chemical shift		DEPT	Interpretation
	(Actual)	(Calculated)		
A	22.33	13.6	2 CH$_3$	Alkyl
B	25.47	18.1	CH	Alkyl
C	43.19	31.6	CH$_2$	Alkyl
D	179.98	160–185	C	Carboxylic acid

Calculations:

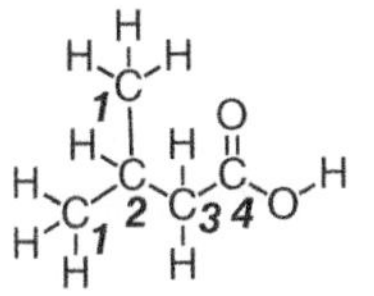

3-methylbutanoic acid

1	2	3	4
15.6	16.1	15.6	
−2	2	16	
13.6	18.1	31.6	160–185

Labeled structure:

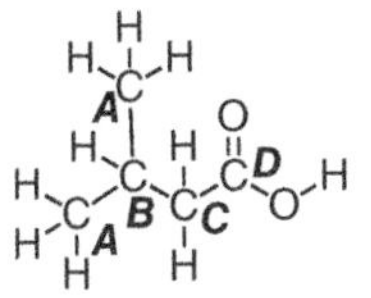

3-methylbutanoic acid

COSY:

H_a is adjacent to **H_b**.

H_b is adjacent to **H_a, H_c**.

H_c is adjacent to **H_b**.

H_d is adjacent to nothing.

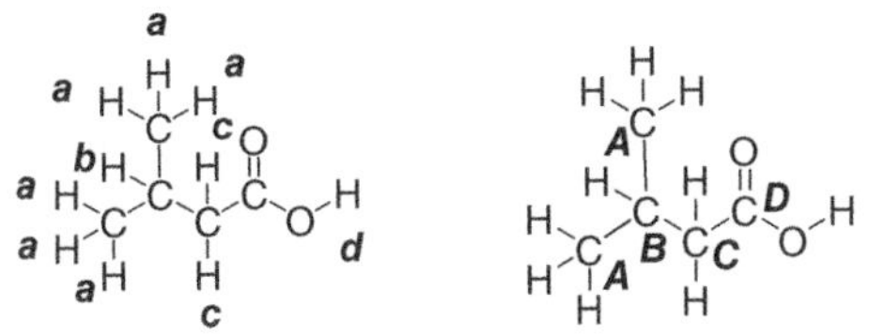

3-methylbutanoic acid

HSQC:

H_a is attached to C_A.

H_b is attached to C_B.

H_c is attached to C_C.

H_d is attached to nothing.

3-Methylbutanoic acid is treated with thionyl chloride to form the 3-methylbutanoyl chloride. This acid chloride reacts with ammonia to form 3-methylbutanamide.

3-methylbutanoic acid → (SOCl₂, thionyl chloride) → 3-methylbutanoyl chloride + SO₂ + HCl → (2 NH₃) → 3-methylbutanamide + NH₄Cl

3. The unknown is in solubility class N, which includes alcohols, aldehydes, ketones, esters with one functional group and more than five but fewer than nine carbons, ethers, epoxides, alkenes, alkynes, and some aromatic compounds with activating groups. The compound is not an aldehyde or ketone (negative 2,4-dinitrophenylhydrazine test), does not have an active hydrogen (negative acetyl chloride test), and does not have a double bond or triple bond (negative potassium permanganate and bromine test). Alcohols, aldehydes, ketones, alkenes, and alkynes are thus eliminated. Anisole matches the boiling point of the sample and the melting point of the derivative.

IR spectrum:

Frequency	Bond	Compound type
691	C—H out-of-plane bend	Monosubstituted aromatic
754	C—H out-of-plane bend	Monosubstituted aromatic
1037	C—O stretch	Aryl alkyl ether
1241	C—O stretch	Aryl alkyl ether
1451	C—H stretch	—CH₃
1494	C=C stretch	Aromatic
1597	C=C stretch	Aromatic
2960	C—H stretch	Alkyl
3004	C—H stretch	Aromatic

¹H NMR spectrum:

	Chemical shift (Actual)	Chemical shift (Calculated)	Splitting	Integration	Interpretation
a	3.84	3.7	s	3H	CH₃ isolated
b	6.94–6.99	7.00	m	3H	Aromatic
c	7.31–7.35	7.19	m	2H	Aromatic

Calculations:

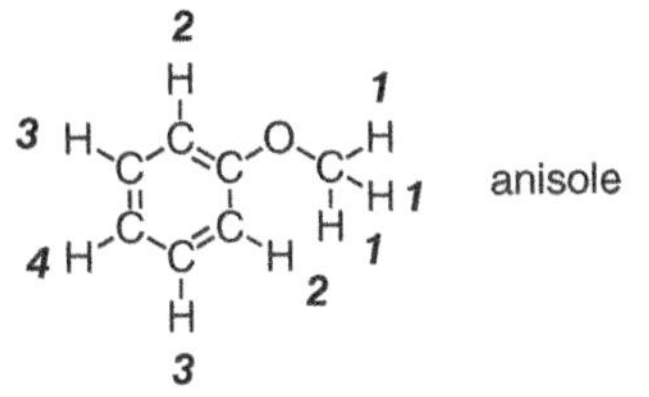

anisole

1	2	3	4
0.9	7.27	7.27	7.27
2.8	−0.27	−0.08	−0.27
3.7	7.00	7.19	7.00

Labeled structure:

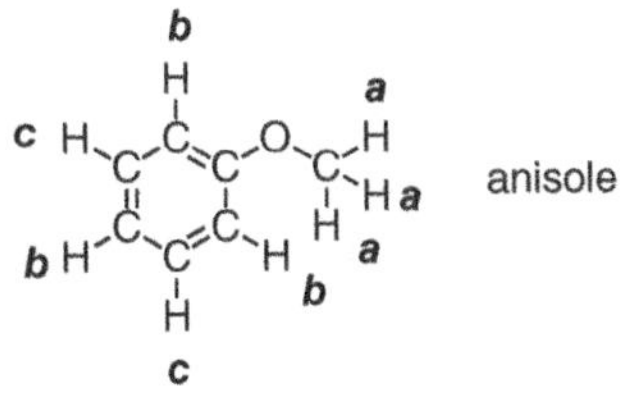

anisole

^{13}C NMR and DEPT spectra:

	Chemical shift		DEPT	Interpretation
	(Actual)	(Calculated)		
A	56.15	55.9	CH_3	Ether
B	113.95	114.3	2CH	Aromatic
C	120.70	121.0	CH	Aromatic
D	129.50	129.7	2CH	Aromatic
E	159.62	160.1	C	Aromatic

Calculations:

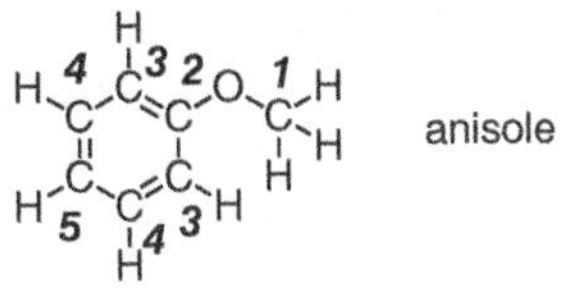

anisole

1	2	3	4	5
−2.1	128.7	128.7	128.7	128.7
58	31.4	−14.4	1.0	−7.7
55.9	160.1	114.3	129.7	121.0

Labeled structure:

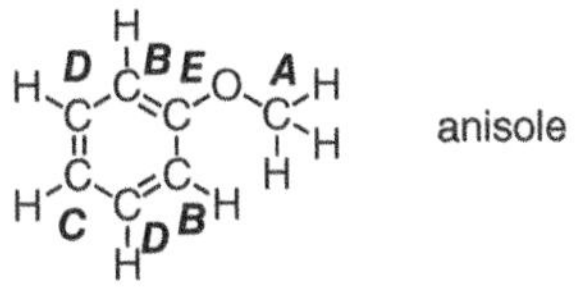

anisole

COSY:

H$_a$ is adjacent to nothing.

H$_b$ is adjacent to *H*$_c$.

H$_c$ is adjacent to *H*$_a$.

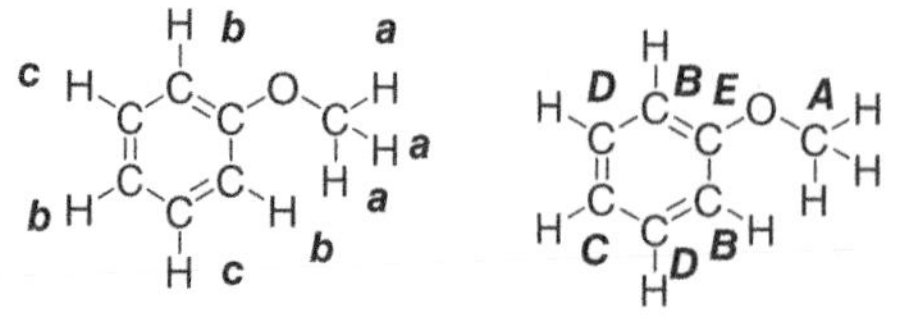

anisole

HSQC:

H$_a$ is attached to *C*$_A$.

H$_b$ is attached to *C*$_B$, *C*$_c$.

H$_c$ is attached to *C*$_D$.

   Anisole is treated with chlorosulfonic acid to yield 4-methoxybenzene-sulfonyl chloride. The sulfonyl chloride is converted to 4-methoxybenzene-sulfonamide with ammonium carbonate or ammonium hydroxide.

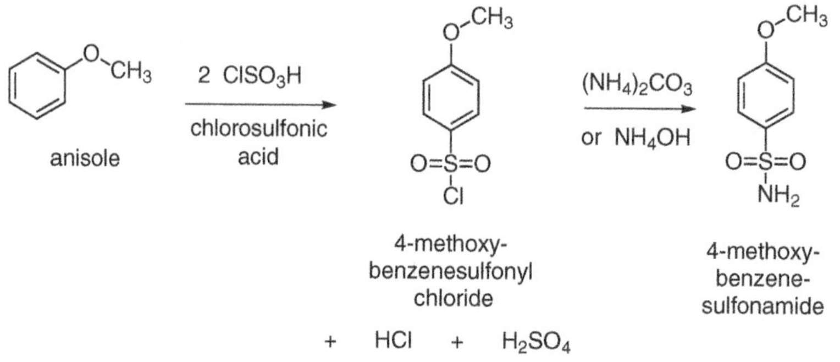

4. The unknown compound is in solubility class A$_2$, which includes weak organic acids: phenols, enols, oximes, imides, sulfonamides, thiophenols, all with more than five carbons, β-ketones, nitro compounds with α-hydrogens. The sample did not contain an aldehyde or ketone (negative 2,4-dinitrophenylhydrazine test) and had an active hydrogen (positive acetyl chloride test). The compound contains a phenol (brown color with ceric ammonium nitrate). The boiling point of the unknown and the melting point of the derivative match guaiacol.

IR spectrum:

Frequency	Bond	Compound type
745	C—H out-of-plane bend	*o*-Disubstituted aromatic
745	O—H bend	Phenol
1022	C—O stretch	Aryl alkyl ether
1207	C—O stretch	Phenol
1241	C—O stretch	Aryl alkyl ether
1363	C—O stretch	Phenol
1456	C=C stretch	Aromatic

Frequency	Bond	Compound type
1456	C—H bend	$—CH_3$
1597	C=C stretch	Aromatic
2970	C—H stretch	Alkyl
3067	C—H stretch	Aromatic
3403	O—H stretch	Phenol

^{1}H NMR spectrum:

	Chemical shift (Actual)	Chemical shift (Calculated)	Splitting	Integration	Interpretation
a	3.89	3.0	s	3H	CH_3 isolated
b	5.85	4.5–7.7	bs	1H	OH isolated
c	6.89–6.90	6.86	m	2H	Aromatic
d	6.92–6.94	6.69	m	1H	Aromatic
e	6.99–7.0	6.90	m	1H	Aromatic

Calculations:

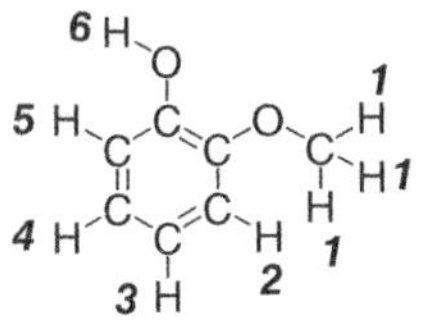

guaiacol

1	*2*	*3*	*4*	*5*	*6*
0.9	7.27	7.27	7.27	7.27	
2.1	–0.27	–0.08	–0.27	–0.08	
	–0.14	–0.40	–0.14	–0.50	
3.0	6.86	6.79	6.86	6.69	4.5–7.7

Labeled structure:

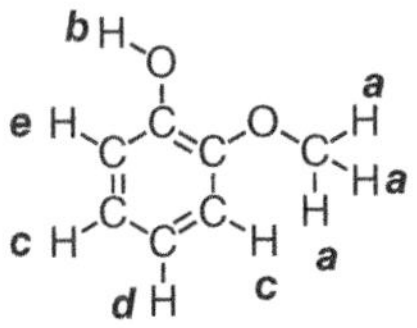

guaiacol

^{13}C NMR and DEPT spectra:

	Chemical shift (Actual)	Chemical shift (Calculated)	DEPT	Interpretation
A	56.91	55.9	CH_3	Ether
B	110.89	115.7	CH	Aromatic
C	114.69	117.0	CH	Aromatic
D	120.24	122.4	CH	Aromatic
E	121.53	122.4	CH	Aromatic
F	145.76	141.2	C	Aromatic
G	146.71	147.4	C	Aromatic

Calculations:

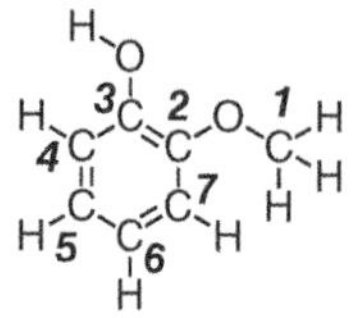

guaiacol

1	2	3	4	5	6	7
−2.1	128.7	128.7	128.7	128.7	128.7	128.7
58	31.4	−14.4	1.0	−7.7	1.0	−14.4
	−12.7	26.9	−12.7	1.4	−7.3	1.4
55.9	147.4	141.2	117.0	122.4	122.4	115.7

Labeled structure:

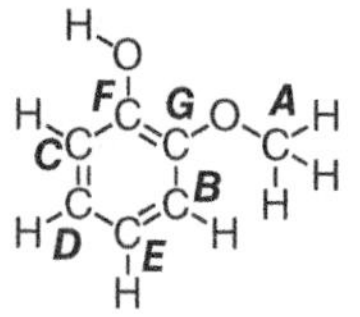

guaiacol

COSY:

H_a is adjacent to nothing.

H_b is adjacent to nothing.

H_c is adjacent to H_d, H_e.

H_d is adjacent to H_c.

H_e is adjacent to H_c.

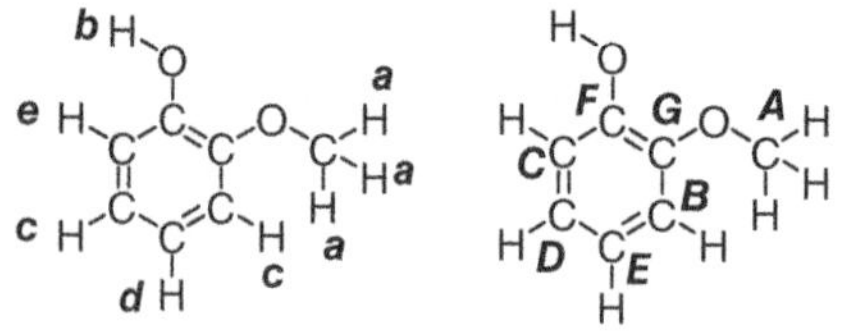

guaiacol

HSQC:

H_a is attached to C_A.

H_b is attached to nothing.

H_c is attached to C_B, C_D.

H_d is attached to C_E.

H_e is attached to C_C.

Guaiacol reacts with phenyl isocyanate to give phenylcarbamic acid 2-methoxyphenyl ester.

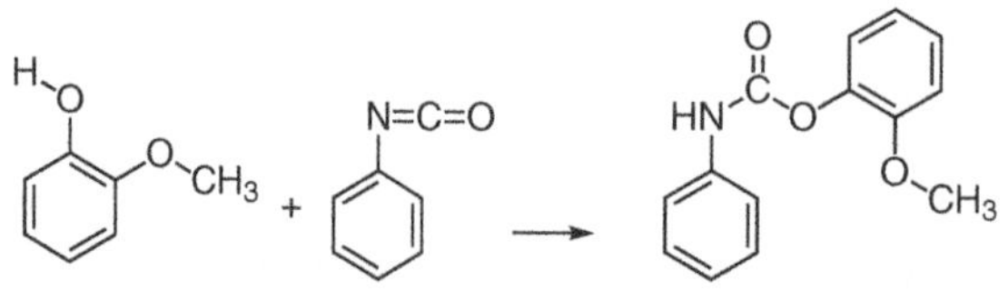

guaiacol   phenyl isocyanate   phenylcarbamic acid
2-methoxyphenyl ester

5. The unknown was in solubility class S_B, which consists of monofunctional amines of six carbons or fewer. The unknown has an active hydrogen (positive sodium and acetyl chloride test). The compound is a primary amine (Hinsberg test results). Butylamine is the only compound that is a primary amine with the boiling point indicated.

IR spectrum:

Frequency	Bond	Compound type
1057	C—N stretch	Primary aliphatic amine
1319	C—H bend	—CH$_2$
1380	C—H bend	—CH$_3$
1465	C—H bend	—CH$_2$
1577	N—H bend	Primary amine
2925	C—H stretch	Alkyl
3288	N—H stretch	Primary amine
3369	N—H stretch	Primary amine

^{1}H NMR spectrum:

	Chemical shift (Actual)	(Calculated)	Splitting	Integration	Interpretation
a	0.75	0.9	t	3H	CH$_3$ adjacent to CH$_2$
b	1.00	1.0–5.0	s	2H	NH$_2$ isolated
c	1.17	1.3	sext	2H	CH$_2$ adjacent to 5H
d	1.27	1.4	quint	2H	CH$_2$ adjacent to 4H
e	2.52	2.7	t	2H	CH$_2$ adjacent to CH$_2$

Calculations:

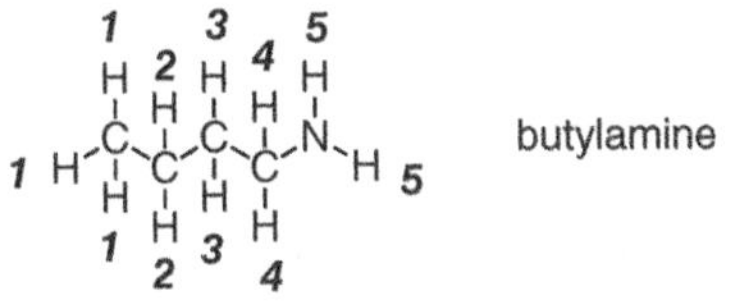

butylamine

1	2	3	4	5
0.9	1.2	1.2	1.2	
	0.1	0.2	1.5	
0.9	1.3	1.4	2.7	1.0–5.0

Labeled structure:

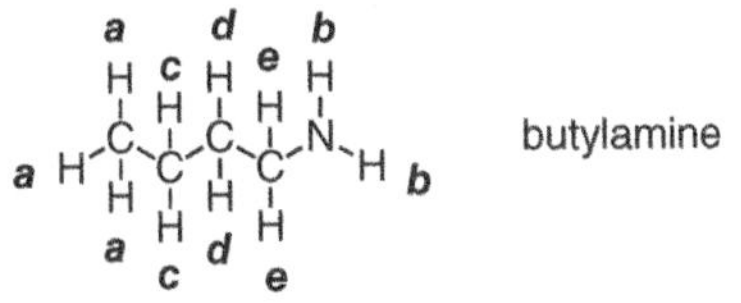

butylamine

^{13}C NMR and DEPT spectra:

	Chemical shift		DEPT	Interpretation
	(Actual)	(Calculated)		
A	13.75	13.2	CH_3	Alkyl
B	19.83	20.0	CH_2	Alkyl
C	36.85	36.0	CH_2	Alkyl
D	47.71	42.2	CH_2	Amine

Calculations:

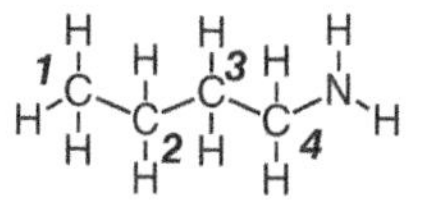

butylamine

1	2	3	4
13.2	25.0	25.0	13.2
	−5	11	29
13.2	20.0	36.0	42.2

Labeled structure:

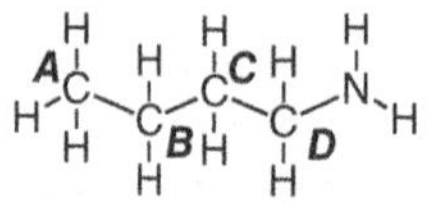

butylamine

COSY:

H_a is adjacent to H_c.

H_b is adjacent to nothing.

H_c is adjacent to H_a, H_d.

H_d is adjacent to H_c.

H_e is adjacent to H_d.

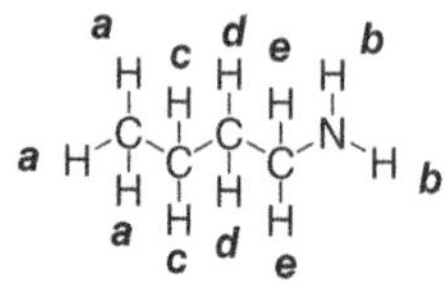

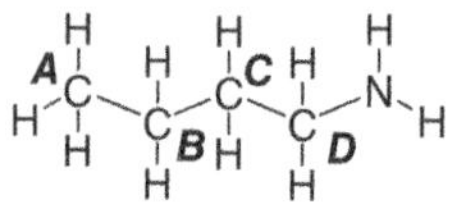

butylamine

HSQC:

H_a is attached to C_A.

H_b is attached to nothing.

H_c is attached to C_B.

H_d is attached to C_C.

H_e is attached to C_D.

The treatment of butyl amine with 4-toluenesulfonyl chloride produces the
N-butyl-4-toluenesulfonamide.

butylamine   4-toluenesulfonyl chloride   (soluble)

+   NaCl   +   2 H_2O

↓ HCl

N-butyl-4-toluenesulfonamide

+   NaCl

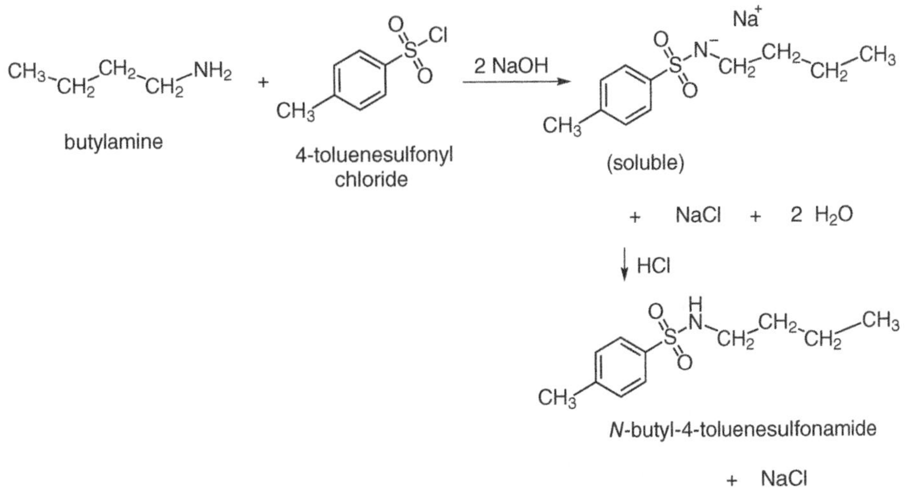

6. The unknown was in solubility class S_B, which consists of monofunctional amines of six
carbons or fewer. The compound has an active hydrogen (positive acetyl chloride test)
and is a secondary amine (Hinsberg test results and positive nitrous acid test). The only
secondary amine that has a boiling point near 84° is diisopropylamine.

IR spectrum:

Frequency	Bond	Compound type
924	C—H bend	—$CH(CH_3)_2$
1178	C—N stretch	Secondary aliphatic amine
1368	C—H bend	—$CH(CH_3)_2$
1380	C—H bend	—$CH(CH_3)_2$
2965	C—H stretch	Alkyl
3308	N—H stretch	Secondary amine

¹H NMR spectrum:

	Chemical shift		Splitting	Integration	Interpretation
	(Actual)	(Calculated)			
a	0.43	1.0–5.0	bs	1H	NH isolated
b	0.90	1.1	d	6H	2 (2 CH_3 adjacent to CH)
c	2.77	3.0	sept	1H	2 (CH adjacent to 2 CH_3)

Calculations:

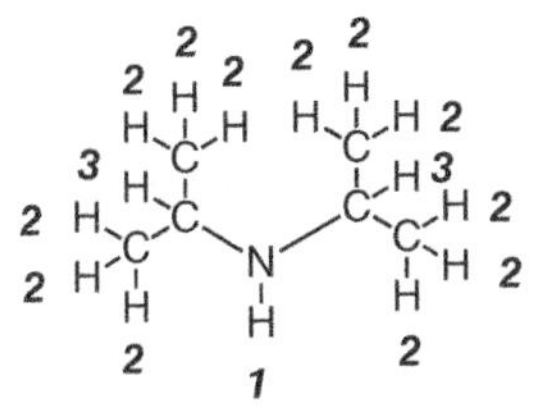

diisopropylamine

	1	2	3
		0.9	1.5
		0.2	1.5
	1.0–5.0	1.1	3.0

Labeled structure:

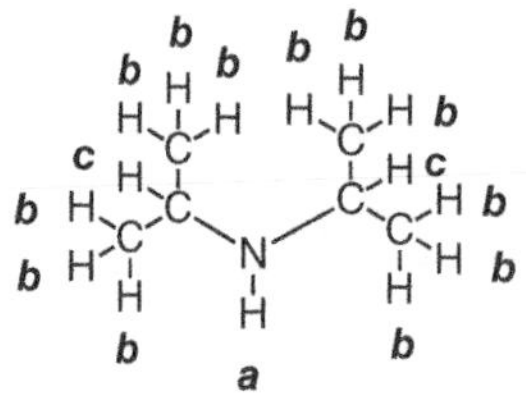

diisopropylamine

¹³C NMR and DEPT spectra:

	Chemical shift		DEPT	Interpretation
	(Actual)	**(Calculated)**		
A	23.28	23.6	4CH₃	Alkyl
B	45.07	53.1	2H	Amine

Calculations:

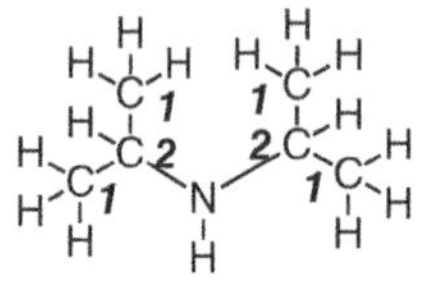

diisopropylamine

1	2
15.6	16.1
8	37
23.6	53.1

Labeled structure:

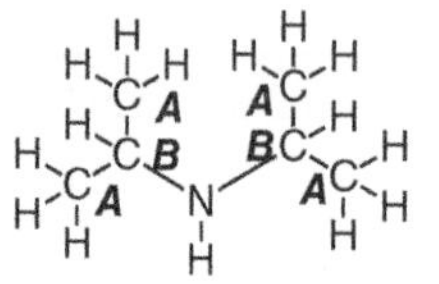

diisopropylamine

COSY:

H$_a$ is adjacent to **H**$_b$.

H$_b$ is adjacent to **H**$_a$.

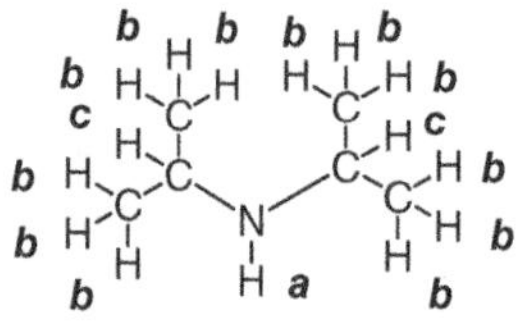

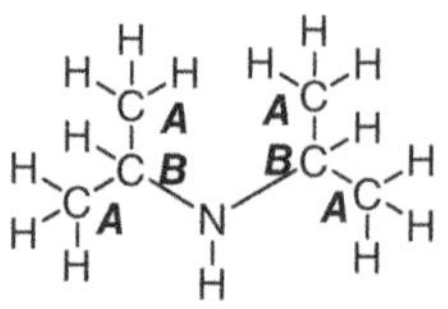

diisopropylamine

HSQC:

H_a is attached to C_A.

H_b is attached to C_B.

7. The unknown is in solubility class S_A, which includes monofunctional carboxylic acids with five carbons or fewer and arylsulfonic acids. The unknown contains a double bond or triple bond (positive bromine and positive potassium permanganate test) and is a carboxylic acid (positive sodium bicarbonate test). 2-Methylpropenoic acid matches the classification tests, the boiling point, and the melting point of the derivative.

IR spectrum:

Frequency	Bond	Compound type
896	C—H out-of-plane bend	$R_2C=CH_2$
1299	C—O stretch	Carboxylic acid
1376	C—H bend	$-CH_3$
1428	O—H bend	Carboxylic acid
1456	C—H bend	$-CH_3$
1638	C=C stretch	$R_2C=CH_2$
1690	C=O stretch	α,β-Unsaturated carboxylic acid
2933	O—H stretch	Carboxylic acid
3078	C—H stretch	Alkene

¹H NMR spectrum:

	Chemical shift (Actual)	(Calculated)	Splitting	Integration	Interpretation
a	1.95	1.7	s	3H	CH_3 isolated
b	5.68	5.76	d	1H	CH adjacent to CH
c	6.25	6.34	d	1H	CH adjacent to CH
d	11.84	10–13	bs	1H	OH isolated

Calculations:

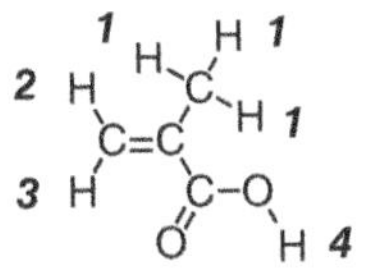

2-methylpropenoic acid

1	2	3	4
0.9	5.28	5.28	
0.8	−0.26	−0.29	
	0.74	1.35	
1.7	5.76	6.34	10–13

Labeled structure:

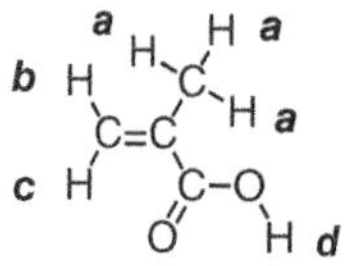

2-methylpropenoic acid

¹³C NMR and DEPT spectra:

	Chemical shift		DEPT	Interpretation
	(Actual)	(Calculated)		
A	17.82	20.9	CH₃	Alkyl
B	127.86	125.7	CH₂	Alkene
C	135.75	141.2	C	Alkene
D	173.25	160–185	C	Carboxylic acid

Calculations:

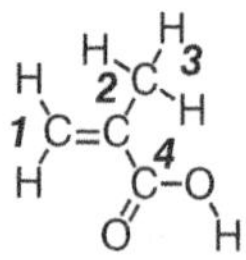

2-methylpropenoic acid

1	2	3	4
123.3	123.3	−2.1	
−7.4	12.9	23	
9.8	5.0		
125.7	141.2	20.9	160–185

Labeled structure:

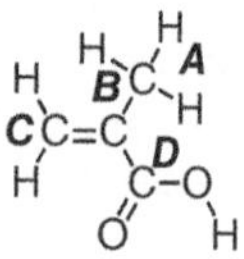

2-methylpropenoic acid

COSY:

H$_a$ is adjacent to nothing.

H$_b$ is adjacent to *H$_c$*

H$_c$ is adjacent to *H$_b$*.

H$_d$ is adjacent to nothing.

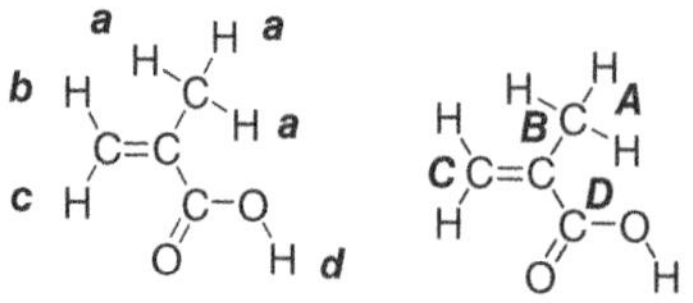

2-methylpropenoic acid

HSQC:

H_a is attached to C_A.

H_b is attached to C_C.

H_c is attached to C_C.

H_d is attached to nothing.

2-Methylpropenoic acid is treated with thionyl chloride to form the 2-methylpropenoyl chloride. This acid chloride then reacts with aniline to form the *N*-phenyl-2-methylpropenamide.

2-methylpropenoic acid → (SOCl₂, thionyl chloride) → 2-methylpropenoyl chloride → (aniline) → *N*-phenyl-2-methylpropenamide

8. The unknown is in solubility class A_1, which includes strong organic acids: carboxylic acids with more than six carbons, phenols with electron-withdrawing groups in the *ortho* and/or *para* positions, and 1,3-diketones. The compound is not an aldehyde or ketone (negative sodium bisulfite test), not an alcohol or phenol (negative ceric ammonium nitrate test), but is a carboxylic acid (positive sodium bicarbonate test). The classification tests, melting point of the unknown, and melting point of the derivative indicate that the unknown is *m*-toluic acid.

IR spectrum:

Frequency	Bond	Compound type
710	C—H out-of-plane bend	*m*-Disubstituted aromatic
746	C—H out-of-plane bend	*m*-Disubstituted aromatic
1214	C—O stretch	Carboxylic acid
1376	C—H bend	—CH₃
1420	C=C stretch	Aromatic
1420	O—H bend	Carboxylic acid
1590	C=C stretch	Aromatic
1678	C=O stretch	Aryl carboxylic acid
2868	C—H stretch	Aromatic
3066	O—H stretch	Carboxylic acid

¹H NMR spectrum:

	Chemical shift (Actual)	Chemical shift (Calculated)	Splitting	Integration	Interpretation
a	2.44	2.3	s	3H	CH₃ isolated
b	7.27–7.45	7.30, 7.32	m	2H	Aromatic
c	7.93–7.95	7.89, 7.90	m	2H	Aromatic
d	10.43	10–13	bs	1H	OH isolated

Calculations:

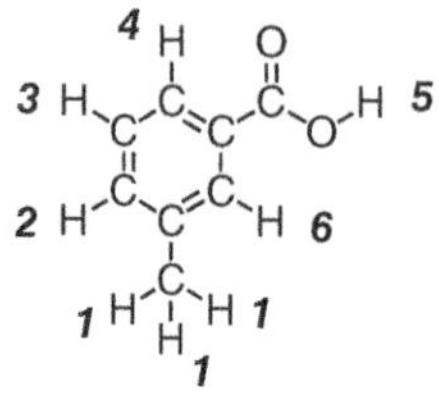

m-toluic acid

1	2	3	4	5	6
0.9	7.27	7.27	7.27		7.27
1.4	−0.17	−0.09	−0.18		−0.17
	0.20	0.14	0.80		0.80
2.3	7.30	7.32	7.89	10−13	7.90

Labeled structure:

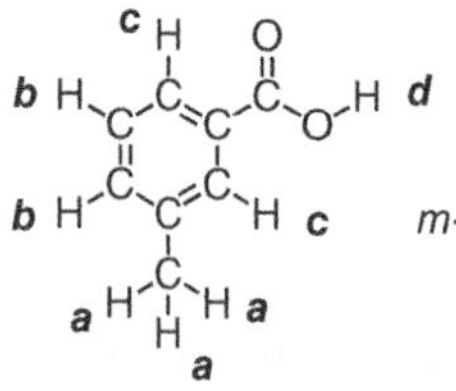

m-toluic acid

^{13}C NMR and DEPT spectra:

	Chemical shift		DEPT	Interpretation
	(Actual)	(Calculated)		
A	21.27	20.9	CH_3	Alkyl
B	127.39	127.3	CH	Aromatic
C	128.40	128.6	CH	Aromatic
D	129.23	130.7	C	Aromatic
E	130.73	130.9	CH	Aromatic
F	134.60	134.5	CH	Aromatic
G	138.33	137.6	C	Aromatic
H	172.29	160−185	C	Carboxylic acid

Calculations:

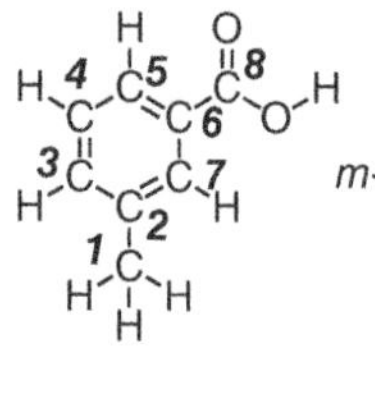

m-toluic acid

1	2	3	4	5	6	7	8
−2.1	128.7	128.7	128.7	128.7	128.7	128.7	
23	8.9	0.7	−0.1	−2.9	−0.1	0.7	
	0.0	5.1	0.0	1.5	2.1	1.5	
20.9	137.6	134.5	128.6	127.3	130.7	130.9	160−185

Labeled structure:

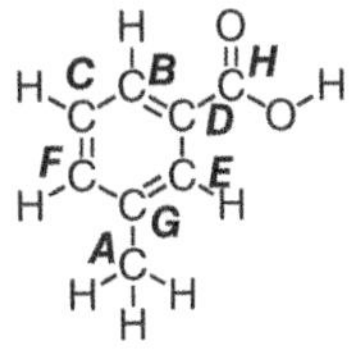

m-toluic acid

COSY:

H_a is adjacent to nothing.

H_b is adjacent to H_c

H_c is adjacent to H_b.

H_d is adjacent to nothing.

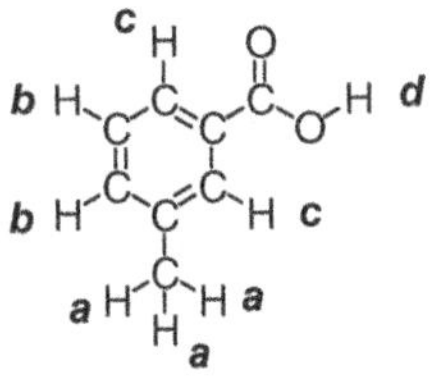

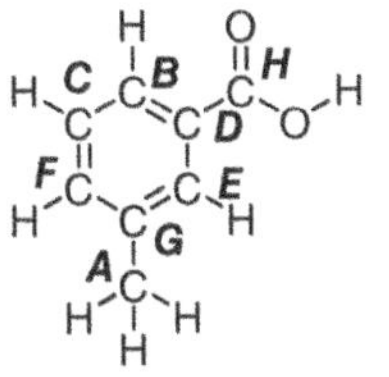

m-toluic acid

HSQC:

H_a is attached to C_A.

H_b is attached to C_C, C_F.

H_c is attached to C_B, C_E.

H_d is attached to nothing.

    m-Toluic acid is treated with sodium hydroxide to form sodium *m*-toluate, which reacts with *S*-benzylthiuronium bromide to give the *S*-benzylthiuronium salt.

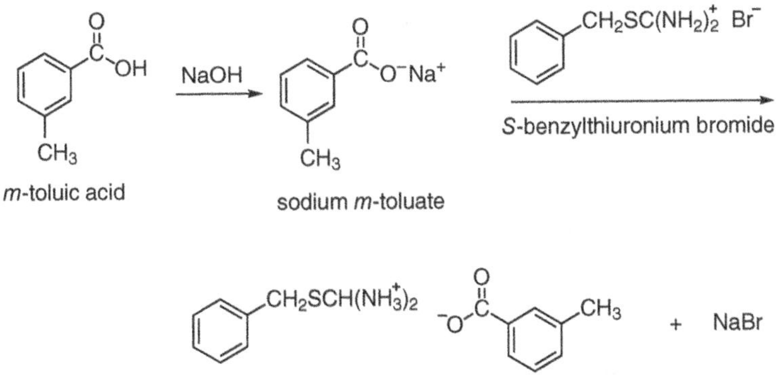

m-toluic acid   sodium *m*-toluate

S-benzylthiuronium salt of *m*-toluic acid

9. The compound is in solubility class S_1, which includes monofunctional alcohols, aldehydes, ketones, esters, nitriles, and amides with five carbons or fewer. The compound is not an aldehyde or ketone (negative sodium bisulfite test), is not a phenol (negative ferric chloride test), but has an active hydrogen (positive sodium test). From the interpretation of the ¹H NMR, the compound is narrowed down to 1-propanol.

IR spectrum:

Frequency	Bond	Compound type
755	O—H bend	Alcohol
1053	C—O stretch	Primary alcohol
1235	C—H bend	—CH_2
1339	O—H bend	Primary alcohol
1380	C—H bend	—CH_3
1456	C—H bend	—CH_3
2876	C—H stretch	Alkyl
3324	O—H stretch	Alcohol

^1H NMR spectrum:

	Chemical shift		Splitting	Integration	Interpretation
	(Actual)	(Calculated)			
a	0.84	1.0	t	3H	CH_3 adjacent to CH_2
b	1.42	1.5	sext	2H	CH_2 adjacent to 5H
c	3.35	3.5	q	2H	CH_2 adjacent to 3H
d	4.38	0.5–6.0	t	1H	OH adjacent to CH_2

Calculations:

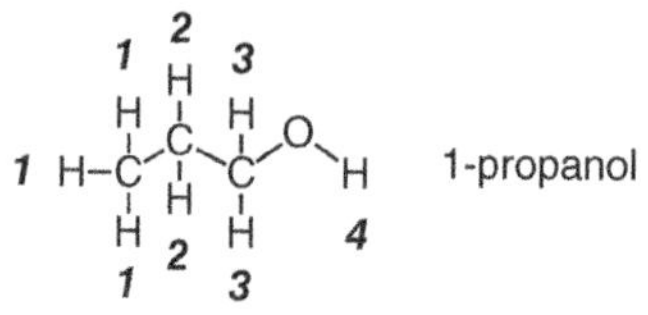

1-propanol

1	*2*	*3*	*4*
0.9	1.2	1.2	
0.1	0.3	2.3	
1.0	1.5	3.5	0.5–6.0

Labeled structure:

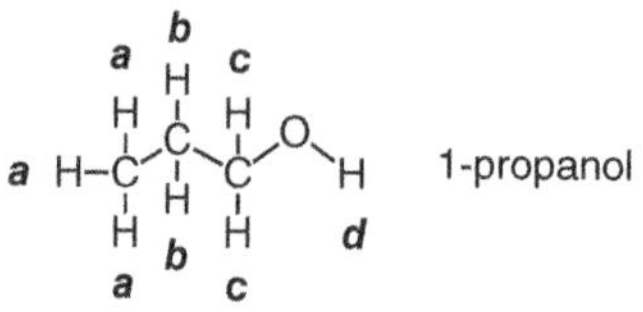

1-propanol

^{13}C NMR and DEPT spectra:

	Chemical shift		DEPT	Interpretation
	(Actual)	(Calculated)		
A	10.70	10.6	CH_3	Alkyl
B	26.05	26.1	CH_2	Alkyl
C	62.95	63.6	CH_2	Alcohol

Calculations:

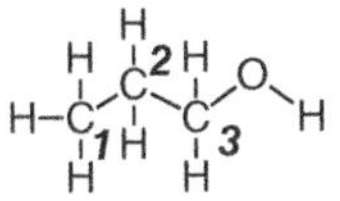

1-propanol

1	2	3
15.6	16.1	15.6
−5	10	48
10.6	26.1	63.6

Labeled structure:

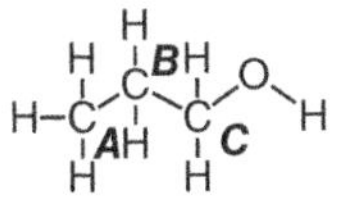

1-propanol

COSY:

H_a is adjacent to H_b.

H_b is adjacent to H_a, H_c.

H_c is adjacent to H_b, H_d.

H_d is adjacent to H_c.

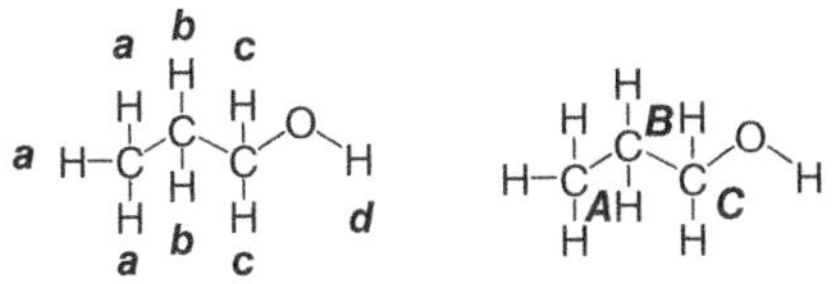

1-propanol

HSQC:

H_a is attached to C_A.

H_b is attached to C_B.

H_c is attached to C_C.

H_d is attached to nothing.

10. The unknown is in solubility class N, which includes alcohols, aldehydes, and ketones, with one functional group and more than five but fewer than nine carbons, ethers, epoxides, alkenes, alkynes, and some aromatic compounds. The compound contains a bromine (pale yellow with silver nitrate test), does not contain an active hydrogen (negative acetyl chloride test), and is an aldehyde (positive purpald and Tollens test). The melting point of the unknown and the derivative match 4-bromobenzaldehyde.

IR spectrum:

Frequency	Bond	Compound type
686	C—Br stretch	Bromide
807	C—H out-of-plane bend	*p*-Disubstituted aromatic
1384	C—H bend	Aldehyde
1481	C=C stretch	Aromatic
1586	C=C stretch	Aromatic
1686	C=O stretch	Aryl aldehyde
2759	C—H stretch	Aldehyde
2816	C—H stretch	Aldehyde
3082	C—H stretch	Aromatic

¹H NMR spectrum:

	Chemical shift		Splitting	Integration	Interpretation
	(Actual)	(Calculated)			
a	7.81–7.86	7.69, 7.72	m	4H	Aromatic
b	10.0	9.0–10.0	s	1H	CH isolated

Calculations:

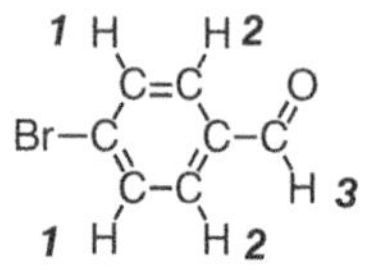

4-bromobenzaldehyde

1	**2**	**3**
7.27	7.27	
0.22	−0.13	
0.21	0.58	
7.69	7.72	9.0–10.0

Labeled structure:

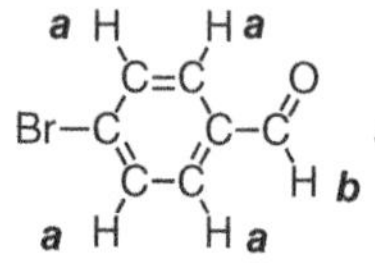

4-bromobenzaldehyde

¹³C NMR and DEPT spectra:

	Chemical shift		DEPT	Interpretation
	(Actual)	(Calculated)		
A	129.15	128.7	C	Aromatic
B	131.70	131.7	2 CH	Aromatic
C	132.77	132.7	2 CH	Aromatic
D	135.59	135.7	C	Aromatic
E	192.80	175–220	CH	Aldehyde

Calculations:

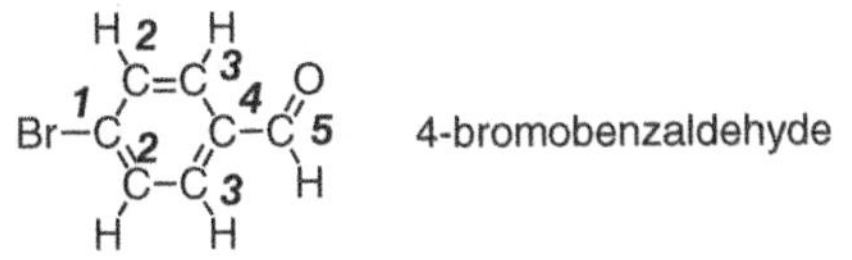

4-bromobenzaldehyde

1	**2**	**3**	**4**	**5**
128.7	128.7	128.7	128.7	
−5.5	3.4	1.7	−1.6	
5.5	0.6	1.3	8.6	
128.7	132.7	131.7	135.7	175–220

Labeled structure:

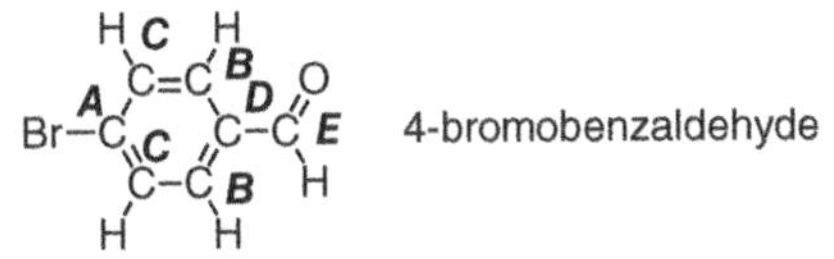

4-bromobenzaldehyde

COSY:

H_a is adjacent to nothing.

H_b is adjacent to nothing.

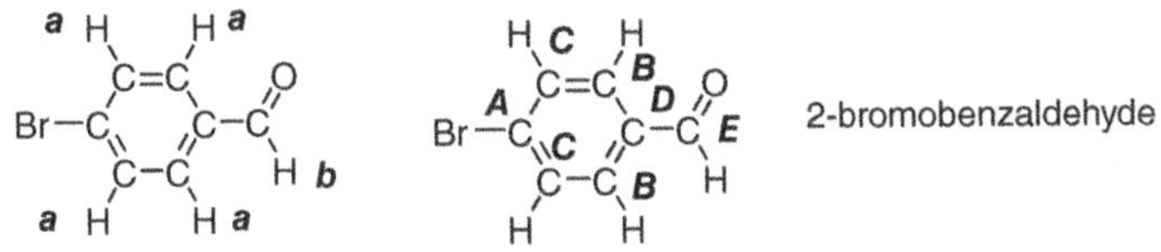

     2-bromobenzaldehyde

HSQC:

H_a is attached to C_B, C_C.

H_c is attached to C_E.

4-Bromobenzaldehyde reacts with semicarbazide hydrochloride to yield the semicarbazone of 4-bromobenzaldehyde.

4-bromobenzaldehyde	semicarbazide hydrochloride

semicarbazone of
4-bromobenzaldehyde

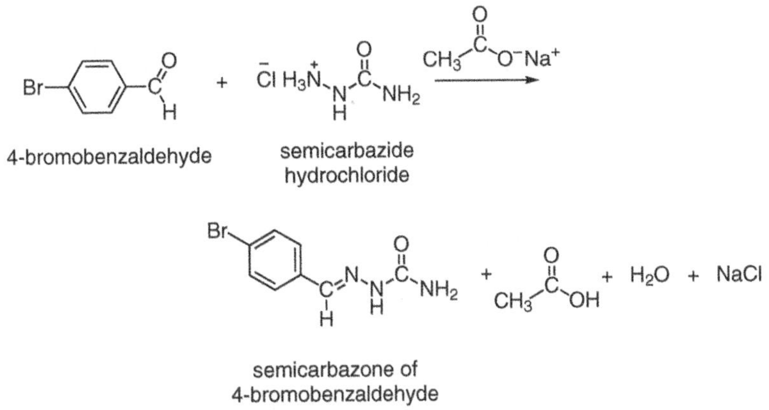

PROBLEM SET 4

1. The unknown is in solubility class N, which consists of alcohols, aldehydes, ketones, esters with one functional group and more than five but fewer than nine carbons, ethers, epoxides, alkenes, alkynes, some aromatic compounds (especially those with activating groups). The unknown does not have an active hydrogen (negative sodium test) and is not an alkene or alkyne (negative potassium permanganate test).

    4-Heptanone matches the criteria above. 4-Heptanone reacts with 2,4-dinitrophenyl-hydrazine to form the 2,4-dinitrophenylhydrazone.

2. The unknown is in solubility class S_1, which consists of monofunctional alcohols, aldehydes, ketones, esters, nitriles, and amides with five carbons or fewer. The unknown contains nitrogen (positive sodium fusion filtrate with 10% ammonium polysulfide, 5% hydrochloric acid, and 5% ferric chloride), does not have an active hydrogen (negative acetyl chloride test), is not an aldehyde or ketone (negative 2,4-dinitrophenylhydrazine test).

    N-Methylmethanamide matches the criteria above and is hydrolyzed to the amine, which then reacts with benzoyl chloride to form the benzamide.

3. The unknown is in solubility class S_1, which consists of monofunctional alcohols, aldehydes, ketones, esters, nitriles, and amides with five carbons or fewer. The unknown has a chlorine (positive silver nitrate test), is not an aldehyde or ketone (negative 2,4-dinitrophenylhydrazine test), and contains an active hydrogen (positive sodium test).

    2-Chloroethanol reacts with 4-nitrobenzoyl chloride to yield the 4-nitrobenzoate derivative.

4. The unknown is in solubility class A$_2$, which consists of weak organic acids; phenols, enols, oximes, imides, sulfonamides, thiophenols, all with more than five carbons; β-diketones; nitro compounds with α-hydrogens. The unknown contains sulfur (positive sodium fusion filtrate with lead acetate test), nitrogen (positive sodium fusion filtrate with 10% ammonium polysulfide, 5% hydrochloric acid, and 5% ferric chloride), and a nitro group (positive ferrous hydroxide test). The unknown does not have an active hydrogen (negative acetyl chloride test).

The unknown must be 3-nitrobenzenesulfonamide. It reacts with phosphorus pentachloride, followed by aniline to form 3-nitro-*N*-phenylbenzenesulfonamide.

| 3-nitrobenzene-sulfonamide | 3-nitrobenzene-sulfonic acid | 3-nitrobenzene-sulfonyl chloride | 3-nitro-*N*-phenyl-benzenesulfonamide |

5. Compound A is in solubility class A$_2$, which consists of weak organic acids: phenols, enols, oximes, imides, sulfonamides, thiophenols, all with more than five carbons; β-diketones (1,3-diketones); and nitro compounds with α-hydrogens. Compound A has an active hydrogen (positive sodium test), contains a phenol or alcohol (positive ceric ammonium nitrate test), and a primary amine (benzenesulfonyl chloride test). Compound A does not contain a halogen (negative silver nitrate test), a nitro group (negative ferrous hydroxide test), or a carboxylic acid. (negative sodium bicarbonate test). Since the unknown reacted with three equivalents of benzoyl chloride, then three NH or OH must be present.

Compound A is 3,4-diaminophenol. Compound A is treated with three equivalents of benzoyl chloride to produce the tribenzoyl derivative of 3,4-diaminophenol, compound B.

Compound A: 3,4-diaminophenol   benzoyl chloride

Compound B: tribenzoyl derivative of 3,4-diaminophenol

6. An unknown is in solubility class A$_2$, which consists of weak organic acids: phenols, enols, oximes, imides, sulfonamides, thiophenols, all with more than five carbons; β-diketones (1,3-diketones); nitro compounds with α-hydrogens. The unknown does not contain sulfur (negative sodium fusion test with lead acetate), or halogen (negative silver nitrate test). The unknown contains a phenol or alcohol (positive ceric ammonium nitrate test); a primary alcohol, a secondary alcohol, or an aldehyde (positive chromium trioxide test); an active hydrogen (positive acetyl chloride test); and an aldehyde or a ketone (positive 2,4-dinitrophenylhydrazine test).

The unknown is vanillin.

4-hydroxy-3-methoxy-
benzaldehyde,
vanillin

7. Compound A is in solubility class N, which consists of alcohols, aldehydes, ketones, esters with one functional group and more than five but fewer than nine carbons, ethers, epoxides, alkenes, alkynes, some aromatic compounds (especially those with deactivating groups). Compound A does not contain an alkene or alkyne (negative potassium permanganate test), an aldehyde or ketone (negative 2,4-dinitrophenylhydrazine test), or an active hydrogen (negative sodium test).

Compound B is in solubility class A_1, which consists of strong organic acids: carboxylic acids with more than six carbons; phenols with electron-withdrawing groups in the ortho and/or para position(s); β-diketones (1,3-diketones). Compound B contains a carboxylic acid (positive sodium bicarbonate test), but does not contain an active hydrogen (negative sodium test), or an aldehyde or ketone (negative 2,4-dinitrophenylhydrazine test).

Compound C is in solubility class A_2, which consists of weak organic acids: phenols, enols, oximes, imides, sulfonamides, thiophenols, all with more than five carbons; β-diketones (1,3-diketones); nitro compounds with α-hydrogens. Compound C does not contain a carboxylic acid (negative sodium bicarbonate test) or a nitro group (negative ferrous hydroxide test). Compound C contains an active hydrogen (positive sodium test). Compound A is an ester.

Compound A is phenyl benzoate, compound B is benzoic acid, and compound C is phenol.

Compound A: Compound B: Compound C:
phenyl benzoate   benzoic acid   phenol

8. The unknown is in solubility class N, which consists of alcohols, aldehydes, ketones, esters with one functional group and more than five but fewer than nine carbons, ethers, epoxides, alkenes, alkynes, some aromatic compound (especially those with deactivating groups). The unknown does not contain a primary alcohol, a secondary alcohol, or an aldehyde (negative Jones reagent),an active hydrogen (negative sodium test), an aldehyde (negative Schiff test), an aldehyde or a ketone (negative 2,4-dinitrophenylhydrazine test), an alkene or alkyne (negative bromine test), or an anhydride, an acyl halide, an ester, a nitrile, or an amide (negative hydroxamic acid and ferric chloride test). The unknown contains oxygen (positive ferric ammonium sulfate test).

The unknown is 1,4-dimethoxybenzene.

1,4-dimethoxybenzene

PROBLEM SET 5

1. The unknown is in solubility class B, which consists of aliphatic amines with eight or more compounds; anilines (only one phenyl group attached to nitrogen); some ethers. *N*-Ethylaniline matches the boiling point and the melting point of the derivative.

    N-Ethylaniline (compound A) reacts with benzoyl chloride to produce *N*-ethyl-*N*-phenylbenzamide (compound B), which is saponified to make sodium benzoate and *N*-ethylaniline (compound A). Acidification of the sodium benzoate yielded benzoic acid (compound C), which is treated with thionyl chloride to form benzoyl chloride (compound D). Benzoyl chloride (compound D) was reacted with 4-toluidine to produce *N*-4-tolylbenzamide (compound E).

2. Compound A is in solubility class S_1, which consists of monofunctional alcohols, aldehydes, ketones, esters, nitriles, and amides with five carbons or fewer. Compound A does not contain nitrogen (negative ammonium polysulfide, hydrochloric acid, and ferric chloride test) but has an active hydrogen (positive acetyl chloride test). Compound A is not an aldehyde or ketone (negative 2,4-dinitrophenylhydrazine test).

    Compound B contains chlorine (positive silver nitrate test).

    Compound C is in solubility class S_1, which consists of monofunctional alcohols, aldehydes, ketones, esters, nitriles, and amides with five carbons or fewer. Compound C does not contain an active hydrogen (negative acetyl chloride test) or a labile halogen (negative silver nitrate test). Compound C is not an aldehyde or a ketone.

    Propyl propanoate matches the boiling point of compound C and the hydrazide derivative.

    1-Propanol (compound A) reacts with propanoyl chloride (compound B) to yield propyl propanoate (compound C). Propyl propanoate (compound C) reacts with hydrazine to produce propanoic acid hydrazide (compound D).

CH₃CH₂CH₂OH + propanoyl chloride → propyl propanoate

Compound A:
1-propanol

Compound B:
propanoyl chloride

Compound C:
propyl propanoate

NH₂NH₂ →

Compound D:
propanoic acid hydrazide

3. Compound A is in solubility class MN, which consists of miscellaneous neutral compounds containing nitrogen or sulfur and having more than five carbon atoms. Compound A contains sulfur (positive lead acetate test), nitrogen (positive ammonium polysulfide, hydrochloric acid, and ferric chloride), and chlorine (positive for silver nitrate).

   4-Methyl-3-nitrobenzenesulfonyl chloride matches the melting points of compound A and both derivatives.

   4-Methyl-3-nitrobenzenesulfonyl chloride (compound A) is hydrolyzed to 4-methyl-3-nitrobenzenesulfonic acid (compound B). 4-Methyl-3-nitrobenzenesulfonic acid (compound B) is treated with phosphorus pentachloride to give 4-methyl-3-nitro-benzenesulfonyl chloride (compound A). 4-Methyl-3-nitrobenzene-sulfonyl chloride (compound A) is reacted with ammonium hydroxide to give 4-methyl-3-nitrobenzenesulfonamide (compound C).

Compound A:
4-methyl-3-nitro-
benzenesulfonyl
chloride

Compound B:
4-methyl-3-
nitrobenzene-
sulfonic acid

Compound C:
4-methyl-3-
nitrobenzene-
sulfonamide

4. The unknown is in solubility class S₂, which consists of salts of organic acids, amine hydrochlorides, amino acids, polyfunctional compounds with hydrophilic functional groups, carbohydrates (sugars), polyhydroxy compounds, and polybasic acids. The unknown contains an active hydrogen (positive acetyl chloride test) but does not contain nitrogen (negative ammonium polysulfide, hydrochloric acid, and ferric chloride test).

   The unknown is sucrose.

sucrose
α-D-glucopyranosyl-β-D-fructofuranose

5. The unknown contains an active hydrogen (positive acetyl chloride), a phenol (positive ferric chloride and pyridine test), a nitro group (positive ferrous hydroxide test), and 1° amine (positive benzenesulfonyl chloride test). The unknown does not contain a 1° alcohol, 2° alcohol, and aldehyde (negative chromium trioxide test).

   The unknown is 2-amino-4,6-dinitrophenol.

2-amino-4,6-dinitrophenol

6. The unknown contains an alkene or alkyne (positive bromine test), a carboxylic acid (positive sodium bicarbonate test), and an aldehyde or ketone (2,4-dinitrophenylhydrazine test). The unknown does not contain an aldehyde (negative Tollens test) or primary alcohol, secondary alcohol, aldehyde (negative chromium trioxide test).

   The unknown is 3-benzoylbenzoic acid.

3-benzoylpropenoic acid

7. The unknown contains an active hydrogen (positive acetyl chloride test); a primary alcohol, a secondary alcohol, or an aldehyde (positive chromium trioxide test); and chlorine (positive silver nitrate test). The unknown does not contain an aldehyde or a ketone (negative 2,4-dinitrophenylhydrazine test) or an alkene or an alkyne (negative bromine test).

   The unknown is 2,3-dichloro-1-propanol.

ClCH$_2$–CH–CH$_2$OH   2,3-dichloro-1-propanol

8. The unknown contains an alkene or alkyne (positive bromine test and positive potassium permanganate test), an aldehyde or a ketone (positive 2,4-dinitrophenylhydrazine test), an aldehyde (positive Tollens test), and a primary alcohol, a secondary alcohol, or an aldehyde (positive chromium trioxide test).

   The unknown is 2-hexenal.

1-hexenal

PROBLEM SET 6

1. Aldehydes are oxidized to an acid by hydrogen peroxide. The peak at δ 9.0 in the ^{1}H NMR spectrum is indicative of an aldehyde. A yellow color indicates conjugation, thereby indicating that the original compound is a α-ketoaldehyde.

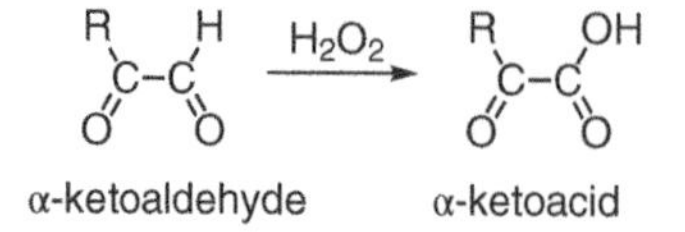

α-ketoaldehyde   α-ketoacid

2. The carbonyl groups in aldehydes and ketones react with phenylhydrazine. Water is eliminated when a phenylhydrazine undergoes reaction with a carbonyl group. Ethanol can be eliminated if the phenylhydrazone of a β-ketoester cyclizes. Thus, the unknown is a β-ketoester.

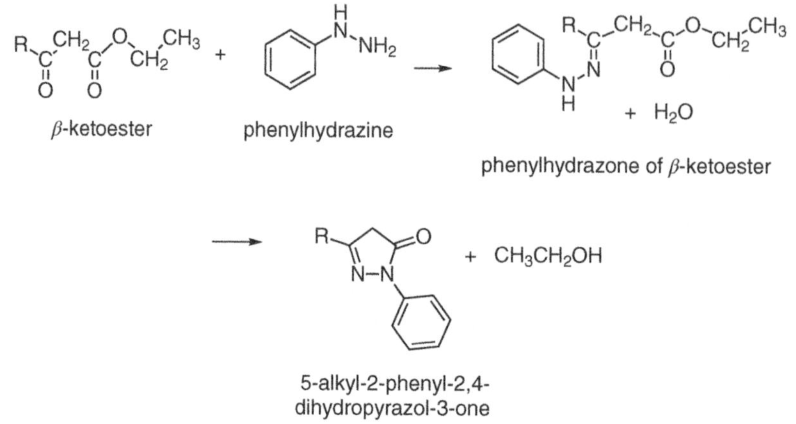

β-ketoester   phenylhydrazine

phenylhydrazone of β-ketoester

5-alkyl-2-phenyl-2,4-dihydropyrazol-3-one

3. Reaction with acetyl chloride indicated the presence of an active hydrogen in such compounds as alcohols and amines. The unknown cannot be an aldehyde or ketone (negative 2,4-dinitrophenylhydrazine test). Periodic acid oxidizes compounds such as 1,2-diols, α-hydroxyketones, or 1,2-diketones.

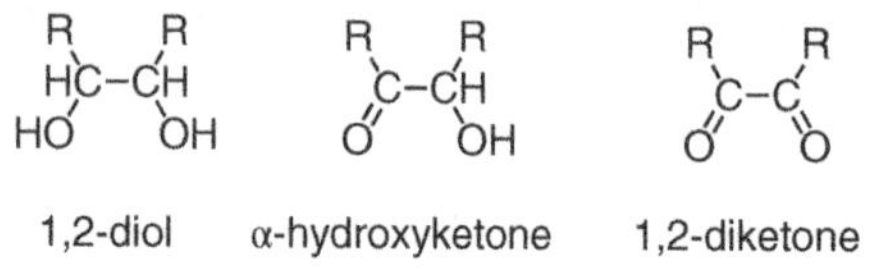

1,2-diol   α-hydroxyketone   1,2-diketone

The original compound must be a 1,2-diol since it must contain an active hydrogen and is not an aldehyde or ketone.

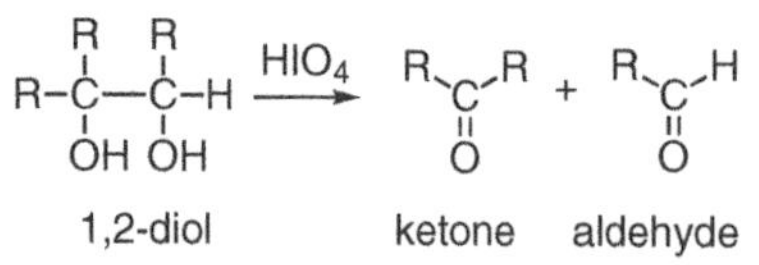

1,2-diol   ketone   aldehyde

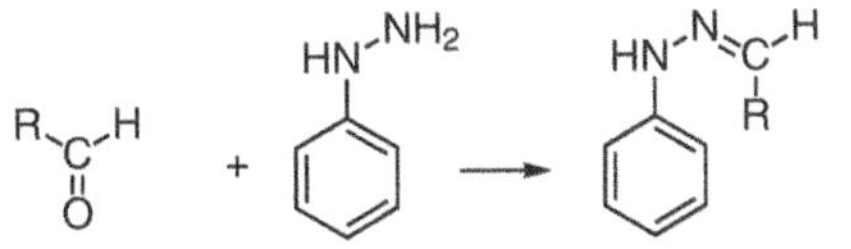

aldehyde   phenylhydrazine   phenylhydrazone

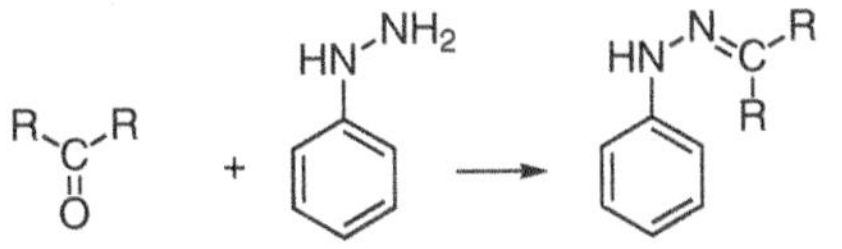

ketone   phenylhydrazine   phenylhydrazone

4. An IR band at $1710\,cm^{-1}$ is near the frequency for a C=O stretching for a ketone. A yellow color indicates conjugation. A α-diketone is a possibility.

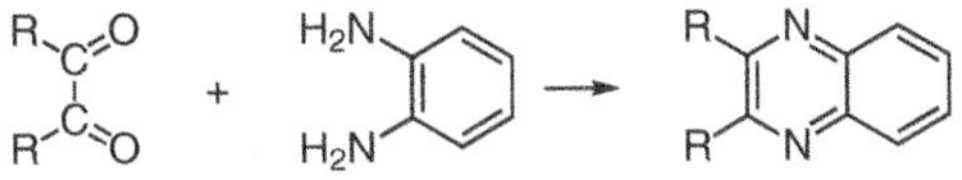

α-diketone   *o*-phenylenediamine   quinoxaline

5. An alcohol that gives a positive test with iodoform indicates a structure of the type shown below.

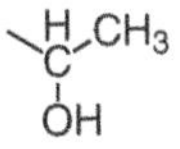

A negative Lucas test indicates a primary alcohol. The only possibility is ethanol.

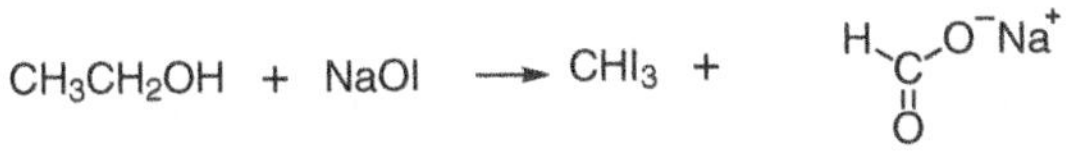

$$CH_3CH_2OH \; + \; NaOI \; \longrightarrow \; CHI_3 \; + \;$$

   ethanol   sodium   iodoform   sodium
           iodate   formate

6. The original compound reacted with acetyl chloride, so it must have an active hydrogen. The original compound yielded a negative test with phenylhydrazine, thereby indicating that it is not an aldehyde or a ketone. Treatment of the unknown with mineral acid yielded a compound which is not an alcohol (negative acetyl chloride test) but is an aldehyde or ketone (positive phenylhydrazine test). Compounds containing carbon-carbon double bonds, methyl ketones, and similarly sized ketones give a positive test with bromine.

The ^1H NMR spectrum of the original compound contains two singlets, indicating that there are no hydrogens on the adjacent carbons. In the presence of mineral acid, 2,3-dimethylbutane-2,3-diol (pinacol) rearranges to 3,3-dimethyl-2-butanone.

Thus, the original compound (2,3-dimethylbutane-2,3-diol) is an alcohol, but not an aldehyde or ketone and the product (3,3-dimethyl-2-butanone) after treatment with mineral acid gave a positive test for a ketone and negative test for an alcohol.

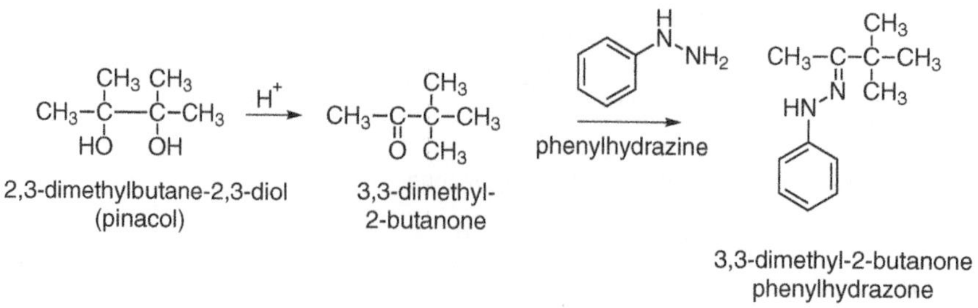

2,3-dimethylbutane-2,3-diol
(pinacol)

3,3-dimethyl-
2-butanone

phenylhydrazine

3,3-dimethyl-2-butanone
phenylhydrazone

7. The secondary amine must also contain another functional group that makes it soluble in base, such as an alcohol group or a carboxyl group. Possibilities would be an amino acid or an amino alcohol.

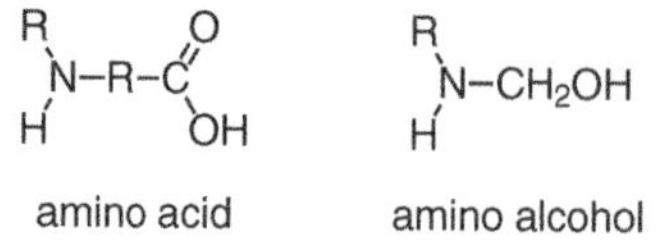

amino acid

amino alcohol

8. Aldehydes and ketones react with ethanol to yield acetals and ketals.

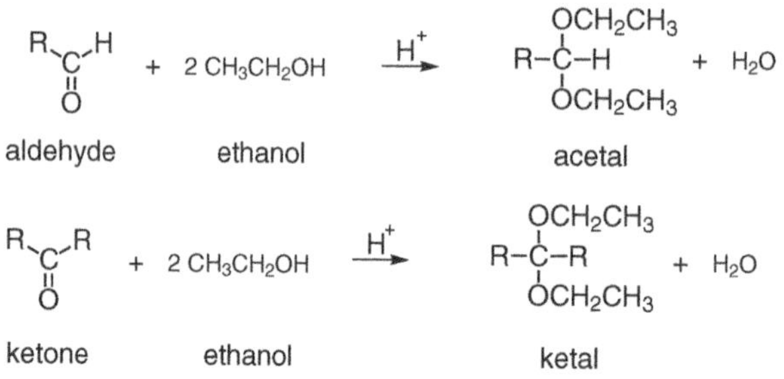

aldehyde   ethanol   acetal

ketone   ethanol   ketal

A signal of δ 9–10 in the ^{1}H NMR spectrum is indicative of an aldehyde group. After the treatment with ethanol and acid, the C–H is moved to δ 5.0.

9. The unknown is benzaldehyde (or a substituted benzaldehyde), which, in the presence of sodium cyanide, undergoes a self-condensation reaction to yield benzoin (or a substituted benzoin).

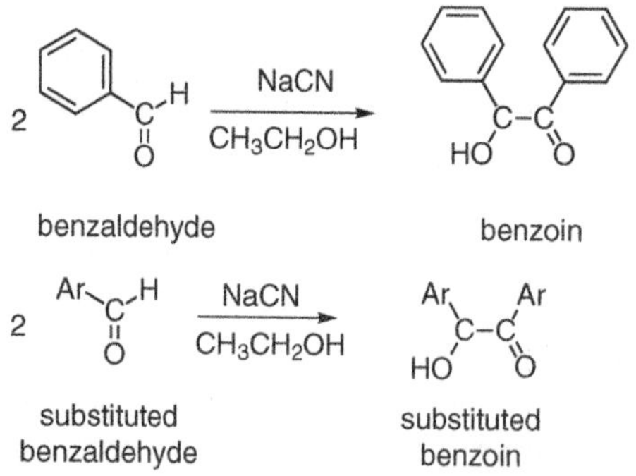

benzaldehyde   benzoin

substituted
benzaldehyde

substituted
benzoin

The IR bands at 2700 and 2800 cm^{-1} correspond to the C–H stretching of an aldehyde group.

10. Tertiary amines react with benzenesulfonyl chloride to form a quaternary ammonium sulfate salt, which yield sodium sulfonate and a water-insoluble tertiary amine in basic solution. Acidification results in the formation of sulfonic acids and water-soluble amine salts.

N,N-dialkyl-   benzenesulfonyl
aniline   chloride

2 NaOH

(soluble)   N,N-dialkyl-   benzenesulfonic   N,N-dialkyl-
                  aniline   acid   anilinium chloride
+ NaCl + H₂O   (insoluble)   (soluble)

+ 2 NaCl

Nitrous acid undergoes reaction with *N,N*-dialkylanilines to form nitroso compounds that precipitate as a green solid.

N,N-dialkylaniline   4-nitroso-N,N-dialkylaniline
                          (green solid)

Another indication that the unknown compound is an *N,N*-dialkylaniline, is the lack of an N–H band near 3333 cm⁻¹. Aromatic nitroso compounds yield N═O stretching bands near 1550 cm⁻¹ in an IR spectrum.

11. An ester with a SE of 59 ± 1 would have a formula of $C_2H_4O_2$. The only possibility that would match the integration ratio of 3 : 1 in the ¹H NMR spectrum would be methyl methanoate.

methyl methanoate
MW = 60, SE = 60

12. The original acid must contain more than one –COOH group in order to have an increase in neutralization equivalent. Since CO_2 is lost with heat, the two –COOH groups must have originally been attached to the same carbon. The following equation can be set up to calculate the number of –COOH groups, with n = the number of original –COOH groups. Carbon dioxide has a molecular weight of 44.

$$(n \times 54) - (n-1)59 = 44$$
$$54n - 59n + 59 = 44$$
$$-5n = -15$$
$$n = 3$$

Thus, there must be three –COOH groups present in the original carboxylic acid. The initial molecular weight must be approximately 162 (3 × 54). To calculate the remainder of the molecule, the three –COOH groups can be subtracted from the molecular weight, leaving a value of 27, which can be a CH_2CH.

$$162 - (3 \times 45) = 27$$

The only possibility is 2-carboxybutanedioic acid.

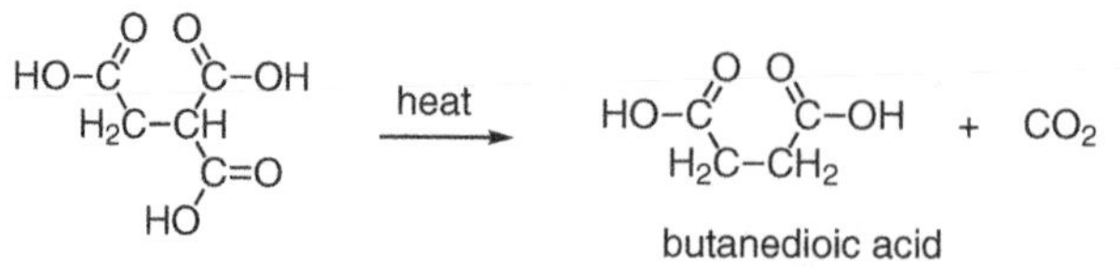

2-carboxybutanedioic acid
MW = 162, NE = 54

butanedioic acid
(succinic acid)
MW = 118, NE = 59

13. An insoluble barium salt is barium sulfate. Amines form hydrochloride salts. Thus, the original compound must contain both a sulfate group and an amine group. The only compound that meets these criteria with a neutralization equivalent of 142 ± 1 is anilinium sulfate. The N–H stretch of the amine salt is in the range of 2500–3333 cm^{-1}.

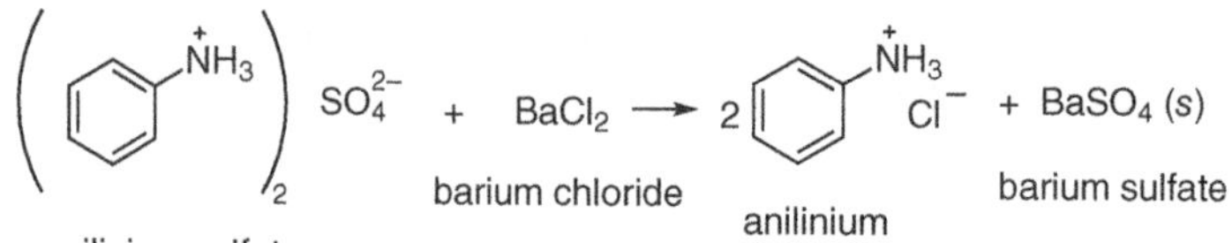

anilinium sulfate
MW = 284,
NE = 142

barium chloride

anilinium
chloride

barium sulfate

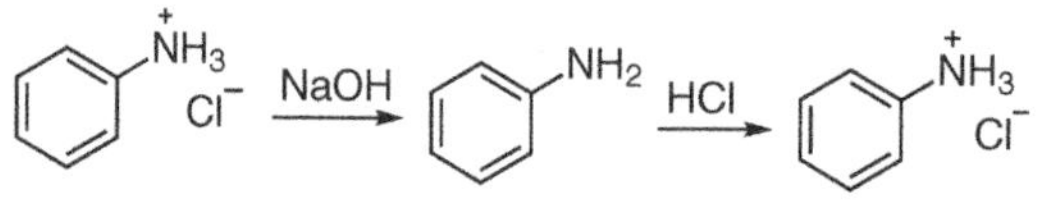

anilinium
chloride

aniline

anilinium
chloride
MW = 129.5,
NE = 129.5

14. Only secondary amines give a precipitate with benzenesulfonyl chloride and sodium hydroxide solution. In the ^1H NMR spectrum, a quartet indicates that there are three hydrogens on the adjacent carbons and a triplet indicates that there are two hydrogens on the adjacent carbons, thus concluding that two ethyl groups are present. A broadened singlet indicates an N–H group. The only compound that fits this data is diethylamine.

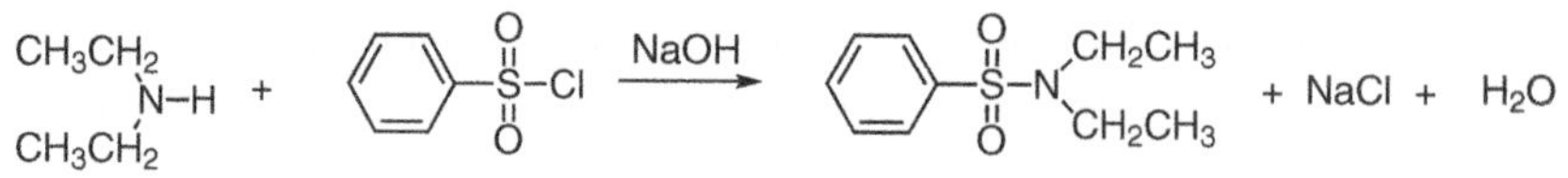

diethylamine

benzenesulfonyl
chloride

N,N-diethylbenzene-
sulfonamide
(insoluble)

15. The following calculations need to be performed first.

$$\frac{12\,g}{120\,g/mol} = 0.100\,mol\ of\ compound\ C_8H_8O$$

$$\frac{60\,g}{160\,g/mol} = 0.375\,mol\ of\ Br_2\ initially$$

$$\frac{16.2\,g}{81\,g/mol} = 0.200\,mol\ of\ HBr\ evolved$$

$$\frac{12\,g}{160\,g/mol} = 0.075\,mol\ of\ Br_2\ unused$$

$$0.375 - 0.075 = 0.300\,mol\ of\ Br_2\ used$$

Therefore, 0.200 mol of Br_2 reacted with the compound to yield 0.200 mol of HBr. Another 0.100 mol of Br_2 were added across a multiple bond in the compound. Bromine substitutes on phenols with the evolution of hydrogen bromide. Additionally, bromine adds across multiple bonds. Thus, the unknown is a phenol containing an aliphatic double bond. With a formula of C_8H_8O, the only possibilities would be 2-hydroxystyrene, 3-hydroxystyrene, or 4-hydroxystyrene. The equation below is given for 4-hydroxystyrene.

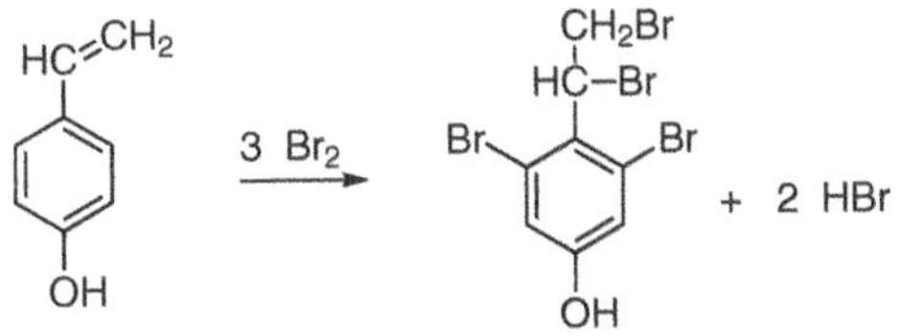

4-hydroxystyrene   3,5-dibromo-4-(1,2-dibromoethyl)phenol

The ^{1}H NMR spectrum of the original compound would include aromatic signals and the protons would be in an integration ratio of 2H (CH_2) : 1H (CH) : 4H (aromatic) : 1H (OH). In the IR spectrum, C=C stretching and bending, O–H stretching and bending, and aromatic frequencies would be present.

The product would have an integration ratio of 2H (CH_2) : 1H (CH) : 2H (aromatic) : 1H (OH) in the ^{1}H NMR spectrum. The IR spectrum would contain the C=C aromatic stretching and bending in addition to the C–Br stretch.

PROBLEM SET 7

I. A negative test with 2,4-dinitrophenylhydrazine indicates that the unknown is not an aldehyde or ketone. A positive test with potassium permanganate and a negative test with bromine in carbon tetrachloride signifies an aldehyde or an alcohol. An active hydrogen is indicated by the positive test for acetyl chloride. A positive periodic acid test indicates that a carbonyl group is adjacent to a carbonyl group, a carbonyl group is adjacent to a hydroxy group, or a hydroxy group is adjacent to a hydroxy group.

Carboxylic acids have neutralization equivalents and 4-nitrobenzyl ester derivatives. Thus, the unknown must contain a carboxylic acid group adjacent to a hydroxy group. In the carboxylic acid derivative tables, (±)-mandelic acid matches the melting point of 117–118 °C and the melting point of 123 °C for the 4-nitrobenzyl ester derivative.

mandelic acid
MW = 152, NE = 152

Mandelic acid reacts with 4-nitrobenzyl chloride to form the 4-nitrobenzyl 1-hydroxy-1-phenylacetate.

mandelic acid   4-nitrobenzyl chloride

4-nitrobenzyl 1-hydroxy-1-phenylacetate

The IR spectrum of mandelic acid contains signals at 2700–3333 cm^{-1} for the O–H stretch of the acid and the alcohol; 1716 cm^{-1} for the C=O stretch of the acid; 1498 and 1588 cm^{-1} for the C=C stretch of aromatic; 1078 cm^{-1} for the C–O stretch of the secondary alcohol; and 1300 cm^{-1} for the C–O stretch of the carboxylic acid.

In the ^{1}H NMR spectrum, the singlet at δ 5.22 signifies the C–H and the multiplet at δ 6.93–7.88 indicates five aromatic protons and both –OH protons.

II. The unknown is in the solubility class of N, indicating an alcohol, aldehyde, ketone, ester, ether, alkene, alkyne, or aromatic.

A negative hydroxylamine hydrochloride test indicates that the unknown is not an aldehyde or a ketone. A negative acetyl chloride test indicates the absence of an active hydrogen. Negative results from the silver nitrate and the sodium iodide tests show that the bromine is not labile. Both the potassium permanganate and bromine tests give negative results, indicating the absence of a multiple bond.

The elemental analysis yielded the presence of bromine, but the tests above indicate that there is no labile halogen present. Therefore, the halogen must be attached to an aromatic ring. From the results of the classification tests, alcohol, aldehyde, ketone, alkene, and alkyne are eliminated as possibilities. Ester, ether, and aromatic compounds are left as possibilities.

Saponification equivalents indicate the presence of an ester. Hydrazine and 3,5-dinitrobenzoic acid react with esters to yield derivatives. From the derivative tables, by comparison of the boiling point of 259–261°C of the original compound and of the melting points of the hydrazine derivative (164°C) and the 3,5-dinitrobenzoate derivative (92°C), the unknown must be ethyl 4-bromobenzoate.

The product from the sodium hydroxide hydrolysis is 4-bromobenzoic acid (mp 250°C).

ethyl 4-bromobenzoate
MW = 229, SE = 229

4-bromobenzoic acid

Ethyl 4-bromobenzoate reacts with 3,4-dinitrobenzoic acid to yield 4-bromobenzoic acid and ethyl 3,5-dinitrobenzoate.

ethyl 4-bromobenzoate

3,5-dinitrobenzoic acid

4-bromobenzoic acid

ethyl 3,5-dinitrobenzoate

Ethyl 4-bromobenzoate reacts with hydrazine to produce the 4-bromobenzoic acid hydrazide.

ethyl 4-bromobenzoate

hydrazine

4-bromobenzoic acid hydrazide

III. The solubility tests give the unknown compound an A_2 solubility class indicating a weak organic acid such as a phenol, an enol, an oxime, an imide, a sulfonamide, a thiophenol, a β-diketone, or a nitro compound.

The unknown has an active hydrogen from a positive acetyl chloride test, but cannot be an aldehyde or a ketone from negative results in the 2,4-dinitrophenylhydrazine test. A precipitate in the bromine in water test and a violet color in the iron (III) chloride test indicate a phenol. A brown color in the cerium ammonium nitrate test indicates a phenol.

The compound must be a phenol from the classification and elemental tests. A compound that matches the boiling point of 198–200°C and the melting point of 102–103°C for the chloroacetic acid derivative is 3-methylphenol.

3-methylphenol

3-Methylphenol reacts with chloroacetic acid to form the aryloxyacetic acid.

| 3-hydroxytoluene | chloroacetic acid | 3-methylphenyloxyacetic acid |

The IR spectrum of this compound indicates an O–H stretching of an alcohol at 3340 cm⁻¹, a C—O stretching of a phenol at 1249 and 1357 cm⁻¹, and C–H out-of-plane bending for m-disubstituted aromatic compounds at 670 and 786 cm⁻¹, O–H bend for phenol at 670 cm⁻¹.

The ¹H NMR spectrum indicates an isolated CH_3 at δ 2.25, the –OH group at δ 5.67, and four aromatic peaks in the range of δ 6.5–7.3.

IV. The solubility class would be I, which consists of saturated hydrocarbons, haloalkanes, aryl halides, diaryl ethers, and deactivated aromatic compounds. The negative classification tests of sulfuric acid with sulfur trioxide and aluminum chloride with azoxybenzene show that the unknown is not an aromatic compound. The negative test with bromine indicates no multiple bond.

The ¹H NMR and the ¹³C NMR indicate one peak. The only compound that fits the boiling point, the molecular formula, and one peak in both NMR spectra is cyclohexane.

cyclohexane

V. The unknown is in solubility class SA, which includes monofunctional carboxylic acids with five carbons or fewer and aryl sulfonic acids. The negative results in the potassium permanganate and the bromine tests show that the unknown does not have a multiple bond. The unknown is not an aldehyde or a ketone due to the negative 2,4-dinitrophenylhydrazine test. A negative acetyl chloride test indicates that an active hydrogen of an alcohol is not present. The neutralization equivalent of 59 confirms the presence of a carboxylic acid.

The melting points of the original compound (187–188 °C) and the 4-bromophenacyl ester derivative (209–210 °C), and the neutralization equivalent of 59 indicate that the compound is butanedioic acid or succinic acid.

Succinic acid reacts with 4-bromophenacyl bromide to form the di-4-bromophenacyl ester of succinic acid.

| butanedioic acid | 4-bromophenacyl bromide | di-4-bromophenacyl ester of butanedioic acid |

The IR spectrum indicates the O–H stretch of a carboxylic acid (2926 cm⁻¹), C=O stretch of an aliphatic carboxylic acid (1696 cm⁻¹), O—H bend of a carboxylic acid (1419 cm⁻¹), and the C–H bend of a CH_2 (1310 and 1457 cm⁻¹), and the C–H stretch of alkyl (2928 cm⁻¹).

In the ¹H NMR spectrum, the signals at δ 2.43 (s) and δ 11.80 (bs) in the integration ratio of 2 : 1, indicate the CH_2 to COOH comparison.

VI. The unknown is in solubility class B, which consists of amines of eight or more carbons, anilines, or ethers. A negative potassium permanganate test shows that the unknown does not contain a multiple bond, an alcohol group, or is a phenol. A precipitate in the bromine in carbon tetrachloride test indicates an aromatic amine. The unknown contains an aldehyde or ketone group from a positive 2,4-dinitrophenylhydrazine test. A tertiary amine is indicated from the benzenesulfonyl chloride test, followed by hydrochloric acid. The negative iron (II) hydroxide test shows the absence of a nitro group. A positive result with the Tollens reagent means that an aldehyde is present.

From the above information, a tertiary amine is present which contains an aldehyde group. Phenylhydrazone and semicarbazone derivatives are prepared from aldehydes. From the melting points of the unknown (72 °C) and the phenylhydrazone (148 °C) and semicarbazone (224 °C) derivatives, the unknown is 4-(N,N-dimethylamino)benzaldehyde.

4-(N,N-Dimethylamino)benzaldehyde (compound A) reacts with base in a Cannizzaro reaction to yield 4-(N,N-dimethylamino)benzoic acid (compound B), (mp 236–240 °C), and 4-(N,N-dimethylamino)benzyl alcohol (compound C).

Compound A:
4-(N,N-dimethyl-amino)benzaldehyde

Compound B:
4-(N,N-dimethyl-amino)benzoic acid

Compound C:
4-N,N-dimethyl-aminobenzyl alcohol

4-(N,N-Dimethylamino)benzaldehyde (compound A) reacts with acetone and sodium hydroxide in a crossed aldol condensation to yield 4-[4-(N,N-dimethylamino)phenyl]-3-buten-2-one (compound D) (mp 134–135 °C).

Compound A:
4-(N,N-dimethylamino)-benzaldehyde

acetone

Compound D:
4-[4-N,N-dimethylamino)-phenyl-3-buten-2-one

4-(N,N-Dimethylamino)benzaldehyde reacts with phenylhydrazine to form the phenylhydrazone.

4-(N,N-dimethylamimo)-benzaldehyde

phenylhydrazine

phenylhydrazone of
4-(N,N-dimethylamino)-benzaldehyde

4-(N,N-Dimethylamino)benzaldehyde reacts with semicarbazide hydrochloride to form the semicarbazone.

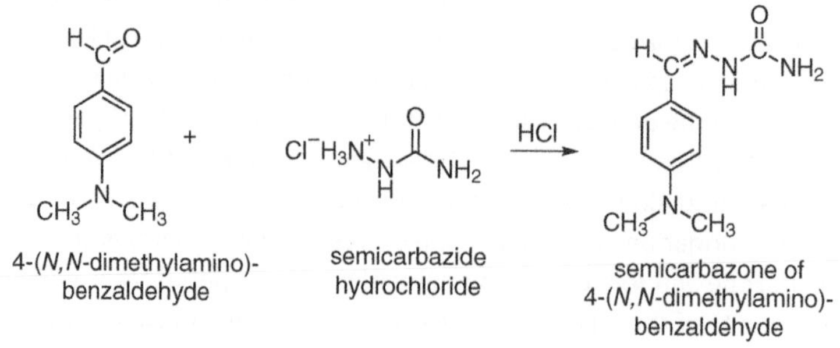

4-(N,N-dimethylamino)-benzaldehyde

semicarbazide hydrochloride

semicarbazone of 4-(N,N-dimethylamino)-benzaldehyde

The IR spectrum has no absorption near 3333 cm^{-1}, showing that it is not a primary or secondary amine. A spectrum indicates the C=O stretch of an aromatic aldehyde (1694 cm^{-1}), a C–H stretch of an aldehyde (2733 and 2818 cm^{-1}), C–H out-of-plane bend of a p-disubstituted aromatic (800 cm^{-1}), C=C stretch of an aromatic (1414 and 1602 cm^{-1}).

The ^{1}H NMR spectrum indicates two CH$_3$ groups at δ 3.05, p-disubstituted aromatic signals at δ 6.69 and δ 7.71, and the aldehyde signal at δ 9.70.

PROBLEM SET 8

1. The solubility class for this unknown is B, which means it is an amine, an aniline, or an ether. A negative acetyl chloride test means that there is not an active hydrogen present. The compound is not a primary or secondary amine, since it did not react with acetyl chloride. A tertiary amine is present (positive nitrous acid test).

   A tertiary amine that fits the boiling point of 193–195 °C is *N,N*-dimethylaniline (compound I).

   This compound is treated with nitrous acid to produce 4-nitroso-*N,N*-dimethylaniline (compound II) (mp 164 °C), which undergoes a reaction with hot sodium hydroxide solution to yield *N,N*-dimethylamine (compound III) and a product which is acidified to produce 4-nitrosophenol (compound V) (mp 125–126 °C). Dimethylamine (compound III) reacts with phenyl isocyanate to form 1,1-dimethyl-3-phenyl-2-thiourea (compound IV) (mp 134–135 °C).

Compound I:
N,N-dimethylaniline

Compound II:
4-nitroso-*N,N*-
dimethylaniline

Compound III:
dimethylamine

sodium
4-nitro-
phenolate

sodium
4-nitrophenolate

Compound V:
4-nitrosophenol

Compound III:
dimethylamine

phenyl
isocyanate

Compound IV:
1,1-dimethyl-3-
phenyl-2-thiourea

   The ¹H NMR spectrum for compound V shows a pair of doublets (4H) in the aromatic region at δ 6.63 and δ 7.67, indicating a *p*-disubstituted aromatic ring. The broad singlet at δ 8.7 is the –OH group.

2. The compound must be in solubility class S_A, S_B, or S_1, which includes carboxylic acids, arylsulfonic acids, amines, alcohols, aldehydes, ketones, esters, nitriles, and amides. Arylsulfonic acids, amines, nitriles, and amides are eliminated as possibilities because of the elemental tests. The compound must contain a multiple bond since it yielded positive potassium permanganate and bromine tests. The presence of an active hydrogen is indicated by the reaction with acetyl chloride and sodium. From a negative

iodoform test, the unknown is neither a methyl ketone nor a secondary alcohol with a methyl group on the carbon adjacent to the carbon bearing the –OH group. A negative 2,4-dinitrophenylhydrazine test indicates that the compound is not an aldehyde or a ketone.

From the boiling point of 94–96 °C of the compound and the melting point of 47–48 °C for the 3,5-dinitrobenzoate derivative, the compound must be 2-propen-1-ol or allyl alcohol.

Allyl alcohol reacts with 3,5-dinitrobenzoyl chloride to produce the 3,5-dinitrobenzoate.

| allyl alcohol | 3,5-dinitrobenzoyl chloride | 3,5-dinitrobenzoate of allyl alcohol |

The ^{1}H NMR spectrum of the original compound can be analyzed as follows: δ 3.58, 1H, s (**a**); δ 4.13, 2H, m (**b**); δ 5.13, 1H, m (**c**); δ 5.25, 1H, m (**d**); and δ 6.0, 1H, 10 lines (**e**). The protons **c** and **d** are in slightly different environments and are observed at different chemical shifts.

2-propen-1-ol
(allyl alcohol)

3. Compound I is in solubility class of B, which contains amines of eight or more carbons, anilines, or ethers. Compound II is in solubility class of A$_2$, which includes phenols, enols, oximes, imides, sulfonamides, thiophenols, and nitro compounds. Compounds I and II must have a nitro group since the products obtained by treatment of I and II with zinc and boiling solution of ammonium chloride readily reduced Tollens reagent.

A compound that is an amine, has a melting point of 113–114 °C, and has a nitro group is 3-nitroaniline. Treatment of 3-nitroaniline (compound I) with nitrous acid produces 3-nitrophenol (compound II) (mp 95–96 °C).

Compound I:
3-nitroaniline

Compound II:
3-nitrophenol

The reaction of 3-nitroaniline (compound I) with benzenesulfonyl chloride in the presence of sodium hydroxide produced N-(3-nitrophenyl)benzene-sulfonamide (compound III) (mp 135–136 °C).

Compound I:
3-nitroaniline

Benzenesulfonyl
chloride

Compound III:
N-(3-nitrophenyl)-
benzenesulfonamide

The ^{1}H NMR spectrum of compound II indicates a disubstituted aromatic compound (4H) in the range of δ 7.0–7.75 and a singlet at δ 9.8 which could be attributed to the –OH.

4. A solubility class of A_1 is indicated from the data for compound I, suggesting a carboxylic acid, a phenol, or a β-diketone.

Compound I does not contain a multiple bond (negative bromine and potassium permanganate tests), an active hydrogen (negative acetyl chloride test), or an aldehyde or ketone (negative 2,4-dinitrophenylhydrazine test).

A carboxylic acid that matches data given for compound I, has a melting point of 186–187°C, a neutralization equivalent of 179, and contains nitrogen is N-benzoylaminoethanoic acid.

Compound II, with its melting point of 120–121°C and neutralization equivalent of 122, corresponds to benzoic acid.

Compound III is in solubility class S_2, which consists of salts of organic acids, amine hydrochlorides, amino acids, and polyfunctional compounds. Compound III contains a labile halogen (positive silver nitrate test) and is a primary amine (nitrous acid test). Compound III is glycine hydrochloride because of the data given and the reaction of N-benzoylaminoethanoic acid (compound I) with boiling hydrochloric acid.

Compound I:
N-benzoylamino-
ethanoic acid
MW = 179, NE = 179

Compound II:
benzoic acid
MW = 122, NE = 122

Compound III:
glycine
hydrochloride

Treatment of glycine hydrochloride (compound III) with 4-toluenesulfonyl chloride produced the 4-toluenesulfonamide of glycine hydrochloride (compound IV) (mp 149°C).

Compound III:
glycine hydrochloride

4-toluenesulfonyl
chloride

Compound IV:
4-toluenesulfonamide
of glycine

The ^{1}H NMR spectrum for *N*-benzoylaminoethanoic acid (compound I) shows an isolated CH$_2$ at δ 3.96, an OH at 5.00, a monosubstituted aromatic ring (5H) at δ 7.35–7.68 and δ 7.82–7.87, and an N–H at δ 8.81. Benzoic acid (compound II) has a monosubstituted aromatic ring (5H) at δ 7.3–7.7 and δ 8.0–8.25, with a singlet for the –OH group at δ 12.8.

5. Compound I is in solubility class of S$_2$, which includes salts of organic acids, amino hydrochlorides, amino acids, and polyfunctional compounds. Compound I contains an active hydrogen (positive acetyl chloride test), has a multiple bond or is an alcohol (positive potassium permanganate test), and is an aldehyde (positive Fehling's solution and Tollens reagent tests). Also, many carbohydrates are optically active. A carbohydrate that has the melting point of 168 °C is D-galactose. Its osazone derivative (mp 199–201 °C) is compound II.

Compound I:
D-galactose

phenylhydrazine

Compound II:
osazone of D-galactose

Nitric acid oxidizes both terminal carbons of D-galactose to –COOH groups to yield mucic or galactaric acid (compound III) (mp 220 °C).

Compound I:
D-galactose

HNO$_3$

Compound III:
mucic acid
MW = 210, NE = 105

Mucic acid gives a negative 2,4-dinitrophenylhydrazine test for aldehydes and ketones, but a positive test for active hydrogen with acetyl chloride.

After prolonged heating mucic acid (III) cyclizes to furoic acid (compound IV) (mp 132–133 °C).

low heat

Compound III:
mucic acid

Compound IV:
furoic acid
MW = 112, NE = 112

Mucic acid is treated with sodium hydroxide to form the sodium salt, which reacts with 4-nitrobenzyl bromide to yield the 4-bromophenacyl ester (compound V) (mp 136 °C).

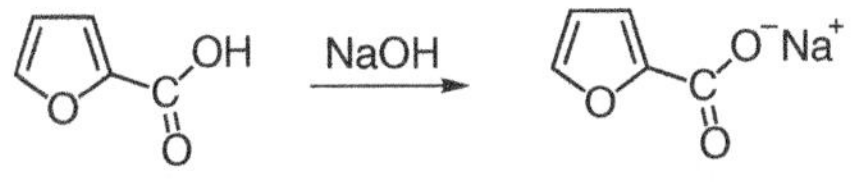

furoic acid   sodium furoate

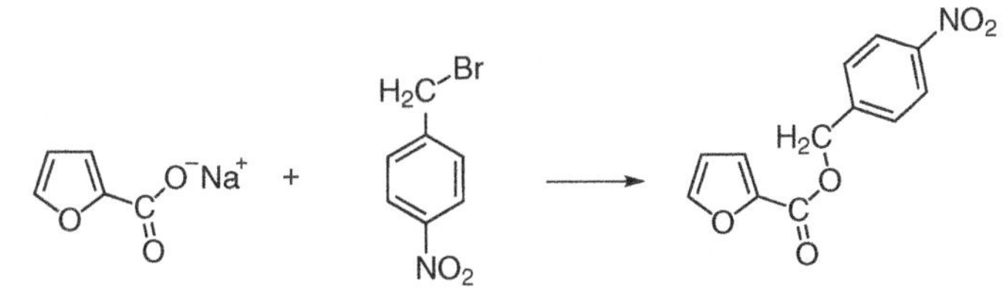

sodium furoate   4-nitrobenzyl bromide   4-nitrobenzyl furoate

The ¹H NMR spectrum of the sodium salt of compound IV indicates a conjugated system. The signals in the range of δ 6.57–7.64 are the hydrogens attached to the carbons in the ring and the signal at δ 12.35 is the OH of the carboxylic acid. The protons are identified as follows: δ 6.57, 1H, m (**a**); δ 7.35, 1H, m (**b**); and δ 7.64, 1H, m (**c**), and δ 12.35, 1H, s (**d**).

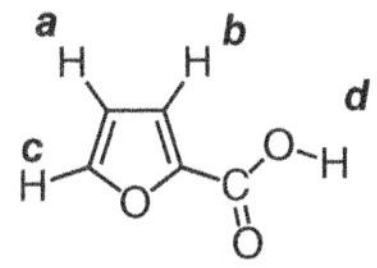

furoic acid

The IR spectrum of IV shows an –OH stretching of a carboxylic acid at 2400–3100 cm⁻¹, and the C=O stretching of an α,β-unsaturated carboxylic acid at 1692 cm⁻¹. Additional bands are the C–O stretch of a carboxylic acid at 1305, O–H bend of a carboxylic acid at 1426 cm⁻¹, C=C stretch of a *cis*-alkene at 1671 cm⁻¹.

PROBLEM SET 9

1. The compound is in the solubility class of MN, which means that the compound is a miscellaneous neutral compound containing nitrogen or sulfur and having more than five carbons. A benzenesulfonamide derivative that is soluble in base is prepared from a primary amine. The ^{1}H NMR spectrum looks like aniline. Nitrobenzene is identified as compound I and is treated with tin and hydrochloric acid to yield aniline (compound II).

nitrobenzene → 1. Sn, HCl / 2. neutralized → aniline

Aniline is treated with benzenesulfonamide to yield *N*-phenylbenzenesulfonamide.

aniline   benzenesulfonyl chloride   *N*-phenylbenzene-sulfonamide

The ^{1}H NMR spectrum confirms the structure of aniline, with the –NH$_2$ appearing as a singlet at δ 3.55, and the 5 aromatic protons at δ 6.62–7.11.

2. The compound is in solubility class S$_A$, which includes monofunctional carboxylic acids with five carbons or fewer and arylsulfonic acids. Since the compound contains sulfur, it could be an arylsulfonic acid. 2-Methylbenzenesulfonic acid matches the melting point of the unknown and of the derivative.

   2-Methylbenzenesulfonic acid reacts with 4-toluidine to form the 4-toludinium salt.

2-methylbenzene-sulfonic acid   +   4-toluidine   →   4-toluidinium salt of 2-methylbenzenesulfonic acid   +   NaCl

3. This compound is in solubility class B, MN, N, or I, which includes amines, anilines, nitrogen or sulfur compounds, alcohols, aldehydes, ketones, esters, ethers, epoxides, alkenes, alkynes, aromatic compounds, saturated hydrocarbons, haloalkanes, aryl halides, or diaryl ethers. The compound does not have a nitroso or nitro group (negative tin and hydrochloric acid test). However, since the compound contains a nitrogen, it is probably an amide or an amine. Amides hydrolyze after prolonged treatment with sodium hydroxide to yield amines and acids. The amine is soluble in dilute hydrochloric acid and reacts with acetyl chloride to produce an acetamide.

   Comparing the melting points of the compound (145–146 °C), the acetamide (111–112 °C), and the carboxylic acid (120–121 °C), the compound must be *N*-benzoyl-2-methylaniline (compound I). *N*-Benzoyl-2-methylaniline undergoes base hydrolysis to 2-methylaniline (compound II) (bp 200 °C) and sodium benzoate.

N-benzoyl-2-
methylaniline (compound I)
NaOH →
2-methylaniline,
(compound II)
+ sodium benzoate

2-Methylacetanilide (compound III) is made from the reaction of 2-methylaniline
(compound II) with acetyl chloride.

2-methylaniline,
(compound II)
acetyl
chloride
2-methylacetanilide
(compound III)

Sodium benzoate is acidified to form benzoic acid (compound IV).

sodium benzoate
HCl →
benzoic acid
MW = 122, NE = 122
(compound IV)

The ^{1}H NMR spectrum of 2-methylaniline shows the methyl peak at δ 2.00. The aromatic
range contains four carbons at δ 6.45–7.15. The $-NH_2$ is indicated by a peak at δ 3.48. A ^{1}H
NMR spectrum of benzoic acid indicates a monosubstituted aromatic at δ 7.41–8.32 (5H)
and the $-OH$ of the carboxylic acid at δ 12.08.

4. This compound is in solubility class I, consisting of hydrocarbons, haloalkanes, aryl
halides, diaryl ethers, or deactivated aromatic compounds. The unknown contains
an aromatic ring since it dissolved in fuming sulfuric acid. The chlorine is attached
to an aromatic ring or an alkene carbon (negative silver nitrate test). Hot potassium
permanganate solution causes oxidation of a $-R$ group attached to the benzene ring,
resulting in a benzoic acid derivative. 2-Chlorobenzoic acid agrees with the melting point
of 138–139 °C and a neutralization equivalent of 156.5.

According to the ^{1}H NMR spectrum the $-R$ group that is oxidized must be a methyl group,
since a singlet of three protons appears at δ 2.37. The aromatic region with four protons at
δ 7.0–7.35 indicates a disubstituted benzene ring. The IR spectrum of the acid shows an O–H
stretch of a carboxylic acid at 2881–3333 cm^{-1}, a C=O stretch of a carboxylic acid at 1678 cm^{-1}
and C–H out-of-plane bend for an aromatic *o*-disubstituted benzene ring at 742 cm^{-1}. For the
carboxylic acid, the C–O stretch is at 1316 cm^{-1} and the O–H bend is at 1400 cm^{-1}. The
aromatic C–H stretch appears at 3050 cm^{-1}. The C–Cl stretch appears at 1100 cm^{-1}.

2-Chlorotoluene (bp 159–161 °C) is the original compound.

2-chlorotoluene
1. hot KMnO$_4$
2. H$^+$
2-chlorobenzoic acid
MW = 156.5, NE = 156.5

5. The compound is in solubility class N, which consists of alcohols, aldehydes, ketones, esters, ethers, epoxides, alkenes, alkynes, or aromatic compounds. A negative test with 2,4-dinitrophenylhydrazine eliminates aldehydes and ketones as possibilities. The compound does not have an active hydrogen (negative acetyl chloride test) or a multiple bond (negative bromine test). Boiling alkalies cause the compound to hydrolyse. Of the choices listed above, only esters, upon treatment with boiling alkali, break into two organic products, alcohols and carboxylic acids.

The distillate of the alkali reaction is an alcohol. An alcohol that matches the boiling point of 130 °C with a 1-naphthylurethane derivative of melting point 70 °C is 3-methyl-1-butanol. The residue is a salt of the carboxylic acid, which upon acidification yields the acid. The acid is treated with thionyl chloride, followed by aniline to form an anilide. 3-Methylbutanoic acid will react with thionyl chloride athen aniline to form an anilide (mp 108–109 °C). The original compound is 3-methylbutyl 3-methylbutanoate or isopentyl isovalerate (bp 188–192 °C).

3-Methylbutyl 3-methylbutanoate undergoes base hydrolysis to form 3-methyl-1-butanol and sodium 3-methylbutanoate.

1-Methylbutyl 3-methylbutanoate reacts with 1-naphthyl isocyanate to produce the 1-naphthylurethane of 3-methyl-1-butanol.

The sodium-3-methylbutanoate is acidified to form the 3-methylbutanoic acid. The carboxylic acid is treated with thionyl chloride to yield the 3-methylbutanoyl chloride, which reacts with aniline to give *N*-phenyl-3-methylbutanamide.

Treatment of the original compound with lithium aluminum hydride produced the same alcohol as from the saponification reaction. The conclusion is reached that the acyl portion and the alkyl portion of the ester each must have five carbons arranged as an isopentyl group.

3-methylbutyl 3-methylbutanoate   3-methyl-1-butanol

The ¹H NMR spectrum agrees with the structure of 3-methyl-1-butanol. The six protons at δ 0.92 that are split into a doublet are the two methyls adjacent to a methine proton. The three protons at δ 1.2–1.7 are the protons attached to carbons two and three. The two hydrogens split into a triplet at δ 3.63 corresponds to the methylene group at carbon one. The –OH is seen as a broad singlet at δ 2.13.

6. A smoky flame indicates that the compound is aromatic. The unknown is in solubility class B, which consists of amines, anilines, and ethers. The compound reacted with benzenesulfonyl chloride to yield a benzenesulfonamide which was soluble in alkali, indicating that the original compound was a primary amine. Comparing the melting point (112–114 °C) of the original compound and the melting points of the benzenesulfonamide (101–102 °C) and acetamide (132 °C) derivatives, the original compound must be 2-aminonaphthalene.

2-Aminonaphthalene reacts with benzenesulfonyl chloride to give N-naphthalen-2-ylbenzenesulfonamide.

2-aminonaphthalene   benzenesulfonyl chloride   N-naphthalen-2-yl-benzenesulfonamide

The acetamide derivative is prepared by reacting 2-aminonaphthalene with acetyl chloride.

2-aminonaphthalene   acetyl chloride   N-naphthalen-2-yl-acetamide

The IR spectrum shows the N–H stretch of a primary amine at 3323 and 3396 cm⁻¹, C–N stretch of a primary aromatic amine at 1282 cm⁻¹, and N–H bend of a primary amine at 1613 cm⁻¹.

PROBLEM SET 10

1. Compound A is not an aldehyde or ketone (negative 2,4-dinitrophenylhydrazine test). Since compound A gave a positive iodoform test, one of the following fragments must be present in the structure. Compound B is an alkene or alkyne (positive bromine and potassium permanganate tests).

   However, it cannot be a ketone according to the negative 2,4-dinitrophenylhydrazine test. If 45 (CH_3CHOH) and 45 (COOH) are subtracted from the neutralization equivalent of 103 ± 1, then 13 ± 1 is left. This is probably a CH_2. Therefore, compound A must be 3-hydroxybutanoic acid.

   3-Hydroxybutanoic acid (compound A) loses water with sulfuric acid to yield 2-butenoic acid (compound B).

Compound A:
3-hydroxybutanoic acid
MW = 104, NE = 104

Compound B:
2-butenoic acid,
crotonic acid
MW = 86, NE = 86

   Treatment of 3-hydroxybutanoic acid (compound A) with iodine and sodium hydroxide produces propanedioic acid (compound C).

Compound A:
3-hydroxybutanoic acid
MW = 104, NE = 104

Compound C:
propanedioic acid
MW = 104, NE = 52

2. The ^1H NMR spectrum of the second acid indicates a symmetrical aromatic dicarboxylic acid. The lack of reactivity for bromination indicates the presence of deactivating groups. Since a 2 : 1 ratio is present between the aromatic region (δ 8.08) and the —OH group (δ 11.0), the second carboxylic acid must be 1,4-benzenedicarboxylic acid (terephthalic acid). The first carboxylic acid cannot contain a labile hydrogen since substitution with bromine does not occur thus must be 2-(4-carboxyphenyl)-2-oxo-ethanoic acid.

4-oxalylbenzoic acid
MW = 194, NE = 97

1,4-benzenedicarboxylic acid,
terephthalic acid
MW = 166, NE = 83

3. This compound is in the solubility class of N, which consists of alcohols, aldehydes, ketones, esters, ethers, epoxides, alkenes, alkynes, or aromatic compounds. Since the compound is a hydrocarbon, all oxygen-containing compounds are eliminated as possibilities. The compound must contain a multiple bond since it decolorized bromine and potassium permanganate solutions.

Since the molecular ion in the mass spectrum was at m/z of 68, thus the formula must be C_5H_8. Therefore, the structure contains two rings, two carbon-carbon double bonds, a carbon–carbon triple bond, or a carbon-carbon double bond and a ring. For oxidation to yield a carboxylic acid, a carbon-carbon double bond must be broken. Both the original compound and the carboxylic acid product contain the same number of carbons. Since the original compound is chiral, it must be 3-methylcyclobutene. The product after oxidation must be a dicarboxylic acid, 2-methylbutanedioic acid.

3-methylcyclobutene
MW = 68

2-methylbutanedioic acid
MW = 132, NE = 66

4. The original compound is not an alkene or alkyne (negative bromine test). For the neutralization equivalent to increase in value from 66 to 88, a carbon dioxide molecule must be lost when the original compound was heated.

The ^{1}H NMR spectrum of the second carboxylic acid agrees with the structure of 2-methylpropanoic acid. The methyl groups appear as a doublet at δ 1.20, the –CH on carbon two appears as a septet at δ 2.57, and the carboxylic acid hydrogen is at δ 12.4. For carbon dioxide to be lost upon heating of the first carboxylic acid, two –COOH groups must be on carbon two. The original acid must be 2,2-dimethylpropanedioic acid.

2,2-dimethylpropane-
dioic acid,
dimethylmalonic acid
MW = 132, NE = 66

2-methylpropanoic acid
MW = 88, NE = 88

5. From the ^{1}H NMR spectrum, both compounds have five protons in the aromatic region (acid-δ 7.52 and δ 8.14, 5H; base-δ 7.23, 5H) and thus contain monosubstituted benzene rings. The spectrum of the base indicates a –NH$_2$ group at δ 0.91, and two –CH$_2$ groups centered at δ 2.78. The spectrum of the acid shows only one other proton at δ 12.82, which is the carboxylic acid proton. The base must be phenylethylamine and the acid is benzoic acid.

phenylethylamine
MW = 121, NE = 121

benzoic acid
MW = 122, NE = 122

6. This compound does not contain a multiple bond since it is unaffected by bromine. A positive iodoform test indicates the presence of one of the following groups.

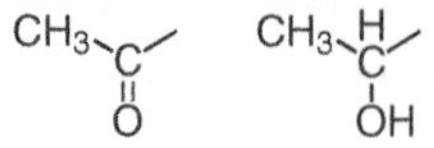

Compounds that fit this data are 2-, 3-, or 4-(1-hydroxyethyl)-benzoic acid.

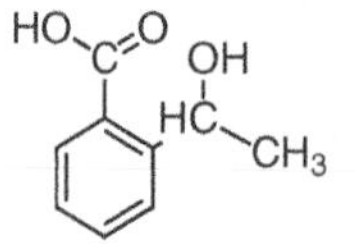

2-(1-hydroxyethyl)-
benzoic acid
MW = 166, NE = 166

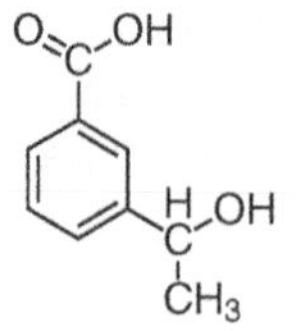

2-(1-hydroxyethyl)-
benzoic acid
MW = 166, NE = 166

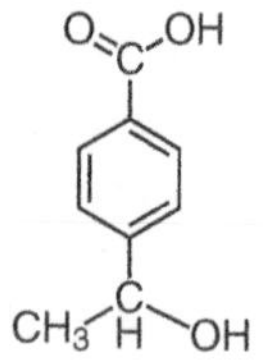

2-(1-hydroxyethyl)-
benzoic acid
MW = 166, NE = 166

7. Compound A belongs to the solubility class of A_1 or A_2, but does not contain sulfur, nitrogen, or halogen. Therefore, the possibilities are carboxylic acids, phenols, enols, or β-diketones. Compound A is not a phenol (negative ferric chloride test) and does not have a multiple bond (negative potassium permanganate test).

The first product from the treatment of compound A with hydrobromic acid is in the solubility class of I, which means that this compound could be a saturated hydrocarbon, haloalkane, aryl halide, diaryl ether, or aromatic compound. This compound contained a labile halogen (positive sodium iodide in acetone test).

Compound B contains a multiple bond as indicated from the positive bromine test and is a phenol as indicated from the positive ferric chloride test. The difference in the neutralization equivalent between compounds A and B is 43 (180−137 = 43), which could be a propyl group. The ^1H NMR spectrum of compound B shows a *p*-disubstituted benzene with doublets of two hydrogens each at δ 6.84 and δ 7.86, and two broad −OH signals at δ 8.2. From the ^1H NMR spectrum and the neutralization equivalent, Compound B must be 4-hydroxybenzoic acid and compound A must be 4-propoxybenzoic acid.

4-Propoxybenzoic acid is treated with hydrogen bromide to yield 4-hydroxybenzoic acid and 1-bromopropane.

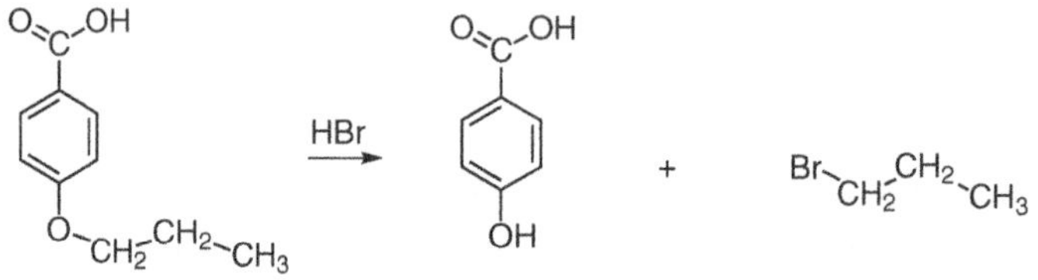

Compound A:
4-propoxybenzoic acid
MW = 180, NE = 180

Compound B:
4-hydroxybenzoic acid
MW = 138, NE = 138

1-bromopropane

8. The acid cannot contain an aromatic ring, due to the value of the molecular weight. The carboxylic acid group has a molecular weight of 45, which leaves 53 (98−45 = 53). The molecular weight of 53 could be C_4H_5. Possibilities from the formula C_4H_5COOH include the following structures.

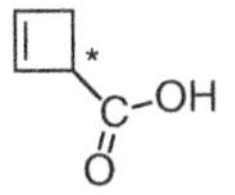

cyclobut-2-enecarboxylic acid

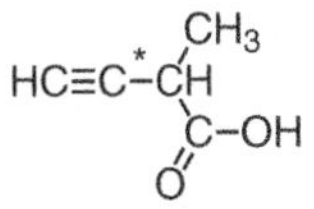

2-methylbut-3-ynoic acid

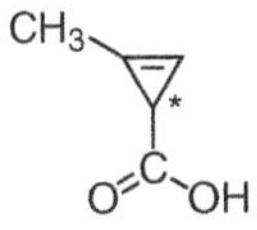

2-methylcycloprop-2-
enecarboxylic acid

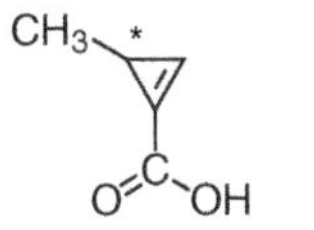

3-methylcycloprop-1-
enecarboxylic acid

9. The original compound I is an aldehyde or a ketone since it yielded a positive 2,4-dinitrophenylhydrazine test. The distillate from the sodium hydroxide treatment contains an active hydrogen (positive sodium test), is a methyl ketone or a methyl secondary alcohol (positive iodoform test), and is not a tertiary or secondary alcohol (negative Lucas test).

The residue from the steam-distillation is a carboxylic acid. The 4-bromophenacyl ester had a saponification equivalent of 257. If the 4-bromophenacyl group is subtracted from this saponification equivalent, a value of 59 (257−198 = 59) is left.

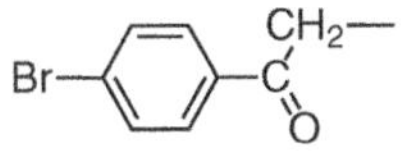

4-bromophenacyl group

The fragment that is left over is an acyl group.

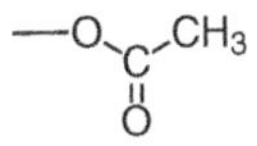

From the mass spectrum, the following peaks can be interpreted as follows.

m/z	Fragment
15	CH_3^+
29	$CH_3CH_2^+$
43	$CH_3-C\equiv O^+$
45	$CH_3CH_2O^+$
85	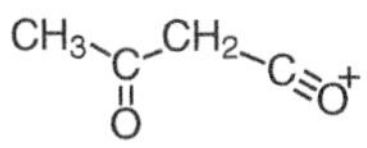
130	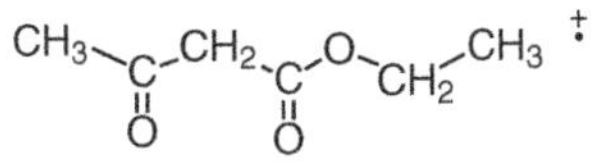

With this information, the original compound must be ethyl acetoacetate.

Ethyl acetatoacetate undergoes base hydrolysis to form ethanol and sodium acetate. The sodium acetate is acidified to form acetic acid.

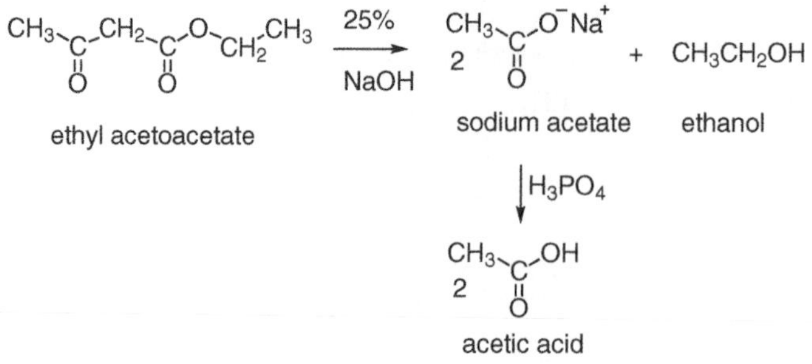

ethyl acetoacetate

sodium acetate   ethanol

H_3PO_4

acetic acid

Sodium acetate reacts with 4-bromophenacyl bromide to yield 2-(4-bromophenyl)-2-oxoethyl acetate.

sodium acetate   4-bromophenacyl bromide   2-(4-bromophenyl)-2-oxoethyl acetate

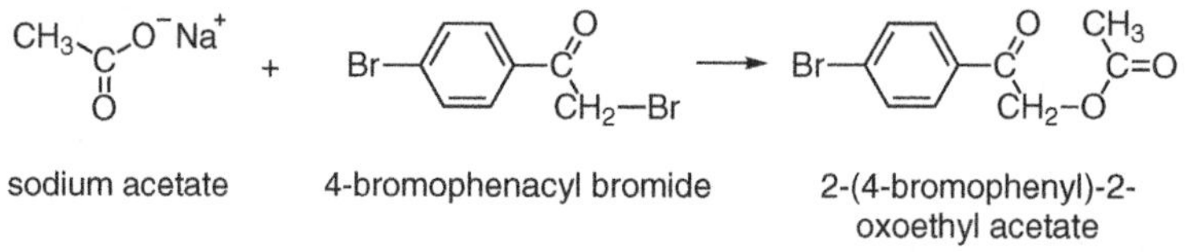

The IR spectrum indicated the C=O stretch of an aliphatic ketone at 1715 cm⁻¹ and the C=O stretch of a β-ketoester at 1652 cm⁻¹. A C–H stretch of an alkyl is at 2969 cm⁻¹ and C–H bend of a –CH₃ is at 1368 and 1448 cm⁻¹. A C–H stretch of a –CH₂ shows up at 1467 cm⁻¹.

PROBLEM SET 11

1. A cyanoacid is treated with thionyl chloride to yield an acyl chloride, which reacts with ammonium hydroxide to give a cyanoamide. The cyanoamide is treated with hypobromite to give a cyanoamine. Then cyano group is oxidized to a carboxylic acid, which is diazotized to a diazonium salt. This diazonium salt reacts with cyanide to give a cyano acid.

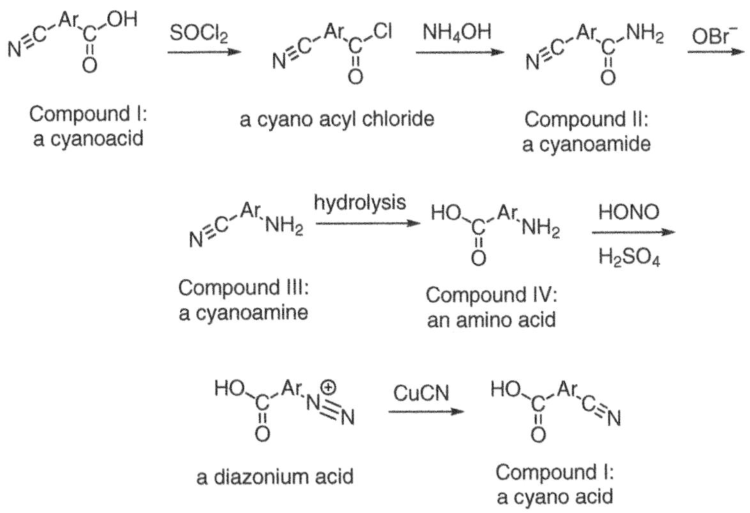

An aromatic ring must be present for the diazotization to occur. Since the final product contains a –CN and a –COOH group, the initial compound also contains these groups. The aryl group can be determined from the neutralization equivalent.

$$197 - (26 + 45) = 126$$
$$-CN, -COOH$$

A value of 126 corresponds to $C_{10}H_6$, which is a disubstituted naphthalene. The original acid (compound I) must be 8-cyano-1-naphthoic acid.

8-Cyano-1-naphthoic acid is treated with thionyl chloride to yield an 8-cyano-1-naphthoyl chloride, which reacts with ammonium hydroxide to give a 8-cyano-1-naphthamide. This cyanoamide is treated with hypobromite to give a 1-amino-8-cyanonaphthalene. Then cyano group is oxidized to give 8-amino-1-naphthoic acid, which is diazotized to a 8-carboxy-2-naphthalene diazonium salt. This diazonium salt reacts with cyanide to give 8-cyano-1-naphthoic acid.

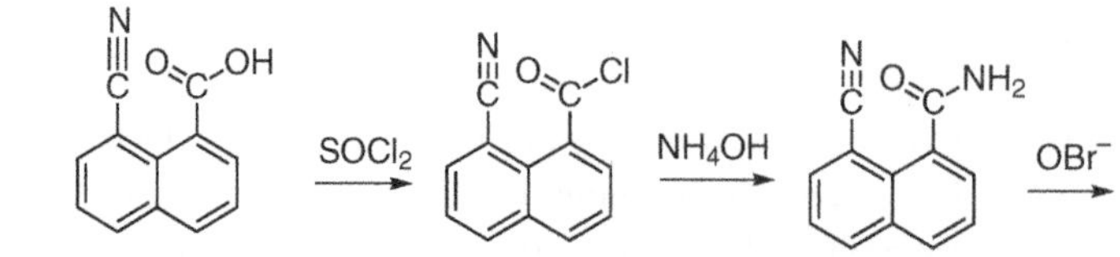

Compound I:
8-cyano-1-
naphthoic acid
MW = 197, NE = 197

8-cyano-1-
naphthoyl
chloride

Compound II:
8-cyano-1-
naphthamide

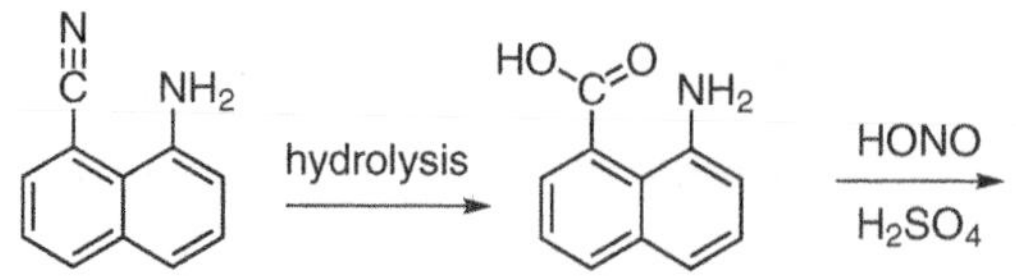

Compound III:
1-amino-8-cyano-
naphthalene

Compound IV:
8-amino-1-naphthoic acid
MW = 187, NE = 187

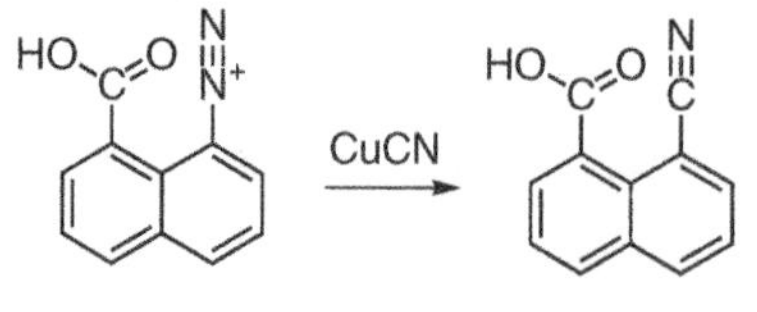

8-carboxy-2-
naphthalene
diazonium salt

Compound I:
8-cyano-1-naphthoic acid
MW = 197, NE = 197

Hydrolysis of 8-cyano-1-naphthoic acid (compound I) produced 1,8-naphthalenedicarboxylic acid (compound V), which is heated to produce 1,8-naphthalenedicarboxylic anhydride (compound VI).

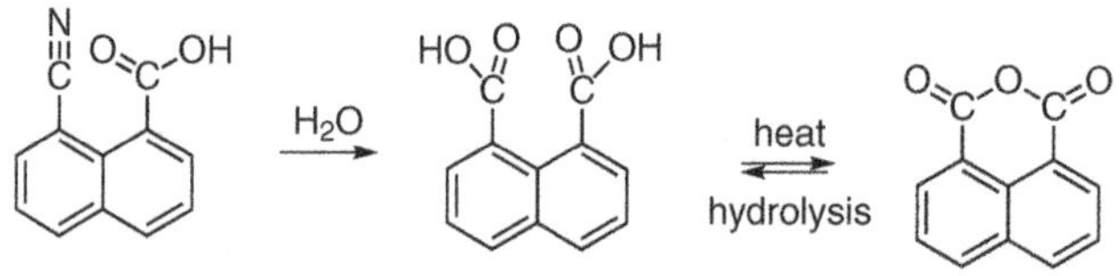

Compound I:
8-cyano-1-naphthoic acid

Compound V:
1,8-naphthalene-
dicarboxylic acid
MW = 216, NE = 108

Compound VI:
1,8-napthalic
anhydride

Oxidation of 8-amino-1-naphthoic acid (compound IV) produced another acid, 1,2,3-benzenetricarboxylic acid (compound VII).

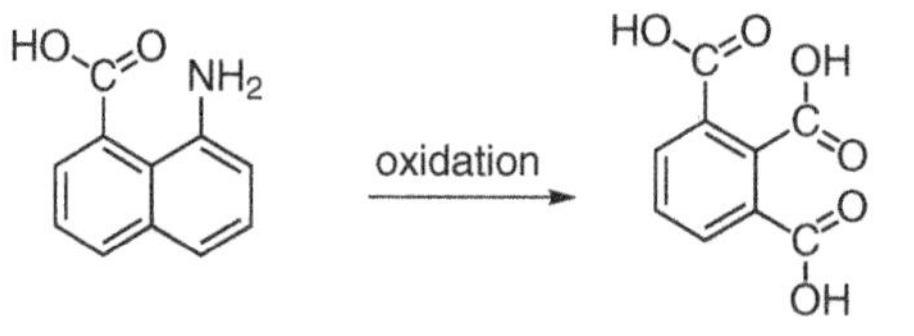

Compound IV:
8-amino-1-naphthoic acid
MW = 187, NE = 187

Compound VII:
1,2,3-benzenetricarboxylic acid
MW = 210, NE = 70

The H NMR signal indicates aromatic peaks from δ 7.6 to 8.28 and the –OH at δ 11.8.

2. The solubility class for this compound is A_1, A_2, or B, consisting of carboxylic acids, phenols, enols, oximes, imides, sulfonamides, thiophenols, nitro compounds, β-diketones, sulfonamides, amines, or anilines. Acetic anhydride reacts with active hydrogens such as those on amines and alcohols. Since nitrogen was evolved when the compound was treated with nitrous acid, a primary amine must be present. From these tests, the original compound (compound I) must contain a –COOH and a –NH$_2$ group, which would also explain the inability to obtain a satisfactory neutralization equivalent.

The ^{1}H NMR spectrum of the oxidation product indicates the presence of a *p*-disubstituted benzene ring (4H) with doublets at δ 7.60 and δ 7.95, and a broad singlet indicating a –OH peak for the carboxylic acid group at δ 12.18. The only bromine-containing compound that matches the ^{1}H NMR spectrum and the neutralization equivalent of 201 is 4-bromobenzoic acid.

Compounds I, II, and III must also be disubstituted with carbon-containing groups in the position *para* to the bromine. By using the neutralization equivalent of 270±2 for compound II, the structure of compound I can be determined.

$$270\pm2 \; - \; \underset{\substack{\text{p-disubstituted} \\ \text{ring}}}{\left(76\right.} \; + \; \underset{\text{Br}}{80} \; + \; \underset{-\text{NH}}{15} \; + \; \underset{-\text{COOH}}{45} \; + \; \underset{\text{CH}_3\text{CO}-}{\left.43\right)} \; = 11\pm2, \text{which is CH}$$

Thus compound I must be amino(4-bromophenyl)acetic acid, which reacts with acetic anhydride to produce acetylamino(4-bromophenyl)acetic acid (compound II).

Compound I:
amino(4-bromo-
phenyl)acetic acid

acetic anhydride

Compound II:
acetylamino(4-bromo-
phenyl)acetic acid
MW = 272, NE = 272

2-Amino-2-(4-bromophenyl)ethanoic acid (compound I) is treated with nitrous acid to produce 2-(4-bromophenyl)-2-hydroxyethanoic acid (compound III).

Compound I:
amino(4-bromo-
phenyl)acetic acid

Compound III:
(4-bromophenyl)-
hydroxyacetic acid

Vigorous oxidation of amino-(4-bromophenyl)acetic acid (compound I), acetylamino-(4-bromophenyl)acetic acid (compound II), or (4-bromophenyl)-hydroxyethanoic acid (compound III) produces 4-bromobenzoic acid.

Compounds I, II, or III $\xrightarrow[\text{oxidation}]{\text{vigorous}}$ 4-bromobenzoic acid

4-bromobenzoic acid
MW = 201, NE = 201

3. Esters hydrolyze to acids and alcohols. Compound II contains an active hydrogen (positive acetyl chloride test) and is a phenol (positive ferric chloride test). The ¹H NMR spectrum for compound II indicates an isolated –CH$_3$ at δ 2.25, an –OH at δ 5.67, and a disubstituted aromatic ring (4H) at δ 6.5–7.3. Therefore, compound II is a methylphenol. The ¹H NMR spectrum for compound III shows a ratio of 2 : 1 between the aromatic signals at δ 7.4–7.9 and the –COOH groups at δ 12.08. Therefore, compound III is 1,2-benzenedicarboxylic acid (phthalic acid), since it undergoes a reversible dehydration.

The original ester (compound I) is di-3-methylphenyl 1,2-benzenedicarboxylate, which is hydrolyzed to 1,2-benzenedicarboxylic acid (phthalic acid) (compound III) and 3-methylphenol (compound II).

Compound I:
di-3-methylphenyl-
1,2-benzenedicarboxylate
MW = 346, SE = 173

1. NaOH
2. H⁺

Compound II:
3-methylphenol

Compound III:
phthalic acid
MW = 166, NE = 83

When heated, phthalic acid (compound III) loses water and becomes phthalic anhydride (compound IV).

Compound III:
phthalic acid

heat / HCl

Compound IV:
phthalic anhydride

4. Solid I is in the solubility class of S$_2$, consisting of salts of an organic acid, amine hydrochlorides, amino acids, or polyfunctional compounds. Treatment of an amine salt of a carboxylic acid with base liberates the salt of a carboxylic acid and an amine (compound II). The reaction of the amine (compound II) with benzenesulfonyl chloride produces a benzenesulfonamide. The amine must be a secondary amine since the benzenesulfonamide is insoluble in base.

The salt of the carboxylic acid from the cold alkali reaction is acidified to liberate the carboxylic acid (compound III). The acid must also contain a nitro group, since it reacts with zinc dust and ammonium chloride to form a product (hydrazine, hydroxylamine, or aminophenol) that reduces Tollens reagent.

The ¹H NMR spectrum of the amine (compound II) corresponds to dibutylamine. The amino is at δ 0.53 (1H, s); the two methyls are at δ 0.90 (6H, t); the four methylenes on carbons two and three are at δ 1.2–1.7 (8H, m); and the two methylenes on carbon one are at δ 2.52 (4H, t). The ^{1}H NMR spectrum of the carboxylic acid (III) is a nitrobenzoic acid; it is *meta* because of the splitting pattern in the aromatic region (δ 7.77–8.96, 4H). The peak at δ 12.0 is the –OH of the acid.

The original compound (compound I) is dibutylammonium 3-nitrobenzoate, which is treated with base to form sodium 3-nitrobenzoate and dibutylamine (compound II).

Compound I: dibutylammonium 3-nitrobenzoate

Compound II: dibutylamine   sodium 3-nitrobenzoate

The dibutylammonium 3-nitrobenzoate (compound I) is acidified to give the dibutyl-ammonium salt and 3-nitrobenzoic acid (compound III).

Compound I: dibutylammonium 3-nitrobenzoate

dibutylammonium salt   Compound III: 3-nitrobenzoic acid
MW = 167, NE = 167

3-Nitrobenzoic acid (compound III) is reduced to 3-hydroxyaminobenzoic acid (compound IV).

Compound III: 3-nitrobenzoic acid   Compound IV: 3-hydroxyamino-
MW = 167, NE = 167   benzoic acid

5. From the formula of $C_5H_{10}O$, it is obvious that the compound contains a double bond. Additional proof is seen in the decolorization of bromine and potassium permanganate solutions. The compound is not an aldehyde or a ketone (negative 2,4-dinitrophenylhydrazine test). A negative Lucas test indicates that the compound is not a tertiary or secondary alcohol. The compound is not a methyl ketone nor does it have a methyl next to a carbon with a hydroxy group (negative iodoform test). The compound contains an active hydrogen (positive acetyl chloride test).

The original compound is oxidized with hot acidified potassium permanganate to a carboxylic acid, with a neutralization equivalent of 59 ± 1. For this acid to be able to lose carbon dioxide upon heating and increase in neutralization equivalent from 59 ± 1 to 73 ± 1, it must be a *gem* dicarboxylic acid. The second acid has a neutralization equivalent of 73 ± 1. The original compound must be 2-methyl-3-buten-1-ol, which is oxidized to 2-methylpropanedioic acid. When heated, the acid is converted to propanoic acid and carbon dioxide.

2-methyl-3-buten-1-ol   2-methylpropanedioic acid   propanoic acid
                                MW = 118, NE = 59   MW = 74, NE = 74

Compound I is in the solubility class of A_2 or B, which include phenols, enols, oximes, imides, sulfonamides, thiophenols, β-diketones, nitro compounds, amines, anilines, or ethers. Compound I was treated with excess acetic anhydride to produce compound II. Compound II is in solubility class of MN, N, or I, which includes miscellaneous compounds containing nitrogen or sulfur and having more than five carbons, alcohols, aldehydes, ketones, esters with one functional group and more than five and less than nine compounds, ethers, epoxides, alkenes, alkynes, some aromatic compounds, saturated hydrocarbons, haloalkanes, aryl halides, other deactivated compounds, and diaryl ethers. Compound I is a phenol since it decolorized bromine water. Compound I reacts with nitrous acid to form an *N*-nitroso (compound III) without the evolution of nitrogen, indicating the presence of a secondary amine.

From the ^{1}H NMR spectrum of compound I, a $-CH_3$ group is present at δ 3.65, a *p*-disubstituted benzene is indicated from the two doublets in the aromatic region (4H) at δ 7.12 and δ 7.49, and the broad singlets corresponding to the amino and hydroxy groups are at δ 3.31 and δ 5.51.

From the spectrum and the classification tests, compound I is *N*-methyl-4-aminophenol. 4-Acetoxy-*N*-methylacetanilide (compound II) is the diacetyl derivative.

Compound I: Compound II:
N-methyl-4-aminophenol   4-acetoxy-*N*-methylacetanilide

Treatment of *N*-methyl-4-aminophenol (compound I) with nitrous acid yielded 4-hydroxy-*N*-methyl-*N*-nitrosoaniline (compound III).

Compound I: Compound III:
N-methyl-4-aminophenol   4-hydroxy-*N*-methyl-*N*-nitrosoaniline

PROBLEM SET 12

1. From the formula the compound must be aromatic with a cycloalkene, in order to obtain the correct number of carbons and hydrogens. As compound I ($C_9H_6O_2$) is oxidized, the aliphatic ring opens to form the potassium salt of a carboxylic acid (compound II) ($C_9H_8O_3$). Compound I must contain a double bond, since it reacts with N-bromosuccinimde to form a bromo structure (compound III) ($C_9H_9O_2Br$). When compound I was heated with sodium hydroxide, followed by acidification, compound IV ($C_9H_6O_3$) was formed, which involved the loss of a hydrogen and bromine and the formation of a carboxylic acid. The best place to start is to look in any handbook that lists formulas and structures and draw out a few possibilities for compounds I and IV. After scrutinizing the various choices, the only one choice that meets the criteria is coumarin for compound I.

    Coumarin (compound I) ($C_9H_6O_2$) is oxidized to the potassium 2-cinnamate (compound II) ($C_9H_8O_3$).

Compound I:
coumarin

potassium salt of
Compound II:
potassium
2-hydroxycinnamate

2-hydroxycinnamate

Treatment of coumarin (compound I) with bromine yields the 3,4-dibromocoumarin (compound III) ($C_9H_8O_2Br_2$). This product undergoes reaction with base, followed by acid to yield coumarilic acid (compound IV) ($C_9H_6O_3$).

Compound I:
coumarin

Compound III:
3,4-dibromocoumarin

Compound IV:
benzofuran-2-carboxylic acid

The ^1H NMR spectrum of 2-hydroxycinnamic acid (compound II) gives an integration ratio of 3:1 between the aromatic and alkene protons at δ 6.5–7.8 and the –OH protons at δ 10.2 and 12.2.

2. From the formula, the structure must be aromatic. For the nitrogen to be lost upon hydrolysis with no loss of carbon(s), a primary amide must be present. For the acid to be optically active, the benzene ring must be monosubstituted. 2-Bromo-2-phenylethanamide (C_8H_8ONBr) reacts with potassium hydroxide to produce the potassium salt of 2-hydroxy-2-phenylethanoic acid ($C_8H_8O_3$).

    The ^1H NMR spectrum of the 2-hydroxy-2-phenylethanoic acid indicates the singlet –CH group at δ 5.22 and the aromatic ring and the –OH groups (7H) at δ 7.2–7.7.

2-bromo-2-phenylethanamide

potassium salt of 2-hydroxy-
2-phenylethanoic acid

3. This compound must be aromatic as deduced from the formula. This is a very difficult problem to solve. When compound I ($C_{11}H_{12}O_4$) is treated with hot sodium hydroxide solution, the ketone group is reduced and the aldehyde group is oxidized to form compound II ($C_{11}H_{13}O_5Na$). Compound II is protonated to form compound III ($C_9H_8O_4$), along with the loss of two carbons. In the last step, compound III dimerizes to form compound IV ($C_{18}H_{12}O_6$). The *o*, *m*, or *p* structures are all possible answers.

Compound I:
dimethoxymethylphenyl-
2-oxoethanol

Compound II:
sodium dimethoxy-
methylphenyl-
2-oxoethanoic acid

Compound III:
carboxyphenyl-2-hydroxy-
ethanoic acid

Compound IV:
3,6-di(carboxyphenyl)-
1,4-dioxane-2,5-dione

4. From the formula of $C_5H_8O_2$, compound I must contain two double bonds, two rings, or one double bond and a ring. Hydrogen chloride is added across a double bond or a ring producing compound II ($C_5H_9O_2Cl$). With potassium hydroxide, the chlorine in compound II is replaced by –OH and the salt of the carboxylic acid is formed to yield compound III ($C_9H_5O_3K$). In the presence of potassium permanganate, compound III is oxidized from a secondary alcohol to a ketone and the anion of the acid is protonated to produce compound IV ($C_5H_8O_3$). Hypochlorite ion oxidizes the methyl ketone in compound IV to give compound V. With a formula of $C_4H_6O_4$ and a neutralization equivalent of 58 ± 1, compound V must be a dicarboxylic acid. The dicarboxylic acid (compound V) cyclizes with loss of water to yield compound VI ($C_4H_4O_3$).

With all of these reactions in mind, the original compound (I) must be either 4-pentenoic acid or γ-methyl-γ-butyrolactone.

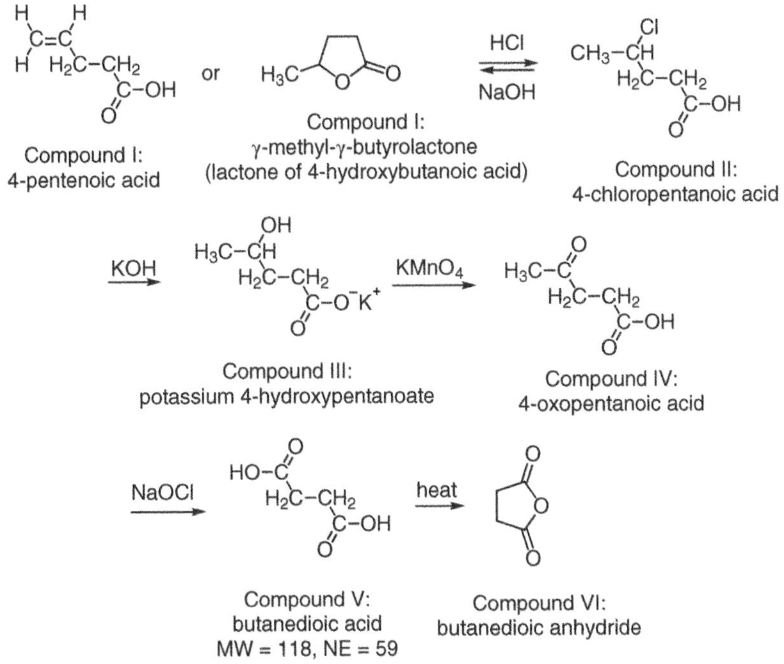
HCl
NaOH
or
Compound I:
4-pentenoic acid
Compound I:
γ-methyl-γ-butyrolactone
(lactone of 4-hydroxybutanoic acid)
Compound II:
4-chloropentanoic acid
KOH
KMnO4
Compound III:
potassium 4-hydroxypentanoate
Compound IV:
4-oxopentanoic acid
NaOCl
heat
Compound V:
butanedioic acid
MW = 118, NE = 59
Compound VI:
butanedioic anhydride

The ^{1}H NMR spectrum of butanedioic acid (compound V) agrees with the structure of butanedioic acid. The integration ratio of 2 : 1 for the signal at δ 2.43 compared to the δ 11.8 correlates to the ratio of $-CH_2$ to –OH peaks.

5. The singlet at δ 3.89 indicates an isolated methyl group. The pair of doublets, with four protons, at δ 6.91 and δ 8.12 indicates a *p*-disubstituted aromatic ring. The structure must be 4-nitroanisole ($C_7H_7NO_3$).

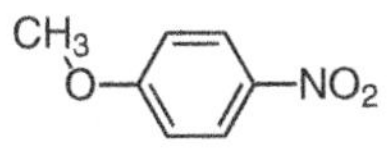
4-nitroanisole

6. The compound must be aromatic as indicated from the formula of $C_{16}H_{13}N$ and has an unsaturation number of 11. The ^{1}H NMR spectrum shows a broad singlet at δ 5.61 which indicates a –NH. The remainder of the peaks are aromatic (12H) (δ 6.60–δ 7.55, δ 7.80). Two structures are possible, 1-phenylaminonaphthalene or 2-phenylaminonaphthalene.

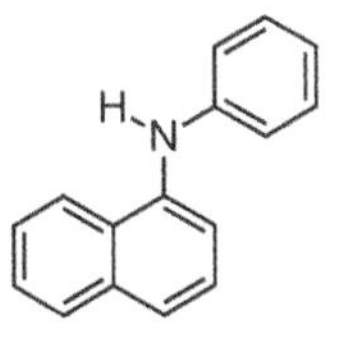
1-phenylaminonaphthalene

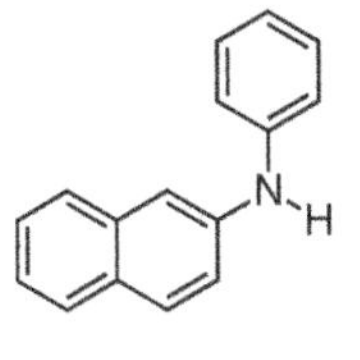
2-phenylaminonaphthalene

7. Potassium permanganate oxidizes alkenes to 1,2-diols, which can be further oxidized to ketones. Potassium permanganate also oxidizes alkynes to carboxylic acids. Sodium tests for the presence of an active hydrogen attached to oxygen, nitrogen, sulfur, or a carbon-carbon triple bond. From the formula, the compound must be aromatic. From drawing various structures with a formula of $C_{14}H_{10}O$, two benzene rings must be present in the structure and these rings are not fused together. Possible structures are 2-phenoxyphenylethyne, 3-phenoxyphenyl-ethyne, or 4-phenoxyphenylethyne.

Oxidation of 2-phenoxyphenylethyne, 3-phenoxyphenylethyne, or 4-phenoxyphenylethyne ($C_{14}H_{10}O$) produces 2-phenoxybenzoic acid, 3-phenoxybenzoic acid, or 4-phenoxybenzoic acid ($C_{13}H_{10}O$).

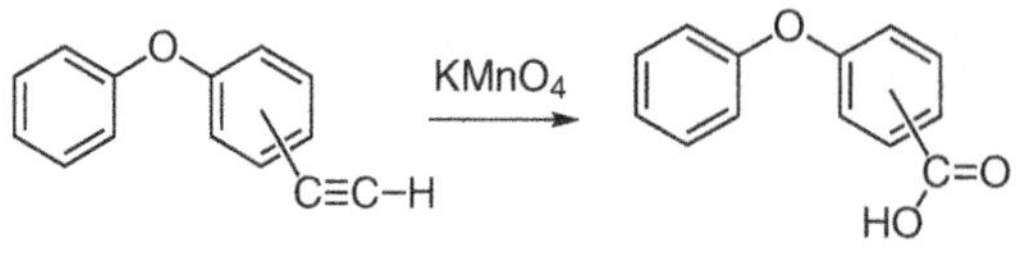

2-phenoxyphenylethyne, 3-phenoxyphenylethyne, or 4-phenoxyphenylethyne → KMnO₄ → 2-phenoxybenzoic acid, 3-phenoxybenzoic acid, or 4-phenoxybenzoic acid

Treatment of 2-phenoxyphenylethyne, 3-phenoxyphenylethyne, or 4-phenoxy-phenylethyne ($C_{14}H_{10}O$) with sodium produces the sodium salt of 2-phenoxyphenylethyne, 3-phenoxyphenylethyne, or 4-phenoxyphenylethyne ($C_{14}H_9ONa$).

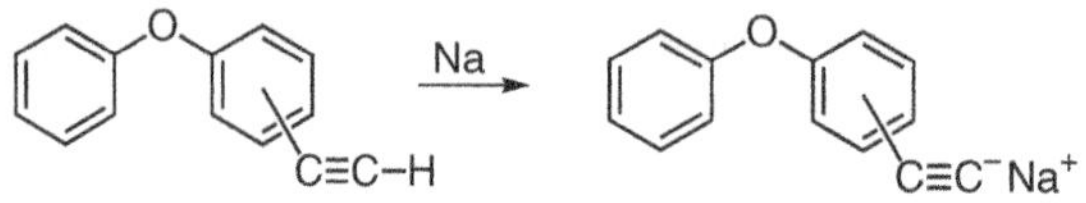

2-phenoxyphenylethyne, 3-phenoxyphenylethyne, or 4-phenoxyphenylethyne → Na → sodium salt of 2-phenoxyphenylethyne, 3-phenoxyphenylethyne, or 4-phenoxyphenylethyne

8. This structure cannot have double bonds that can accept bromine. A possible structure would be 1,2,3,4,5,6-benzenehexacarboxylic acid.

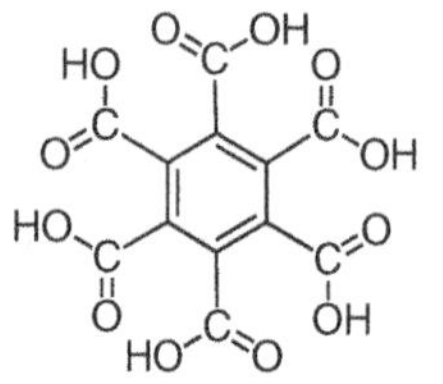

1,2,3,4,5,6-benzene-
hexacarboxylic acid
MW = 342, NE = 57

PROBLEM SET 13

1. Compound I (C$_5$H$_4$O$_2$) has an unsaturation number of 4. The compound has a double bond (positive potassium permanganate test) and is an aldehyde (positive Tollens test). Compound I is dimerized with heat and alkali cyanide to produce compound II. Compound I reacts with excess ethanol to form an acetal (compound III) (C$_9$H$_{14}$O$_3$). Because of the formula, compound I must be cyclic. The original compound (compound I) must be furfural. Furfural (compound I) (C$_5$H$_4$O$_2$) is dimerized with heat and alkali cyanide to form furoin (compound II).

| Compound I: | Compound II: |
| furfural | furoin |

Furfural (compound I) (C$_5$H$_4$O$_2$) reacts with excess ethanol to form the acetal (compound III) (C$_9$H$_{14}$O$_3$) of furfural.

Compound I: Compound III:
furfural   acetal of furfural

The ^{1}H NMR spectrum of furfural (compound I) (C$_5$H$_4$O$_2$) indicates a conjugated system. The three protons in the range of δ 6.63–7.72 are the hydrogens attached to the carbons in the ring. The singlet at δ 9.67 is indicative of an isolated aldehyde. Thus, the protons are identified as follows: δ 6.63, 1H, d of d (***a***); δ 7.28, 1H, d (***b***); δ 7.72, 1H, m (***c***); and δ 9.67, 1H, s (***d***).

furfural

2. Compound I is in solubility class A$_1$ or A$_2$, consisting of carboxylic acids, phenols, β-diketones, enols, oximes, imides, sulfonamides, thiophenols, or nitro compounds. From the formula of C$_4$H$_4$O$_4$, the compound must contain three double bonds, three rings, two double bonds and one ring, two rings and a double bond, a triple bond and one ring, or a triple bond and a double bond. Treatment of compound I with bromine yields compound II (C$_4$H$_4$O$_4$Br$_2$), thus indicating the presence of one carbon–carbon double bond. Compound I is regenerated by the reaction of compound II with zinc dust, confirming the presence of a carbon-carbon double bond in I.

Compound I is hydrogenated to produce compound III. Compound III has a neutralization equivalent of 59 and a molecular weight of 118 (probably C$_4$H$_6$O$_4$), thus indicating the presence of two –COOH groups. Compound III lost a molecule of water when heated, forming a cyclic anhydride (compound IV). The anhydride (compound IV) reacted with aluminum chloride and benzene in a Friedel-Crafts acylation to form a ketone (compound V) (C$_{10}$H$_{10}$O$_3$). The presence of a ketone is confirmed by a positive test with 2,4-dinitrophenyl-hydrazine and vigorous oxidation to produce benzoic acid.

The original compound (compound I) must be *cis* or *trans* butenedioic acid ($C_4H_4O_4$). Butenedioic acid (compound I) ($C_4H_4O_4$) is brominated to yield 2,3-dibromobutanedioic acid (compound II) ($C_4H_4Br_2O_4$), which is debrominated with zinc.

Compound I:
butenedioic acid

Compound II:
2,3-dibromobutanedioic acid

Butenedioic acid (compound I) ($C_4H_4O_4$) is hydrogenated to yield butanedioic acid (compound III), which loses water to produce butanedioic anhydride (compound IV). Butanedioic anhydride (compound IV) reacts with aluminum chloride and benzene to yield 4-oxo-4-phenylbutanedioic acid (compound V) ($C_{10}H_{10}O_3$), which is oxidized to benzoic acid and propanedioic acid.

Compound I:
butenedioic acid

Compound III:
butanedioic acid
MW = 118, NE = 59

Compound IV:
butanedioic anhydride

Compound V:
4-oxo-4-phenylbutanoic acid

benzoic acid

propanedioic acid

The ^1H NMR spectrum of 4-oxo-4-phenylbutanoic acid (compound V) ($C_{10}H_{10}O_3$), shows two triplets of two hydrogens each at δ 2.8 and δ 3.3 indicating the two $-CH_2-$ groups adjacent to each other. Additionally, a monosubstituted benzene (5H) is indicated at δ 7.2–8.0, and the broad singlet at δ 11.7 is the $-OH$.

3. From the formula of $C_{11}H_{10}N_2$, the original compound must be aromatic. Possible answers would be a dipyridylmethane oxidizing to a dipyridyl ketone. The methylene group may be attached to the pyridine ring in the 2, 3, or 4 positions of either ring.

dipyridylmethane

dipyridyl ketone

4. The formula, $C_5H_{10}O$, indicates the presence of a ring, a carbon–carbon double bond, a carbon–oxygen double bond, or an aliphatic ring. A compound that decolorizes a potassium permanganate solution, but is not affected by bromine in carbon tetrachloride contains a carbonyl group.

The only aldehyde that has a boiling point of 75 °C and has a formula of $C_5H_{10}O$ is 2,2-dimethylpropanal.

2,2-dimethylpropanal
(pivalaldehyde)

The following fragments can be assigned in the mass spectrum.

m/z	Fragment
29	$^{+}O\equiv C-H$
57	$H_3C-\underset{+}{\overset{CH_3}{\underset{\vert}{\overset{\vert}{C}}}}-CH_3$ (base peak, 100%)
86	$H_3C-\overset{CH_3}{\underset{CH_3}{\overset{\vert}{\underset{\vert}{C}}}}-\overset{H}{\underset{O}{\overset{\vert}{C}}}\overset{\cdot}{\underset{..}{+}}$

5. From the formula of $C_{14}H_{12}O$, the structure must be aromatic. To produce a carboxylic acid from chromic acid oxidation, the original compound must be a primary alcohol, an aldehyde, or an alkyl group attached to an aromatic ring. By calculating the difference in the molecular weight of the original compound ($C_{14}H_{12}O$) and the neutralization equivalent of the product (226–196 = 30), a value of 30 is obtained, indicating that the original compound must be gaining two oxygens and losing two hydrogens upon oxidation.

   In the ^{1}H NMR spectrum of the oxidation product, the aromatic peaks are seen as nine hydrogens in the range of δ 7.45–8.19. The –OH of the carboxylic acid appears as a singlet at δ 10.8. This compound must consist of two nonfused benzene rings. Possibilities for this structure include 2, 3, or 4-benzoylbenzoic acid. The original compound must be 2, 3, or 4-methylbenzophenone.

2, 3, or 4-methylbenzophenone   2, 3, or 4-benzoylbenzoic acid
MW = 226, NE = 226

6. Compound I must be highly conjugated because of the formula $C_{10}H_6O_4$ and the ^{1}H NMR spectrum. The unsaturation number is 8. Based upon this information, compound I would be 1,2-difuryl-1,2-ethanedione. This compound, when treated with sodium hydroxide, can rearrange in a manner similar to the benzoin rearrangement to benzilic acid to yield 2,2-difuryl-2-hydroxyethanoic acid (compound II).

Compound I: Compound II:
1,2-difuryl-1,2-ethanedione   2,2-difuryl-2-hydroxyethanoic acid
MW = 208, NE = 208

   Treatment of 1,2-difuryl-1,2-ethanedione (compound I) ($C_{10}H_6O_4$) with hydrogen peroxide can cause cleavage between the two carbonyl groups to produce two molecules of furoic acid (compound III).

Compound I: Compound III:
1,2-difuryl-1,2-ethanedione   furoic acid
MW = 112, NE = 112

The ^{1}H NMR spectrum of 1,2-difuryl-1,2-ethanedione (compound I) ($C_{10}H_6O_4$) indicates a conjugated system. The six hydrogens in the range of δ 6.64–7.78 are the hydrogens attached to the carbons in the ring. The protons are identified as follows: δ 6.64, 2H, d of d (**a**); δ 7.63, 2H, d (**b**); and δ 7.78, 2H, d (**c**).

1,2-difuryl-1,2-ethanedione

7. The reaction of compound I with iron and hydrochloric acid reduced the nitro group in structure I to an amino group in structure II. In both compounds, the rest of the formula, $C_{10}H_7$, must be a naphthalene ring. Thus, compound I is 1-nitronaphthalene and compound II is 1-aminonaphthalene.

Compound I:
1-nitronaphthalene

Compound II:
1-aminonaphthalene

Both 1-nitronaphthalene (compound I) ($C_{10}H_7NO_2$) and 1-aminonaphthalene (compound II) ($C_{10}H_9N$) are oxidized, with loss of two carbons each to form 1,2-benzenedicarboxylic acids ($C_8H_5O_6N$ and $C_8H_6O_4$). In 1-nitronaphthalene (compound I) ($C_{10}H_7NO_2$), the nitro group makes that ring less reactive than the other ring, so carbons five and eight become the –COOH groups.

Compound I:
1-nitronaphthalene

3-nitro-1,2-benzene-
dicarboxylic acid

In 2-aminonaphthalene (compound II) ($C_{10}H_9N$), the amino group makes that ring more reactive than the other ring, so carbons one and four become the –COOH groups. When phthalic acid is heated, phthalic anhydride is formed.

Compound II:
1-aminonaphthalene

1,2-benzenedicarboxylic
acid
(phthalic acid)

1,2-benzenedicarboxylic
anhydride
(phthalic anhydride)

The ^{1}H NMR spectrum of any of these compounds will only show protons in the aromatic region.

PROBLEM SET 14

1. Compound I ($C_{10}H_{10}O_2$) does not have an active hydrogen (negative acetyl chloride test) and is not an aldehyde or a ketone (negative 2,4-dinitrophenylhydrazine test). Compound II also does not have an active hydrogen (negative acetyl chloride test) and is not an aldehyde or a ketone (negative 2,4-dinitrophenylhydrazine test). Both compounds contain a multiple bond because they decolorized solutions of bromine and of potassium permanganate. Compound III is an aldehyde or a ketone (positive 2,4-dinitrophenylhydrazine test). The formula ($C_8H_6O_4$) for compound IV indicates that it is a substituted benzoic acid.

   The naturally occurring compound that fits this data is safrole (compound I) which isomerizes to isosafrole (compound II). Isosafrole (compound II) is then converted to piperonal (compound III).

| Compound I: safrole, 1-allyl-3,4-methylenedioxybenzene | heat KOH → | Compound II: isosafrole, 1-propenyl-3,4-methylenedioxybenzene | O₃ → | Compound III: piperonal, 3,4-methylene-dioxybenzaldehyde |

   Isosafrole (compound II) and piperonal (compound III) are both oxidized with alkaline potassium permanganate to piperonylic acid (compound IV).

| Compound II: isosafrole, 1-propenyl-3,4-methylenedioxybenzene | or | Compound III: piperonal, 3,4-methylene-dioxybenzaldehyde | KMnO₄ → | Compound IV: piperonylic acid, 3,4-methylene-dioxybenzoic acid |

   The structure for safrole (compound I) ($C_{10}H_{10}O_2$) is supported by the following ^1H NMR spectrum: the methylene outside of the ring is at δ 3.30, 2H, d of m (***a***); the alkene protons are at δ 4.90, 1H, m (***b***); δ 5.15, 1H, m (***c***); and δ 5.6–6.2, 1H, m (***e***); the methylene that is between two oxygens is at δ 5.88, 2H, s (***d***); and the aromatic protons are located at δ 6.67, 3H, s (***f***).

Compound I:
safrole,
1-allyl-3,4-methylene-
dioxybenzene

2. From the formula ($C_8H_5ClO_2$), compound I has to be aromatic. Treatment of compound I with ethanol resulted in the addition of three molecules of ethanol to form compound II. Thus, an aldehyde, ketone, or acid halide must be present.

Compound II was oxidized with potassium permanganate to yield a dicarboxylic acid ($C_8H_6O_4$), which was then treated with excess thionyl chloride to produce a diacid chloride (compound III) ($C_8H_4Cl_2O_2$).

The reaction of compound I with aniline yielded a compound ($C_{20}H_{16}N_2O$) that had close to two aniline molecules added; aniline adds to aldehydes, ketones, and acid halides.

Since all of the ^{1}H NMR signals for compound III are in the aromatic range of δ 7.72–8.82, the functional groups must be attached to a benzene ring. The benzene ring is disubstituted since there are only 4H. Since there are only three sets of equivalent hydrogens, the groups must be *meta* to each other. If the groups were *para*, then all the aromatic protons would be equivalent and only one aromatic signal would be present. If the groups were *ortho*, there would be two sets of aromatic protons, each with two protons.

Therefore, compound III must be 1,3-benzenedicarboxylic dichloride. Working backwards, compound I must be 2-formylbenzoyl chloride. 2-Formylbenzoyl chloride (compound I) ($C_8H_5ClO_2$) is treated with excess ethanol to form the acetal (compound II) ($C_{14}H_{20}O_4$).

The acetal (compound II) ($C_{14}H_{20}O_4$) is oxidized to 1,3-benzenedicarboxylic acid (isophthalic acid) ($C_8H_6O_4$), which is converted with thionyl chloride to 1,3-benzenedicarboxylic chloride (compound III) ($C_8H_4Cl_2O_2$).

Compound I:
3-formylbenzoyl chloride

Compound II:
ethyl 3-diethoxymethylbenzoate

1,3-benzenedicarboxylic acid,
isophthalic acid

Compound III,
1,3-benzenedicarboxylic dichloride
isophthaloyl dichloride

2-Formylbenzoyl chloride (compound I) ($C_8H_5ClO_2$) is reacted with excess aniline to produce *N*-phenyl-3-phenyliminomethyl-benzamide ($C_{20}H_{16}N_2O$).

Compound I:
2-formylbenzoyl chloride

excess aniline

N-phenyl-3-phenyl-
iminomethyl-
benzamide

3. From the formula of $C_8H_{14}O_3$, the compound contains two rings, two double bonds, one triple bond, or a double bond and a ring. The compound is in the solubility class of A_1 or A_2, indicating a carboxylic acid, phenol, β-diketone, enol, oxime, imide, sulfonamide, thiophenol, or nitro compound.

Compound III gave a positive test with semicarbazide indicating an aldehyde or a ketone, but a negative test with iodoform indicating that a methyl ketone or methyl secondary alcohol is not present. From the ^1H NMR spectrum of the semicarbazone of carbonyl compound III, evidence is seen that compound III is 3-pentanone.

Working backwards, compound I ($C_8H_{14}O_3$) is ethyl 2-methyl-3-oxopentanoate. When ethyl 2-methyl-3-oxopentanoate (compound I) is treated with phenylhydrazine, the initial phenylhydrazone is formed. The reaction continues as the ethoxy group is lost and the chain cyclizes to form ethyl 2-methyl-3-(phenylhydrazono)pentanoate (compound II) ($C_{12}H_{14}N_2O$).

Compound I:
ethyl 2-methyl-3-
oxopentanoate

phenylhydrazine

ethyl 2-methyl-3-
(phenylhydrazono)-
pentanoate

Compound II:
5-ethyl-4-methyl-2-phenyl-
3-pyrazolone

β-Keto acids undergo decarboxylation to yield ketones. Since ethyl 2-methyl-3-oxopentanoate (compound I) ($C_8H_{14}O_3$) is an β-keto ester, it hydrolyzes to produce a β-keto acid. The acid then decarboxylates to give 3-pentanone (compound III) ($C_5H_{10}O$).

Compound I: ethyl 2-methyl-
3-oxopentanoate

2-methyl-3-oxopentanoic acid

Compound III: 3-pentanone

3-Pentanone (compound III) ($C_5H_{10}O$) reacts with semicarbazide to form the semicarbazone.

Compound III:
3-pentanone

semicarbazide

semicarbazone
of 3-pentanone

The ^{1}H NMR spectrum of the semicarbazone shows as a triplet with six hydrogens at δ 1.09, indicating two CH_3 adjacent to a CH_2; a quartet with four hydrogens at δ 2.26, indicating two CH_2 adjacent to a CH_3; a broad singlet with two hydrogens at δ 5.89, indicating a NH_2; and a broad singlet with one hydrogen at δ 8.59, referring to the NH.

4. From the formula of C_9H_7N, a fused ring system is indicated, with nitrogen being a member of one of the rings. Two possibilities of a structure with this formula are quinoline or iso-quinoline. However, later oxidation steps eliminate quinoline as a possibility.

Isoquinoline (compound I) (C_9H_7N), is catalytically reduced to 1,2,3,4-tetrahydroisoquinoline (compound II) ($C_9H_{11}N$). 1,2,3,4-Tetrahydroisoquinoline (compound II) ($C_9H_{11}N$) is treated with excess methyl iodide to form a quaternary ammonium salt, N,N-dimethyl-1,2,3,4-tetrahydroisoquinolinium iodide. With silver oxide, this salt is changed to the hydroxide salt, N,N-dimethyl-1,2,3,4-tetrahydroisoquinolinium hydroxide. This hydroxide salt, when heated, decom-poses to water and 2-(N,N-dimethylaminomethyl)styrene (compound III) ($C_{11}H_{15}N$). Vigorous oxidation of 2-(N,N-dimethylaminomethyl)styrene (compound III) ($C_{11}H_{15}N$) produces 1,2-benzenedicarboxylic acid (phthalic acid) (compound IV) ($C_8H_6O_4$).

Compound I: isoquinoline

Compound II: 1,2,3,4-tetrahydro-isoquinoline

N,N-dimethyl-1,2,3,4-tetrahydro-isoquinoline iodide

N,N-dimethyl-1,2,3,4-tetrahydro-isoquinoline hydroxide

Compound III: 2-(N,N-dimethylamino-methyl)styrene

Compound IV: 1,2-benzene-dicarboxylic acid (phthalic acid)

The ozonolysis of N,N-dimethylaminomethyl) styrene (compound III) ($C_{11}H_{15}N$), followed by hydrolysis, oxidized the alkene group to yield 2-(N,N-dimethylaminomethyl) benzaldehyde (compound V) ($C_{10}H_{13}NO$). When treated with a solution of potassium cyanide, 2-(N,N-dimethylaminomethyl) benzaldehyde (compound V) ($C_{10}H_{13}NO$) dimerized to form 2,2'-(N,N-dimethylaminomethyl) benzoin (compound VI) ($C_{20}H_{26}N_2O_2$), which is then oxidized back to 1,2-benzenedicarboxylic acid (compound IV) ($C_8H_6O_4$).

Compound III:
2-(N,N-dimethylamino-
methyl)styrene

Compound V:
2-(N,N-dimethylamino-
methyl)benzaldehyde

Compound VI:
2,2′-(N,N-dimethylamino-
methyl)benzoin

Compound IV:
1,2-benzenedicarboxylic acid
(phthalic acid)

The ^{1}H NMR spectrum agrees with the structure of isoquinoline (compound I) (C_9H_7N), with seven protons in the aromatic region (δ 7.25–δ 9.26).

5. The very last product with the formula of $C_8H_6O_4$ has a ^{1}H NMR spectrum that shows only two singlets at δ 8.08 and δ 11.00 with an integration ratio of 2 : 1. The structure of this compound is 1,4-benzenedicarboxylic acid (terephthalic acid) and the ratio agrees with the comparison of aromatic protons to –OH protons. This is the *para* product since *ortho* and *meta* would produce more splitting in the aromatic range. Compound I must have an active hydrogen (positive sodium test).

Sulfuric acid and mercuric sulfate react with terminal alkynes to form methyl ketones. Methyl ketones are oxidized to carboxylic acids with sodium hypochlorite. Since treatment of compound III with hydrogen bromide resulted in the loss of a carbon, the CH_3OCH_2 group must be attached to the benzene ring. Therefore, compound I must be 4-methoxymethylphenylethyne ($C_{10}H_{10}O$).

Treatment of 4-methoxymethylphenylethyne (compound I) ($C_{10}H_{10}O$) with sodium produced the sodium salt of 4-methoxymethylphenylethyne ($C_{10}H_9ONa$), which gave 4-methoxymethylphenylethyne (compound I) when treated with water.

Compound I:
4-methoxymethylphenylethyne

Sodium salt of 4-methoxy-
methylphenylethyne

Sulfuric acid and mercuric sulfate reacted with 4-methoxymethylphenylethyne (compound I) ($C_{10}H_{10}O$), to yield 4-methoxymethylacetophenone (compound II) ($C_{10}H_{12}O_2$). Sodium hypochlorite oxidized 4-methoxymethylacetophenone (compound II) ($C_{10}H_{12}O_2$) to 4-methoxymethylbenzoic acid (compound III; $C_9H_{10}O_3$). Treatment of 4-methoxymethylbenzoic acid (Compound III) ($C_9H_{10}O_3$) with hydrogen bromide produced 4-bromomethylbenzoic acid (compound IV) ($C_8H_7BrO_2$), which was oxidized to terephthalic acid (compound V) ($C_8H_6O_4$).

Compound I:
4-methoxymethyl-
phenylethyne

HgSO₄
H₂SO₄

Compound II:
4-methoxymethyl-
acetophenone

NaOCl

Compound III:
4-methoxy-
methylbenzoic acid

HBr

Compound IV:
4-bromomethylbenzoic acid

vigorous
oxidation

Compound V:
1,4-benzene-
dicarboxylic acid
(terephthalic acid)

6. From the formula of $C_{14}H_{12}$, the compound obviously contains two benzene rings that are not fused together since the unsaturation number is 9. Inferred from the oxidation reaction is that both benzene rings are attached to the same carbon. The only possible structure for $C_{14}H_{12}$ is 1,1-diphenylethene. This compound is then oxidized to benzophenone ($C_{13}H_{10}O$). Benzophenone cannot be oxidized any further.

1,1-diphenylmethane

KMnO₄

benzophenone

KMnO₄

No reaction

The ^1H NMR spectrum of 1,1-diphenylmethane ($C_{14}H_{12}$) shows the $-CH_2$ peak at δ 5.35 and 10 aryl protons at δ 7.21.

7. In the ^1H NMR spectrum of the isomer, the singlet at δ 0.95 indicates a $-NH_2$ group, the singlet at δ 2.30 indicates an isolated $-CH_3$, the multiplet at δ 2.5–3.0 signifies two $-CH_2$ groups, and the four protons at δ 7.05 indicates a p-disubstituted benzene. The structure that matches this ^1H NMR spectrum is α-methyl-4-methylbenzylamine. The only optically active p-disubstituted compound that agrees with the formula ($C_9H_{13}N$) is 4-(aminoethyl)-toluene. Vigorous oxidation of either amine produces 1,4-benzenedicarboxylic acid.

α-methyl-4-methyl-
benzylamine
(original compound)

oxidation

1,4-benzenedicarboxylic
acid (phthalic acid)
MW = 166, NE = 83

oxidation

4-(aminoethyl)-
toluene
(isomer)

8. The compound is not a phenol (negative bromine water test) but has two active hydrogens (positive sodium test to give a disodium compound). The first carboxylic acid contains only one –COOH group since the neutralization equivalent of 152 is equal to the molecular weight (probably 152 from $C_8H_8O_3$). The second carboxylic acid must be a dicarboxylic acid since the neutralization equivalent of 82 ± 2 is half of the molecular weight (166, calculated from $C_8H_6O_4$). Since there is only one aromatic peak on the ^1H NMR spectrum, all protons on the ring must be equivalent. Thus, the second acid must be 1,4-benzenedicarboxylic acid (terephthalic acid). Working backwards, the original compound must be 4-hydroxymethylbenzoic acid.

Thus, 4-hydroxymethylbenzoic acid reacts with sodium to give the disodium salt of 4-hydroxymethylbenzoic acid ($C_8H_6O_3Na_2$).

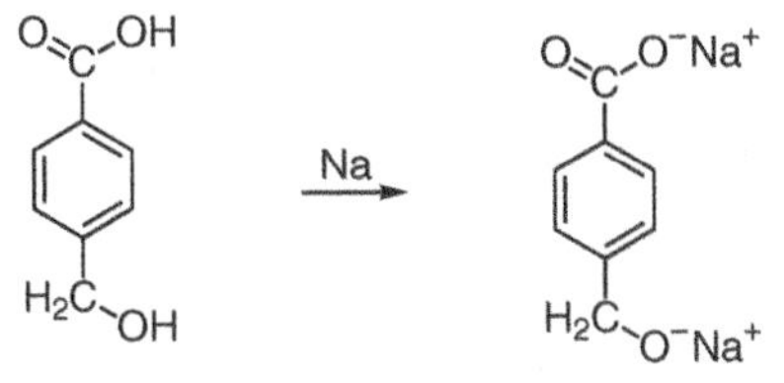

4-hydroxymethylbenzoic
acid (original compound)
MW = 152, NE = 152

disodium salt of
4-hydroxymethyl-
benzoic acid

4-Hydroxymethylbenzoic acid is vigorously oxidized to 1,4-benzenedicarboxylic acid (terephthalic acid) ($C_8H_6O_4$).

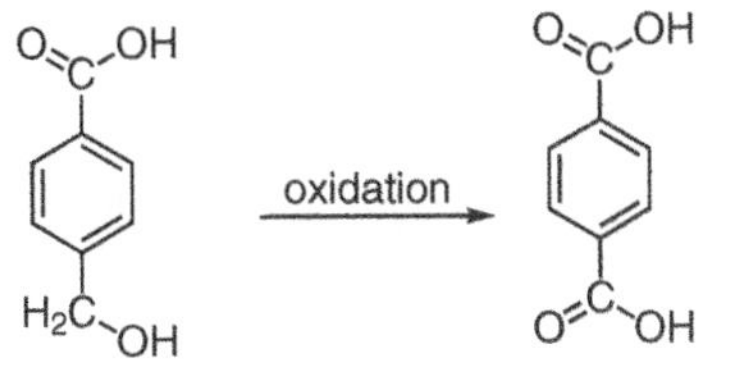

4-hydroxymethylbenzoic
acid (original compound)
MW = 152, NE = 152

1,4-benzene-
dicarboxylic acid
(terephthalic acid)
MW = 166, NE = 83

PROBLEM SET 15

1. Solid I is in solubility class of S_2 and thus is a salt of an organic acid, an amine hydrochloride, an amino acid, or a polyfunctional compound. Evolution of a gas, probably nitrogen, from a reaction with nitrous acid (sodium nitrite and hydrochloric acid) indicates a primary amine; however, another part of this gas also dissolved in potassium hydroxide solution which shows that an acid is present. The second gas is carbon dioxide. The presence of sodium nitrate in the residue shows that the nitrate ion must have been in the original compound.

   When the original compound is treated with dilute sodium hydroxide solution, ammonia is evolved. Upon acidification of this solution, another gas (probably carbon dioxide) is evolved. The only possibility for solid I is urea nitrate.

   Treatment of urea nitrate (compound I) with sodium nitrite and hydrochloric acid produces carbon dioxide, nitrogen, and inorganic salts.

$$H_2N\overset{+}{\underset{\underset{OH}{|}}{C}}NH_2 \quad NO_3^- \xrightarrow[\text{HCl}]{\text{NaNO}_2} CO_2\ (g)\ +\ N_2\ (g)$$

$$+\ \text{inorganic salts}$$

   Compound I:
   urea nitrate
   MW = 123, NE = 123

   The carbon dioxide is soluble in potassium hydroxide.

$$CO_2 \xrightarrow{\text{KOH}} CO_3^{2-}$$

   carbon dioxide   carbonate ion

   Treatment of urea nitrate (compound I) with sodium hydroxide produced carbonate ion, ammonia, and inorganic salts.

$$H_2N\overset{+}{\underset{\underset{OH}{|}}{C}}NH_2 \quad NO_3^- \xrightarrow{\text{NaOH}} CO_3^{-2}\ (g)\ +\ NH_3\ (g)$$

$$+\ \text{inorganic salts}$$

   Compound I:
   urea nitrate
   MW = 123, NE = 123

   Carbonate ion reacts with hydrochloric acid to produce carbon dioxide.

$$CO_3^{-2} \xrightarrow{\text{HCl}} CO_2\ (g)$$

   carbonate ion   carbon dioxide

   Urea (mp 135 °C) is the neutralized compound.

$$H_2N\overset{+}{\underset{\underset{OH}{|}}{C}}NH_2 \quad NO_3^- \xrightarrow[\text{with dilute NaOH}]{\text{neutralized}} H_2N\underset{\underset{O}{||}}{C}NH_2\ +\ \text{NaNO}_3$$

$$+\ H_2O$$

   Compound I:
   urea nitrate
   MW = 123, NE = 123

   urea

The mass spectrum can be interpreted as shown below.

m/e	Fragment

44    $H_2N-C{=\!\!=\!\!=}O^+$ (with $\stackrel{||}{O}$)

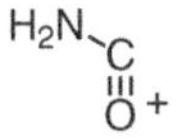

60    $H_2N-C(NH_2){=\!\!=}O^{+\bullet}$

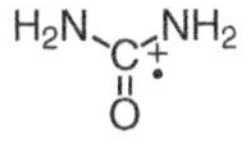

In the IR spectrum, the doublet at 3350 and 3450 cm^{-1} indicates the presence of N–H stretching for a primary amine. The signals at 1640 and 1690 cm^{-1} indicate the C=O stretching of the carbonyl group.

2. From the solubilities given, the compound is in the solubility class of N or I. This means that the compound could be an alcohol, aldehyde, ketone, ester, ether, epoxide, alkene, alkyne, aromatic, saturated hydrocarbon, haloalkane, or aryl halide. Saturated hydrocarbon, haloalkane, and aryl halide are eliminated as possibilities since the compound contains carbon, hydrogen, oxygen, and nitrogen. The compound does not contain an active hydrogen (negative acetyl chloride test) and it is not an aldehyde or ketone (negative 2,4-dinitrophenylhydrazine test).

From a positive iodoform test, the alkali distillate compound is either a methyl ketone or has a methyl adjacent to a hydroxyl group. A negative reaction with zinc chloride and hydrochloric acid indicates that this compound is not a secondary or tertiary alcohol. The alkali distillate could be ethanol since it gives a negative Lucas test. The ^{1}H NMR spectrum showed a peak that was concentration dependent, which indicates a –OH.

The alkali residue, after acidification and distillation, yielded a volatile carboxylic acid. This volatile acid reduced potassium permanganate which indicates the presence of a multiple bond, a primary alcohol, a secondary alcohol, or an aldehyde. Analysis of the ^{1}H NMR spectrum for the volatile acid shows a multiplet in the range of δ 5.8–6.75 with only three hydrogens and a singlet with one proton at δ 12.4. The multiplet must be an $H_2C{=}CH-$ group and the singlet would be from a –COOH group, which would mean that the volatile acid is propenoic acid.

The neutralized residue from the second distillation is a non-volatile carboxylic acid. The non-volatile acid contains a *para*-disubstituted aromatic ring, as indicated from the pair of doublets (4H) at δ 6.75 and δ 7.83. The singlet (3H) at δ 6.55 is the –COOH and the –NH$_2$ protons. Thus, the nonvolatile acid is 4-aminobenzoic acid.

The original compound must have been an ester since esters hydrolyze slowly with base. Based upon the above information, the original compound must be ethyl 4-[*N*-(1-oxo-2-propenyl)]aminobenzoate.

Ethyl 4-[*N*-(1-oxo-2-propenyl)]aminobenzoate reacts with hot aqueous alkali, followed by acidification, to produce 4-[*N*-(1-oxo-2-propenyl)]aminobenzoic acid and ethanol.

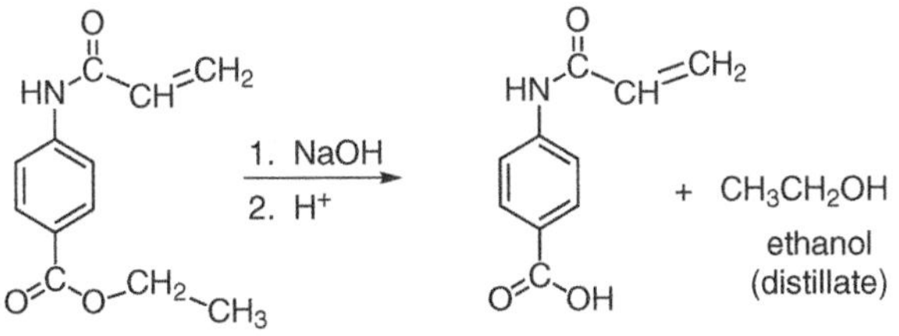

ethyl 4[*N*-(1-oxo-2-propenyl)]amino benzoate

1. NaOH
2. H$^+$

4-[*N*-1-oxo-2-propenyl]-aminobenzoic acid (residue)

+ CH$_3$CH$_2$OH

ethanol (distillate)

The original alkaline solution containing the 4-[N-(1-oxo-2-propenyl)]-aminobenzoic acid was acidified and distilled to yield 4-aminobenzoic acid and propenoic acid.

4-[N-1-oxo-2-propenyl] -   4-aminobenzoic acid   propenoic acid
amino benzoic acid   (nonvolatile acid)   (volatile acid)
                                MW = 137, NE =137

3. Compound I ($C_6H_8O_4$) must be a diacid, since it lost 2 —OH groups and gained two —Cl groups from the treatment of I with phosphorus pentachloride to produce the diacid chloride (compound II) ($C_6H_6O_2Cl$). The diacid chloride (compound II) yielded a diketone (compound III) ($C_{18}H_{16}O_2$) from the Friedel-Crafts acylation reaction. The structure is confirmed as a diketone (negative potassium permanganate test, positive bromine test). The formation of a dioxime is in agreement that III is a diketone. For the last reaction to occur, compound IV ($C_{18}H_{18}O_2N_2$) must be able to be easily hydrolyzed to a diacid. Thus, compound IV ($C_{18}H_{18}O_2N_2$) is a diamide.

From the formula of $C_6H_8O_4$ for compound I, a hexenedioic acid or a cyclobutanedicarboxylic acid can be drawn. However, a double bond is absent as indicated from the negative reaction of potassium permanganate solution with III. Thus, compound I must be a cyclobutanedicarboxylic acid.

In the ^{1}H NMR spectrum of compound I, a multiplet of four protons exists in the range of δ 2.2–δ 2.6 indicating that the carboxylic acid groups are *trans* and are on the ring on adjacent carbons; these peaks are the —CH$_2$ protons. The triplet at δ 3.90 indicates the protons on the same carbons as the —COOH. The broad singlet at δ 11.9 indicates the —OH of the acid. Therefore, compound I ($C_6H_8O_4$) is *trans*-1,2-cyclobutanedicarboxylic acid.

Trans-1,2-cyclobutanedicarboxylic acid (compound I) ($C_6H_8O_4$) is converted with phosphorus pentachloride to *trans*-1,2-cyclobutanedicarboxylic dichloride (compound II) ($C_6H_6O_2Cl_2$). *Trans*-1,2-cyclobutanedicarboxylic dichloride then undergoes a Friedel-Crafts acylation with benzene and aluminum chloride to produce *trans*-1,2-dibenzoylcyclobutane (compound III) ($C_{18}H_{16}O_2$). *Trans*-1,2-dibenzoylcyclobutane (compound III) ($C_{18}H_{16}O_2$) reacts with hydroxylamine to form *trans*-1,2-cyclobutanedicarboxylic acid dioxime, which rearranges in the presence of phosphorus pentachloride to yield *trans*-N,N-diphenyl-1, 2-cyclobutanedicarboxylic acid diamide (compound IV) ($C_{18}H_{18}O_2N_2$). *Trans*-N,N-diphenyl-1,2-cyclobutanedicarboxylic acid diamide (compound IV) ($C_{18}H_{18}O_2N_2$) hydrolyzes to *trans*-1,2-cyclobutanedicarboxylic acid (compound I) ($C_6H_8O_4$).

Compound I:
trans-cyclobutane-
1,2-dicarboxylic acid

PCl₅ →

Compound II:
trans-cyclobutane-
1,2-dicarboxylic dichloride

AlCl₃ →

Compound III: *trans*-1,2-
dibenzoylcyclobutane

NH₂OH →

trans-1,2-cyclobutane-
dicarboxylic acid dioxime

PCl₅ →

Compound IV:
trans-N,N-diphenyl-
1,2-cyclo-butanedicarboxylic
acid diamide

H₂O / H⁺ →

Compound I:
trans-1,2-
cyclobutanedicarboxylic
acid

+ 2 aniline

4. Compound I is in solubility class S₂, indicating a salt of an organic acid, an amine hydrochloride, an amino acid, or a polyfunctional compound. A reaction with silver nitrate shows that the chlorine is labile. When compound I was exactly neutralized to form compound II, the chlorine was released which means that chloride ion was initially present.

The reaction of compound II with acetic anhydride signifies the presence of a primary or secondary amine. The reaction of compound II with benzenesulfonyl chloride and with nitrous acid without the evolution of gas confirms the identity of this compound as a secondary amine.

Compound III is a carboxylic acid since it gave a neutralization equivalent. Since compounds II and III were soluble in base, but not soluble in acid, the presence of a carboxylic acid group is confirmed.

The oxidation of compounds I, II, or III to produce a nitrogen-free acid, which is insoluble in water, shows that the nitrogen is not directly attached to the aromatic ring.

The ¹H NMR spectrum of the nitrogen-free acid shows three sets of aromatic signals (4H) at δ 7.62, δ 8.25, and δ 8.70 indicating *meta* disubstitution. The two protons in a singlet at δ 12.28 correspond to the –OH of the two –COOH. Therefore, the final product is 1,3-benzenedicarboxylic acid (isophthalic acid). Thus, working backwards, compound I must be 3-carboxy-*N*-methylbenzylammonium chloride.

3-Carboxy-*N*-methylbenzylammonium chloride (compound I) is reacted with sodium hydroxide to produce an amine, 3-carboxy-*N*-methylbenzylamine (compound II). Treatment of 3-carboxy-*N*-methylbenzylamine (compound II) with acetic anhydride yields an amide, 3-carboxy-*N*-acetyl-*N*-methylbenzylamine (compound III).

Compound I:
3-carboxy-*N*-methyl-
benzyl ammonium chloride

Compound II:
3-carboxy-*N*-methyl-
benzylamine

acetic anhydride

Compound III:
3-carboxy-*N*-acetyl-
N-methylbenzylamine
MW = 207, NE = 207

Oxidation of 3-carboxy-*N*-methylbenzylammonium chloride (compound I), 3-carboxy-*N*-methylbenzylamine (compound II), or 3-carboxy-*N*-acetyl-*N*-methylbenzylamine (compound III) produces a dicarboxylic acid, 1,3-benzenedicarboxylic acid (isophthalic acid).

Compounds I, II, or III $\xrightarrow{\text{oxidation}}$

1,3-benzenedicarboxylic acid,
isophthalic acid
MW = 166, NE = 166

5. Compound I ($C_{10}H_6O_3$) contains a multiple bond (positive potassium permanganate test) and is an aldehyde or a ketone (positive hydroxylamine test). Compound II ($C_9H_6O_2$) contains an active hydrogen (positive sodium test). Compound III ($C_8H_6O_4$) is a carboxylic acid which decarboxylates with soda lime to form Compound IV ($C_7H_6O_2$). Compound IV ($C_7H_6O_2$) is treated with hydrochloric acid under pressure to yield a weakly acidic compound ($C_6H_6O_2$).

For compound III, an isolated –CH$_2$– is at δ 6.00, a trisubstituted aromatic ring (3H) is in the range of δ 6.8–7.55, and a –COOH is present as a broad singlet at δ 7.6. The ^{1}H NMR spectrum of compound IV indicates an isolated singlet indicating a –CH$_2$– at δ 5.90 and a disubstituted aromatic ring (4H) at δ 6.83.

The only structure that can be drawn for compound IV that matches the formula of $C_7H_6O_2$ and the ^{1}H NMR spectrum is 1,2-methylenedioxybenzene. By working backwards, the other structures can be determined. Compound I is 3-(3,4-methylenedioxyphenyl)propynal ($C_{10}H_6O_3$).

In the presence of heat, 3-(3,4-methylenedioxyphenyl)propynal (compound I) ($C_6H_{10}O_3$) is converted to 1-(3,4-methylenedioxyphenyl)ethyne (compound II) ($C_9H_6O_2$), which is then oxidized to 3,4-methylenedioxybenzoic acid (compound III) ($C_8H_6O_4$). 3,4-Methylenedioxybenzoic acid (compound III) ($C_8H_6O_4$) decarboxylates with heat and soda lime to produce 1,2-methylenedioxybenzene (compound IV) ($C_7H_6O_2$), which is then decomposed by heating with hydrochloric acid and pressure to 1,2-dihydroxybenzene (weak acid).

Compound I:
3-(3,4-methylenedioxy-phenyl)propynal

heat →

Compound II:
1-(3,4-methylene-dioxyphenyl)ethyne

oxidation →

Compound III:
3,4-methylene-dioxybenzoic acid

soda lime →

Compound IV:
1,2-methylene-dioxybenzene

heat / HCl →

1,2-dihydroxybenzene (weak acid)

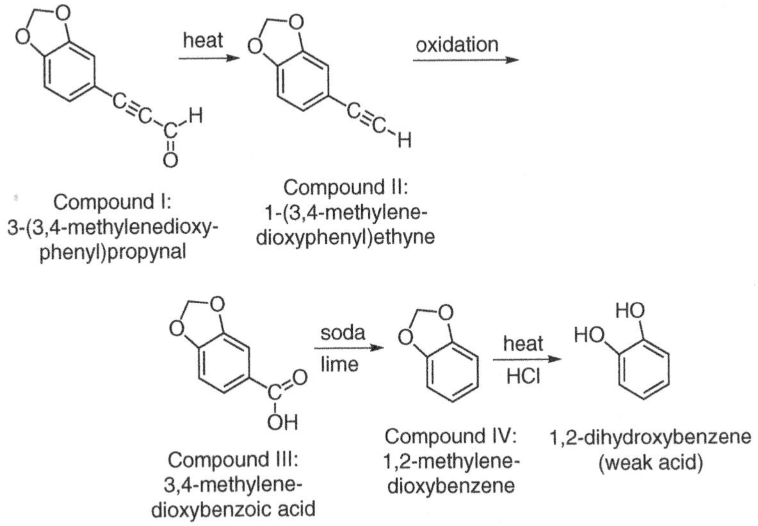

PROBLEM SET 16

1. Compound A belongs to the solubility class of N which indicates that it is an alcohol, aldehyde, ketone, ester, ether, epoxide, alkene, alkyne, or aromatic. Compound A is not an aldehyde or a ketone (negative 2,4-dinitrophenylhydrazine test) and does not contain an active hydrogen (negative acetyl chloride test). The volatile acid, from the base hydrolysis of compound A, must be acetic acid due to the low neutralization equivalent. Compound B contains an active hydrogen (positive acetyl chloride test), but is not an aldehyde or ketone (negative 2,4-dinitrophenylhydrazine test). Compound C contains a labile halogen (positive silver nitrate test) and is a phenol (positive ferric chloride test and positive bromine water test). Compound D was esterified to produce compound E.

   The ^1H NMR spectrum of compound E can be interpreted to have a $-CH_3$ group adjacent to a $-CH_2-$ at both δ 1.35 and δ 1.40. The signal at δ 3.8–4.5 is two overlapping quartets and shows a $-CH_2-$ adjacent to $-CH_3$ for each quartet. These two ethyl groups are in slightly different environments. The pair of doublets (4H) in the aromatic region at δ 6.79 and δ 7.89 indicate a p-disubstituted benzene ring.

   Compound E must be ethyl 4-ethoxybenzoate. This structure matches the ^1H NMR spectrum and is an ester. From this structure, the other structures can be determined. 4-Ethoxybenzyl acetate (compound A) is heated with alkali and neutralized to produce 4-ethoxybenzyl alcohol (compound B) and acetic acid. 4-Ethoxybenzyl alcohol (compound B) is treated with hydrogen bromide to produce 4-ethoxybenzyl bromide (compound C).

| Compound A: 4-ethoxybenzyl acetate | Compound B: 4-ethoxybenzyl alcohol | Compound C: 4-ethoxybenzyl bromide |

Oxidation of 4-ethoxybenzyl alcohol (compound B) yielded 4-ethoxy-benzoic acid (compound D), which is esterified to produce ethyl 4-ethoxybenzoate (compound E).

| Compound B: 4-ethoxybenzyl alcohol | Compound D: 4-ethoxybenzoic acid ME = 166, NE = 166 | Compound E: 4-ethoxybenzoate |

2. Compound I is in the solubility class of A_2, indicating that it is a phenol, enol, oxime, imide, sulfonamide, thiophenol, β-diketone, or nitro compound. Since compound I tests positive for nitrogen, phenol, enol, thiophenol, and β-diketone are eliminated

as possibilities. Sulfonamide is also eliminated since sulfur is not present. Compound I contains an active hydrogen (positive acetyl chloride test) and the chlorine is not labile (negative silver nitrate test). Since boiling alkali liberated ammonia from compound I to produce carboxylic acid III, then compound I must be an imide.

The ^{1}H NMR spectrum of compound IV indicates a benzenetricarboxylic acid. There are three protons present in the aromatic region at δ 7.57 and 8.18 and three carboxylic acid protons at δ 11.2. Since all of the protons are split, the answer must be 1,2,3-benzenetricarboxylic acid. 1,2,3-Benzenetricarboxylic acid (compound IV) has a molecular weight of 210 and a neutralization equivalent of 70, matching the neutralization equivalent given. 1,2,3-Benzenetricarboxylic acid (compound IV) was chlorinated to produce a compound with a neutralization equivalent of 81.5.

Working backwards, compound I is 5-chloro-3-hydroxyphthalimide. Chlorine can actually be on carbons 4, 5, or 6, based upon the information given. 5-Chloro-3-hydroxyphthalimide (compound I) reacts with acetyl chloride to produce 5-chloro-3-hydroxymethylphthalimide acetate (compound II).

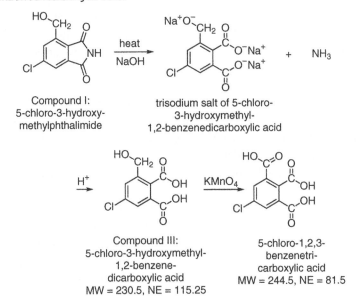

Compound I:
5-chloro-3-hydroxy-
methylphthalimide
MW = 211.5, NE = 211.5

acetyl chloride

Compound II:
5-chloro-3-hydroxy
methylphthalimide acetate
MW = 253.5, NE = 253.5

5-Chloro-3-hydroxymethylphthalimide (compound I) reacts with boiling alkali to liberate ammonia. Acidification of the residue produced a carboxylic acid, 5-chloro-3-hydroxymethyl-1,2-benzenedicarboxylic acid (compound III), which is oxidized to 5-chloro-1,2,3-benzenetricarboxylic acid.

Compound I:
5-chloro-3-hydroxy-
methylphthalimide

heat
NaOH

trisodium salt of 5-chloro-
3-hydroxymethyl-
1,2-benzenedicarboxylic acid

+ NH_3

H$^+$

KMnO$_4$

Compound III:
5-chloro-3-hydroxymethyl-
1,2-benzene-
dicarboxylic acid
MW = 230.5, NE = 115.25

5-chloro-1,2,3-
benzenetri-
carboxylic acid
MW = 244.5, NE = 81.5

1,2,3-Benzenetricarboxylic acid (compound IV) is chlorinated to give 5-chloro-1,2,3-benzenetricarboxylic acid.

Compound IV:
1,2,3-tricarboxylic acid
MW = 210, NE = 70

5-chloro-1,2,3-benzene-
tricarboxylic acid
MW = 244.5, NE = 81.5

The IR spectrum shows the C=O stretch of 5-chloro-3-hydroxyphthalimide (compound I) at 1712 and 1719 cm^{-1}.

3. Compound A is an aldehyde or a ketone (positive 2,4-dinitrophenylhydrazine test) and contains an active hydrogen (positive acetyl chloride test). Compound B is an aldehyde or ketone (positive 2,4-dinitrophenylhydrazine test) but does not have an active hydrogen (negative acetyl chloride test). Compound C is benzoic acid.

The ^{1}H NMR spectrum of compound A shows ten hydrogens in the aromatic region of δ 7.2–7.85, indicating two nonfused benzene rings. The singlet as δ 5.9 is a –CH and the broad singlet at δ 4.5 is an –OH. From this information, compound A is deduced to be benzoin.

Benzoin (compound A) is oxidized to benzil (compound B). Vigorous oxidation of benzil (compound B) produced benzoic acid (compound C).

Compound A: benzoin

Compound B: benzil

Compound C:
benzoic acid
MW = 122, NE = 122

4. Liquid I is in the solubility class N, which includes alcohols, aldehydes, ketones, esters, alkenes, alkynes, ethers, epoxides, or aromatic compounds. Liquid I does not contain a labile halogen (negative silver nitrate test), does not have an active hydrogen (negative acetyl chloride test), but could be an aldehyde or a ketone (positive 2,4-dinitrophenylhydrazine test). A negative Tollens test indicates that liquid I is not an aldehyde. Liquid I has a saponification equivalent of 226 ± 1 and is probably an ester. Liquid I is heated with sodium hydroxide; the distillate from this reaction is a methyl ketone or has a methyl adjacent to a secondary alcohol (positive iodoform test). The residue was acidified to produce a solid II with a neutralization equivalent of 156 ± 1 and an acidic filtrate with a neutralization equivalent of 61 ± 1, indicating possible carboxylic acids.

The ^{1}H NMR spectrum of compound II indicates a possible *ortho* disubstituted benzene ring (4H) in the aromatic region of δ 7.1–7.82 and a –COOH group at δ 9.0. 2-Chlorobenzoic acid matches the ^{1}H NMR spectrum and the neutralization equivalent for compound II. With a neutralization equivalent of 61 ± 1, the acidic filtrate is acetic acid. Working backwards, the initial compound is ethyl 3-(2-chlorophenyl)-3-oxopropanoate.

Ethyl 3-(2-chlorophenyl)-3-oxopropanoate (compound I) is saponified to give ethanol (distillate) and a salt, which is acidified to yield 2-chlorobenzoic acid (compound II) and acetic acid (filtrate).

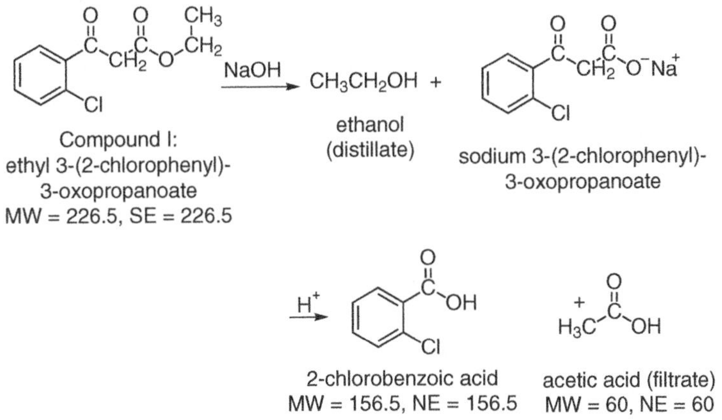

Compound I:
ethyl 3-(2-chlorophenyl)-
3-oxopropanoate
MW = 226.5, SE = 226.5

ethanol
(distillate)

sodium 3-(2-chlorophenyl)-
3-oxopropanoate

2-chlorobenzoic acid
MW = 156.5, NE = 156.5

acetic acid (filtrate)
MW = 60, NE = 60

5. The unknown is an ester since a saponification equivalent was determined. Saponification of an ester yields an alcohol and a carboxylic acid. The alcohol is a phenol (positive ferric chloride test). The acid is acetic acid, because of the neutralization equivalent of 60.

The ^{1}H NMR spectrum indicated two isolated methyl groups in the same environment at δ 2.18, the –OH group at δ 5.73, and three aromatic ring protons at δ 6.33 and δ 6.45. The aromatic protons are all isolated since they are singlets. The benzene ring must be trisubstituted since only three aromatic protons are present. The original compound is 3,5-dimethylphenyl acetate. 3,5-Dimethylphenyl acetate is saponified to yield 3,5-dimethylphenol and acetic acid.

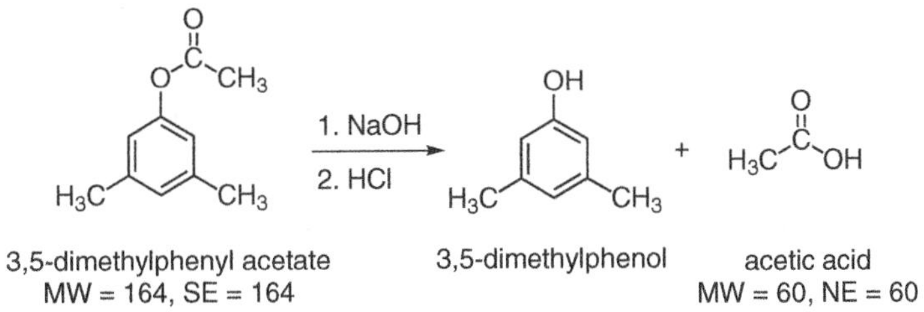

3,5-dimethylphenyl acetate
MW = 164, SE = 164

3,5-dimethylphenol

acetic acid
MW = 60, NE = 60

PROBLEM SET 17

1. Compound I is in the solubility class of S_2, indicating that this compound is an amino acid, a salt of an organic acid, an amine hydrochloride, or a polyfunctional compound. Treatment of compound I with nitrous acid liberated a gas, indicating that this compound is a primary amine. With a neutralization equivalent of 37 ± 1, the structure is small. The ^1H NMR spectrum for compound I shows isolated amino protons at δ 1.08, a methylene adjacent to a methylene at δ 2.75, and a methylene between two methylenes at δ 1.58. 1,3-Diaminopropane agrees with the neutralization equivalent and the ^1H NMR spectrum.

   Treatment of 1,3-diaminopropane (compound I) with sodium hydroxide and benzoyl chloride yielded dibenzamide of 1,3-diaminopropane (compound II).

Compound I:
1,3-diaminopropane
MW = 74 , NE = 37

benzoyl chloride

Compound II: dibenzamide
of 1,3-diaminopropane
MW = 282

   Treatment of 1,3-diaminopropane (compound I) with sodium nitrite and hydrochloride acid, followed by benzoic anhydride produced 1,3-propanediol dibenzoate (compound III).

Compound I:
1,3-diaminopropane
MW = 74, NE = 37

1,3-propanediol

benzoic anhydride

Compound III: 1,3-propanediol dibenzoate
MW = 284, SE = 142

2. Liquid I is in the solubility class of N, thus indicating an alcohol, an aldehyde, a ketone, an ester, an epoxide, an alkene, an alkyne, an aromatic compound, or an ether. This compound does not contain an active hydrogen (negative acetyl chloride test) and is not an aldehyde or a ketone (negative 2,4-dinitrophenylhydrazine test). Compound I was treated with phosphoric acid to yield compound II as an oil and compound VI.

   Compound II is an aldehyde or a ketone (positive 2,4-dinitrophenylhydrazine and sodium bisulfite tests), but does not contain an active hydrogen (negative acetyl chloride test). Compound II was treated with strong alkali to yield compound III and the anion of compound IV.

Compound III contains an active hydrogen (positive acetyl chloride test), but is not an aldehyde or a ketone (negative 2,4-dinitrophenylhydrazine test). Compounds IV and V are carboxylic acids.

Compound VI contains an active hydrogen (positive sodium and acetyl chloride tests), is a methyl ketone or has a methyl next to a –CHOH (positive iodoform test), and is not a tertiary or secondary alcohol (negative Lucas test). Compound VI is ethanol.

The ^{1}H NMR spectrum for compound II shows an isolated methyl at δ 2.42, a p-disubstituted benzene ring (4H) with signals at δ 7.18 and δ 7.66, and an aldehyde signal at δ 9.81. From this spectrum, compound II must be 4-methylbenzaldehyde. Since compound VI is ethanol, compound I is an acetal.

The acetal of 4-methylbenzaldehyde (compound I) is hydrolyzed to 4-methylbenzaldehyde (compound II) and ethanol (compound VI).

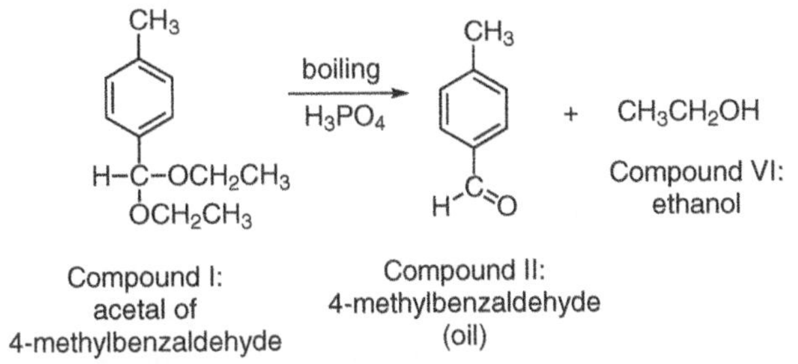

4-Methylbenzaldehyde (compound II) undergoes a Cannizzaro reaction to yield 4-methylbenzyl alcohol (compound III) and sodium 4-methylbenzoate.

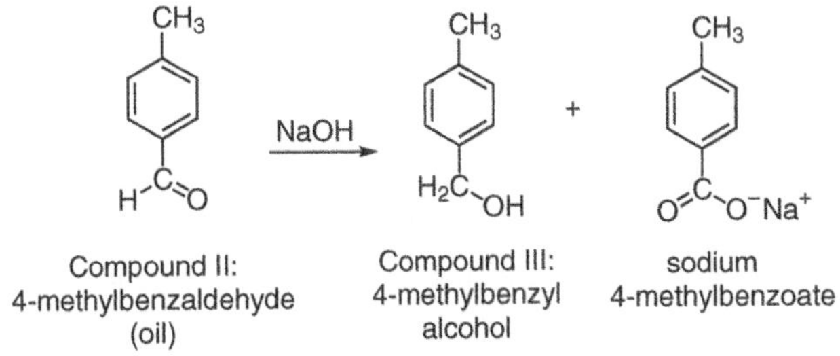

The sodium 4-methylbenzoate was acidified to yield 4-methylbenzoic acid (compound IV). Oxidation of 4-methylbenzoic acid (compound IV) produces 1,4-benzenedicarboxylic acid (terephthalic acid) (compound V).

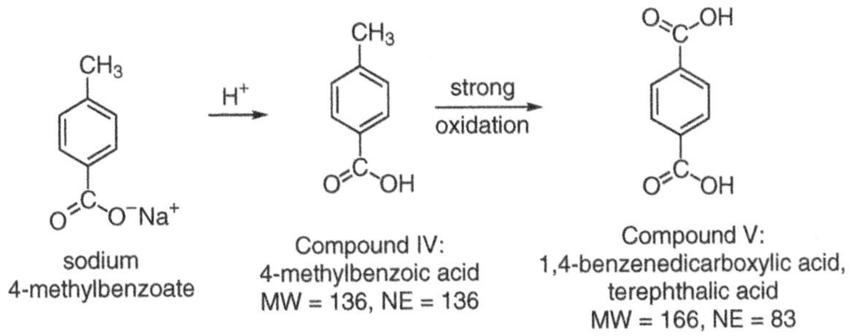

3. Liquid I is in either the S_A, S_B, or S_1 solubility class, indicating a monofunctional carboxylic acid with five carbons or fewer, an arylsulfonic acid; a monofunctional amine with six carbons or fewer; or an alcohol, an aldehyde, a ketone, an ester, a nitrile, or amide with five carbons or fewer. Arylsulfonic acids, amines, nitriles, and amides are eliminated as possibilities since the compound does not contain sulfur or nitrogen. Liquid I does not contain an active hydrogen (negative sodium and acetyl chloride tests), is not an aldehyde or a ketone (negative 2,4-dinitrophenylhydrazine test), and does not have a multiple bond (negative potassium permanganate and bromine tests).

Liquid I reacts with hydrobromic acid to yield oil II. Oil II is in the solubility class of I, which indicates a saturated hydrocarbon, an haloalkane, an aryl halide, a diaryl ether, or an aromatic compound. This compound contains a labile bromine (positive silver nitrate test). Oil II undergoes a reaction with magnesium to produce a gas III, as well as a reaction with alcoholic potassium hydroxide to liberate a gas IV. Compounds III and IV both decolorize potassium permanganate and bromine solutions, thus indicating the presence of a multiple bond.

The ^{1}H NMR spectrum indicates only one signal at δ 3.69, showing that all of the protons are equivalent. One structure that fits the above criteria for compound I is 1,4-dioxane. The m/z value for the parent peak for 1,4-dioxane is 88.

1,4-Dioxane, I, is treated with excess hydrobromic acid to form 1,2-ethanediol (ethylene glycol), 2-bromoethanol (ethylene bromohydrin), and 1,2-dibromoethane (compound II). 1,2-Ethanediol and 2-bromoethanol are infinitely soluble in the aqueous solution; 1,2-dibromoethane (compound II) is only slightly soluble. Therefore, 1,2-dibromoethane (compound II) will separate as an oil.

Compound I:	1,2-ethanediol	2-bromoethanol	Compound II:
1,4-dioxane	(ethylene glycol)	(ethylene bromohydrin)	1,2-dibromoethane
	(water soluble)	(water soluble)	(slightly water soluble)

Treatment of 1,2-dibromoethane (compound II) with magnesium produces ethene (compound III).

Compound II: Compound III:
1,2-dibromoethane   ethene

The reaction of 1,2-dibromoethane (compound II) with potassium hydroxide yields ethyne (acetylene) (compound IV).

$$H_2C=CH_2 \longrightarrow HC\equiv CH$$

Compound III: Compound IV:
ethene   ethyne

4. Compound I is in the solubility class of I, indicating a saturated hydrocarbon, a haloalkane, an aryl halide, a diaryl ether, or a deactivated aromatic compound. When compound I was heated with hydrochloric acid, acid II was formed. Oxidation of compound I with potassium dichromate in sulfuric acid yielded another acid III. The reaction of I with benzaldehyde produced a benzal derivative, thus indicating that I has an acidic proton. Since tin (II) chloride and hydrogen chloride reduces a nitro group to an amino group or a nitrile group to an aldehyde group, then compound I must have either a nitro group or a nitrile group and compound IV has an amino group or an aldehyde group.

Using the ^{1}H NMR spectrum, compound I is a disubstituted aromatic compound with four hydrogens at δ 7.5–8.25. An isolated –CH$_2$– is also present as a singlet at δ 4.25. Compound II also contains the isolated –CH$_2$– group as a singlet at δ 4.02 and is a disubstituted aromatic compound with four hydrogens at δ 7.3–8.0. A –COOH group is present as a broad singlet at δ 11.2. Compound III contains a disubstituted aromatic ring since four hydrogens are present in the aromatic range of δ 7.5–8.0. The broad singlet at δ 12.15 is due to the –COOH. Compound IV shows seven aromatic and alkene protons at δ 6.38–7.65. Based on the information above, compound I is 2-nitrobenzyl cyanide. 2-Nitrobenzyl cyanide (compound I) is hydrolyzed to 2-nitrophenylacetic acid (compound II).

Compound I:
2-nitrobenzyl cyanide

Compound II:
2-nitrophenylacetic acid
MW = 181, NE = 181
(solid)

2-Nitrobenzyl cyanide (compound I) is oxidized to 2-nitrobenzoic acid (compound III).

Compound I:
2-nitrobenzyl cyanide

Compound III:
2-nitrobenzoic acid
MW = 167, NE = 167 (solid)

2-Nitrobenzyl cyanide (compound I) is treated with benzaldehyde to produce a benzal derivative, 2-(2-nitrophenyl)-3-phenylacrylonitrile.

Compound I:
2-nitrobenzyl cyanide

benzaldehyde

2-(2-nitrophenyl)-3-
phenylacrylonitrile

2-Nitrobenzyl cyanide (compound I) is treated with tin (II) chloride and hydrochloric acid, which converts the nitrile group to an acid group and the nitro group to an amino group, to yield (2-aminophenyl)ethanoic acid. Cyclization of this compound occurs to form indole (compound IV) (C$_8$H$_7$N).

Compound I:
2-nitrobenzyl cyanide

(2-aminophenyl)ethanoic
acid

Compound IV:
indole

PROBLEM SET 18

1. Liquid I is an aldehyde or a ketone (positive 2,4-dinitrophenylhydrazine test), does not contain an active hydrogen (negative acetyl chloride test), and has a multiple bond, is an alcohol, or is an aldehyde (positive potassium permanganate test). Compound I reacts with base in a Cannizzaro reaction to yield an alcohol II (positive acetyl chloride test) and an acid III. Upon heating, compound III loses carbon dioxide to produce a compound with a formula of C_4H_4O. This compound contains a multiple bond, is an alcohol, or is an aldehyde (positive potassium permanganate test), does not have an active hydrogen (negative sodium test), and is not an aldehyde or a ketone (negative phenylhydrazine test). Compound I dimerized to produce compound IV, which has an active hydrogen (positive acetyl chloride test) and is an aldehyde or a ketone (positive phenylhydrazine test).

    The only possible structure for the formula of C_4H_4O is furan. Thus, compound III must be 2-furancarboxylic acid, since carbon dioxide is lost from compound III to form furan. Furfural (compound I) reacts with base in a Cannizzaro reaction to yield 2-furylmethanol (compound II) and 2-furancarboxylic acid (compound III).

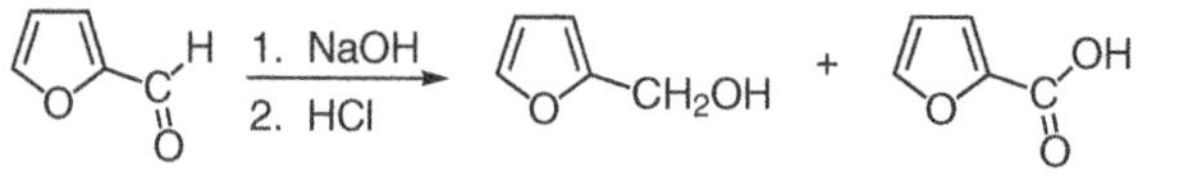

Compound I:
furfural

Compound II:
2-furylmethanol

Compound III:
2-furancarboxylic acid
furoic acid

2-Furancarboxylic acid (compound III) decarboxylates to yield furan (C_4H_4O).

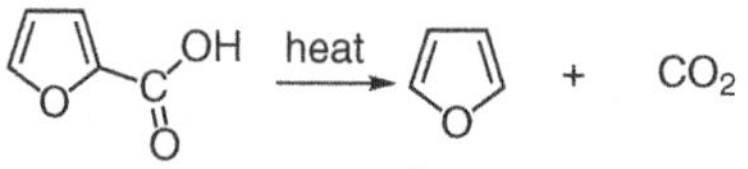

Compound III:
2-furancarboxylic acid

furan

Furfural (compound I) dimerizes to form 1,2-di(2-furyl)-1-oxo-2-ethanol (compound IV) in a reaction similar to the benzoin condensation.

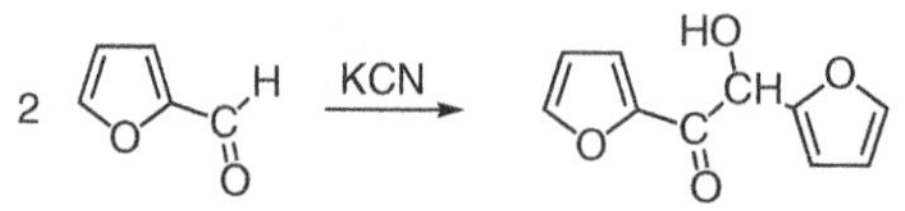

Compound I:
furfural

Compound IV:
1,2-di(2-furyl)-1-oxo-2-ethanol

1,2-Di(2-furyl)-1-oxo-2-ethanol (compound IV) reacts with periodic acid to form furfural (compound I) and 2-furancarboxylic acid (compound III).

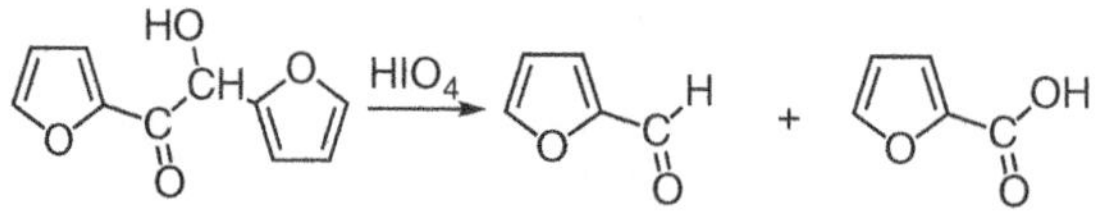

Compound IV:
1,2-di(2-furyl)-1-oxo-2-ethanol

Compound I:
furfural

Compound III:
2-furancarboxylic acid

The ^1H NMR spectra of compounds I, II, III and C_4H_4O do not contain any aromatic protons. The protons are identified in the structures below. Furfural (compound I): δ 6.63, 1H, d of d (*a*); δ 7.28, 1H, d (*b*); δ 7.72, 1H, m (*c*); δ 9.67, 1H, s (*d*). 2-Furylmethanol (compound II): δ 2.83,

1H, s (**a**); δ 4.57, 2H, s (**b**); δ 6.33, 2H, m (**c**); δ 7.44, 1H, m (**d**). 2-Furancarboxylic acid (compound III): δ 6.56, 1H, m (**a**); δ 7.20, 1H, m (**b**); δ 7.65, 1H, m (**c**); δ 12.18, 1H, bs (**d**). Furan (C_4H_4O): δ 6.37, 2H, t (**a**); δ 7.42, 2H, t (**b**).

Compound I: furfural

Compound II: 2-furylmethanol

Compound III:
2-furancarboxylic acid (furan)

furan

2. Compound I contains an active hydrogen (positive acetyl chloride test) and is not an aldehyde or a ketone (negative 2,4-dinitrophenylhydrazine test). Compound I is easily hydrolyzed to compound II. Compound II does not have an active hydrogen (negative acetyl chloride test) and is a ketone (negative Tollens test, positive 2,4-dinitrophenylhydrazine and sodium bisulfite tests). Compound III contains an active hydrogen, since it reacts with benzoyl chloride to produce an ester or an amide IV. The hydrolysis of compound IV produces two compounds, a carboxylic acid V and compound III. Compound VI is an amine salt, which is from a primary amine (liberates a gas with sodium nitrite).

Compound I is cyclopentanone oxime, which is hydrolyzed to cyclopentanone (compound II).

Compound I:
cyclopentanone oxime

Compound II:
cyclopentanone

In the presence of phosphorus pentachloride, cyclopentanone oxime (compound I) rearranges to 5-pentanelactam (compound III). 5-Pentanelactam (compound III) is treated with sodium hydroxide and benzoyl chloride to form (5-benzoylamino)pentanoic acid (compound IV), which is hydrolyzed to benzoic acid (compound V) and 5-pentanelactam (compound III).

Compound I:
cyclopentanone oxime

Compound III:
5-pentanelactam

benzoyl chloride

Compound IV:
(5-benzoylamino)pentanoic acid
MW = 221, NE = 221

Compound V:
benzoic acid
MW = 122, NE = 122

Compound III:
5-pentanelactam

5-Pentanelactam (compound III) is acidified to yield the hydrochloride salt of 5-amino-pentanoic acid (compound VI), which gives off a gas when treated with nitrous acid.

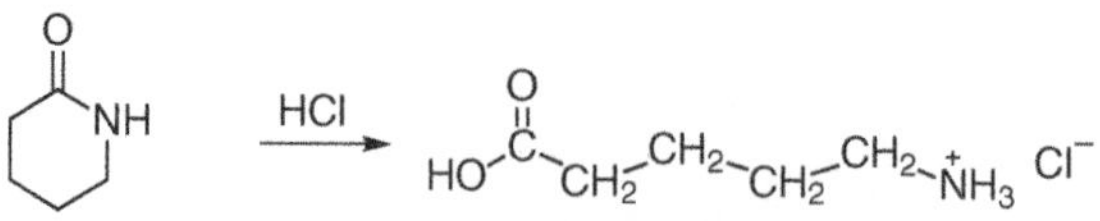

Compound III:
5-pentanelactam

Compound VI:
hydrochloride salt of
5-aminopentanoic acid

The ^{1}H NMR spectrum for cyclopentanone oxime (compound I) shows an −OH at δ 9.12. Cyclopentanone oxime (compound I) has peaks at δ 1.77, 4H, m (**a**); δ 2.40, 4H, m (**b**); δ 9.12, 1H, bs (**c**). For 5-pentanelactam (compound III), the −NH is seen at δ 7.60. 5-Pentanelactam (compound III) has signals at δ 1.80, 4H, m (**a**); δ 2.38, 2H, m (**b**); δ 3.32, 2H, m (**c**); δ 7.60, 1H, bs (**d**). The protons are identified in the structures below.

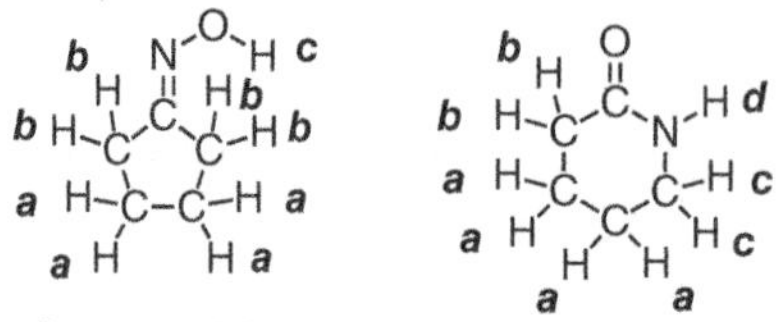

Compound I:
cyclopentanone oxime

Compound III:
5-pentanelactam

3. Compound I is in the solubility class of S_2, which includes salts of organic acids, amine hydrochlorides, amino acids, or polyfunctional structures. Compound II is in the solubility class of I, which includes saturated hydrocarbons, haloalkanes, aryl halides, diaryl ethers, or deactivated aromatic compounds. Both compounds I and II contain a chlorine (positive silver nitrate test with white precipitant). Since compound III did not give a benzenesulfonamide derivative, but did give a nitroso derivative, it must be a tertiary aniline.

Compound I (trimethylanilinium chloride) is decomposed by heat to compound II (methyl chloride) and compound III (*N,N*-dimethylaniline).

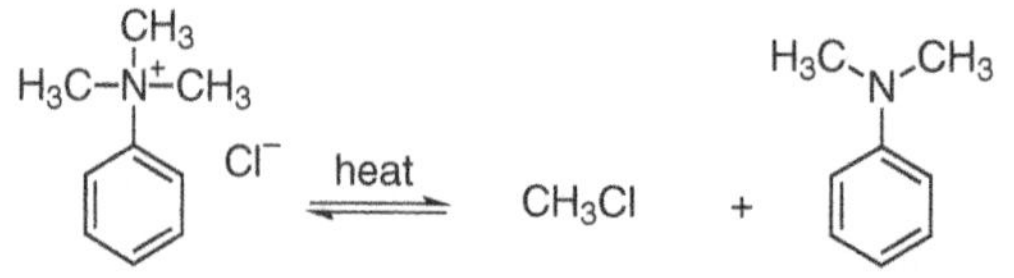

Compound I:
trimethylanilinium
chloride

Compound II:
methyl chloride

Compound III:
N,N-dimethylaniline

The ^{1}H NMR spectrum of I shows a ratio of 9:5 between methyl groups at δ 3.70 (s) and the aromatic range at δ 7.4 (m), indicating three methyls and a monosubstituted aromatic ring. This data corresponds to *N,N,N*-trimethylanilinium chloride. The ^{1}H NMR spectrum for II shows only a methyl at δ 2.20. For compound III, two methyls are seen at δ 2.85 (s) and the monosubstituted ring protons at δ 6.4–δ 7.10 (m). This NMR spectrum corresponds to *N,N*-dimethylaniline.

4. Compound I contains chlorine, bromine, and iodine. Reaction of compound I with silver nitrate to produce a white solid indicates the presence of a labile chlorine. Compound I contains an aldehyde or a ketone (positive 2,4-dinitrophenylhydrazine test), but

does not contain an active hydrogen (negative acetyl chloride test). Compound II contains bromine and iodine and is a carboxylic acid. Since compound II is produced from the hydrolysis of compound I with the loss of chlorine, then I must have been an acid chloride. Compound I is subjected to hot alkali, followed by acidification to yield Compound III as a carboxylic acid. Compound III contains only iodine, indicating the loss of chlorine and bromine. Bromine could be lost from the hydrolysis of an alkyl bromide. Compound III has an aldehyde or a ketone (positive phenylhydrazine test) and has an active hydrogen (positive acetyl chloride test).

Based upon the information given, compound I is a trisubstituted benzene ring, but positions of the various groups on the ring cannot be determined. The example below is given with the groups at the 1, 3, and 5 positions.

3-Bromoacetyl-5-iodobenzoyl chloride (compound I) is treated with cold alkali, followed by acidification, to yield 3-bromoacetyl-5-iodobenzoic acid (compound II). The acid chloride is hydrolyzed to a carboxylic acid.

1. cold $^{-}$OH
2. H^{+}

Compound I:
3-bromoacetyl-5-
iodobenzoyl chloride

Compound II:
3-bromoacetyl-5-
iodobenzoic acid
MW = 369, NE = 369

3-Bromoacetyl-5-iodobenzoyl chloride (compound I) is treated with hot alkali, followed by acidification, to yield 3-hydroxyacetyl-5-iodobenzoic acid (compound III). In this reaction, both the chloride and bromide ions are cleaved.

1. hot $^{-}$OH
2. H^{+}

Compound I:
3-bromoacetyl-5-
iodobenzoyl chloride

Compound III:
3-hydroxyacetyl-5-iodobenzoic acid
MW = 306, NE = 306

With sodium hypochlorite, 3-bromoacetyl-5-iodobenzoyl chloride (compound I) or 3-bromoacetyl-5-iodobenzoic acid (compound II) is oxidized to 5-iodo-1,3-benzenedicarboxylic acid.

or

1. NaOCl
2. H^{+}

Compound I:
3-bromoacetyl-5-
iodobenzoyl chloride

Compound II:
3-bromoacetyl-5-
iodobenzoic acid

Compound IV:
5-iodo-1,3-benzenedicarboxylic acid
MW = 292, NE = 146

5. Compound A does not contain a labile bromine (negative silver nitrate test), an active hydrogen (negative acetyl chloride test), or a multiple bond (negative bromine test). Compound A is not an aldehyde or a ketone (negative 2,4-dinitrophenylhydrazine test). Compound A was treated with alkaline solution, then acidified to produce compounds B, C, and the precursor to D. Compounds B and C are carboxylic acids. The precursor to compound D was treated with alkaline, then treated with benzoyl chloride to yield compound D.

From the ^{1}H NMR spectrum, compound B contains a *p*-disubstituted benzene ring with 4 hydrogens as doublets at δ 7.71 and 7.90. The hydrogen from the –COOH group appears as a broad singlet at δ 7.3. Since compound B also contains bromine, compound B must be 4-bromobenzoic acid.

The ^{1}H NMR spectrum of compound C indicates an aliphatic structure. A CH$_3$– adjacent to a –CH$_2$– is indicated by a triplet with three hydrogens at δ 0.93. Two –CH$_2$– are seen at δ 1.2–δ 1.8, and a –CH$_2$– adjacent to a –CH$_2$– is seen at δ 2.31. At δ 11.7, the hydrogen from the –COOH is indicated. Compound C has a neutralization equivalent of 102 ± 1. From this information, compound C must be pentanoic acid.

The precursor to compound D must be a diol, since two carboxylic acids have already been identified. This diol undergoes reaction with benzoyl chloride to yield compound D, a diester with a saponification equivalent of 135 ± 1. Compound D must be the dibenzoyl ester of 1,2-ethanediol.

Compound A must be 4-bromo-2-pentanoyloxyethyl benzoate, which is hydrolyzed to 4-bromobenzoic acid (compound B), pentanoic acid (compound C), and 1,2-ethanediol (compound D).

Compound A:
4-bromo-2-pentanoyloxyethyl benzoate

Compound B:
4-bromobenzoic acid
MW = 201, NE = 201

Compound C:
pentanoic acid
MW = 102, NE = 102

Compound D:
1,2-ethanediol

1,2-Ethanediol (compound D) undergoes reaction with benzoyl chloride to yield Compound E, the dibenzoyl ester of 1,2-ethanediol.

Compound D:
1,2-ethanediol

benzoyl
chloride

Compound E:
1,2-dibenzoyl 1,2-ethanediol
MW = 270, SE = 135

PROBLEM SET 19

1. Compound I was hydrolyzed into two acids, compounds II and III. Since compound I also has a neutralization equivalent, then compound I contains both an ester group and a carboxylic acid group. Compound II is a phenol, because of the positive test with both bromine water and ferric chloride.

   Since all four of the protons are split in the range of δ 6.7 to 7.75 in the ^{1}H NMR spectrum of compound I, it is an *o*-disubstituted benzene ring. Two hydroxy protons are present at δ 11.55. With a neutralization equivalent of 138 ± 1, Compound II is salicylic acid (2-hydroxybenzoic acid). With a neutralization equivalent of 60 ± 1, Compound III is acetic acid. Therefore, compound I is acetyl salicylic acid (2-acetyl benzoic acid).

   Acetyl salicylic acid (2-acetyl benzoic acid) (compound I) is hydrolyzed to salicylic acid (2-hydroxybenzoic acid) (compound II) and acetic acid (compound III).

Compound I: acetylsalicylic acid, 2-acetylbenzoic acid MW = 180, NE = 180	Compound II: salicylic acid, 2-hydroxybenzoic acid MW = 138, NE = 138	Compound III: acetic acid MW = 60, NE = 60

2. Compound I is in the solubility class B, which contains amines, anilines, and ethers. A primary amine is indicated since the product from treatment with benzenesulfonyl chloride is soluble in acid. Compound I is not an aldehyde or ketone (negative phenylhydrazine test).

   Compound II contains an active hydrogen (positive acetyl chloride and sodium test) and is a primary amine since the product from the reaction with benzenesulfonyl chloride is soluble in acid. Compound II is in solubility class B, which includes amines, anilines, and ethers. Compound IV is the hydrochloride salt of compound II.

   Compound V is an aniline derivative, since it reacts with nitrous acid without the evolution of nitrogen and this solution reacts with 2-sodium naphthoxide to form a red solution. The ^{1}H NMR spectrum of compound V indicates a *p*-disubstituted benzene ring from the doublets with four protons at δ 6.76 and δ 7.83 and three hydroxy and/or amino protons at δ 6.55. Compound V is 4-aminobenzoic acid.

   Methylation of compound V followed by mild reduction produces 4-(*N,N*-dimethylamino)-benzyl alcohol, compound II. Compound III is the sodium salt of compound V. Compound I is 4-(*N,N*-dimethyl)aminobenzyl 4-aminobenzoate.

   4-(*N,N*-Dimethyl)aminobenzyl 4-aminobenzoate (compound I) is decomposed with hot sodium hydroxide to yield 4-(*N,N*-dimethylamino)benzyl alcohol (compound II) and sodium benzoate (compound III).

Compound I: 4-(*N,N*-dimethyl)aminobenzyl 4-aminobenzoate	Compound II: 4-(*N,N*-dimethylamino)-benzyl alcohol	Compound III: sodium benzoate

4-(N,N-Dimethyl)aminobenzyl alcohol (compound II) is treated with hydrochloric acid to yield N,N,N-trimethyl-4-hydroxymethylanilinium chloride (compound IV).

Compound II:
4-(N,N-dimethylamino)-
benzyl alcohol

Compound IV:
N,N,N-trimethyl-4-
hydroxymethyl anilinium chloride
MW = 201.5, NE = 201.5

Sodium 4-aminobenzoate (compound III) is acidified to yield 4-aminobenzoic acid (compound V), which is methylated twice to yield 4-(N,N-dimethylamino)benzyl alcohol (compound II).

Compound III:
sodium 4-aminobenzoate

Compound V:
4-aminobenzoic acid
MW = 137, NE = 137

Compound II:
4-(N,N-dimethyl-
amino)benzyl alcohol

3. Compound I is in the solubility class of S_2, which includes salts of organic acids, salts of sulfonic acids, amine hydrochlorides, amino acids, and polyfunctional compounds.

Compound II is a primary amine since it reacts with benzenesulfonyl chloride and base to yield a soluble compound III, which precipitates with acidification. Compound II does not contain a labile halogen (negative silver nitrate test), contains an activating group on the aromatic ring (positive bromine water test), and contains a multiple bond or is a phenol or an aniline (positive potassium permanganate test).

Compound I reacts with nitrous acid without the evolution of gas, followed by treatment with copper cyanide to produce compound IV. This type of reaction is indicative of the diazotization of an aromatic amine with a subsequent Sandmeyer reaction resulting in a nitrile group attached to the benzene ring. Hydrolysis of the nitrile group to a carboxylic acid group produces compound V. Compound V does not have a labile halogen (negative silver nitrate test) or non-aromatic multiple bond (negative potassium permanganate test).

The ^{1}H NMR spectrum of compound II indicates a p-disubstituted benzene ring with doublets at δ 6.57 and 7.21. The amino protons appear as a broad singlet at δ 3.53. Since this structure also contains bromine, the structure must be 4-bromoaniline. Compound III is the benzenesulfonamide derivative of compound II.

Compound V must be 4-bromobenzoic acid, since it is an acid with a neutralization equivalent of 200±2. Working backwards, compound IV is 4-bromobenzonitrile. Using the neutralization equivalent and eliminating other possibilities, compound I is 4-bromoanilinium sulfate.

4-Bromoanilinium sulfate (compound I) was treated with alkali to yield 4-bromoaniline (compound II), which is then reacted with benzenesulfonyl chloride and alkali to produce *N*-bromophenylbenzenesulfonamide (compound III).

Compound I:
4-bromoanilinium
sulfate
MW = 442, NE = 221

Compound II:
4-bromoaniline

benzenesulfonyl
chloride

Compound III:
N-bromophenyl-
benzenesulfonamide

Treatment of 4-bromoanilinium sulfate (compound I) with nitrous acid produced the diazonium salt, which reacted with copper cyanide to produce 4-bromobenzonitrile (compound IV). Acid hydrolysis of 4-bromobenzonitrile (compound IV) yielded 4-bromobenzoic acid (compound V).

Compound I:
4-bromoanilinium
sulfate
MW = 442, NE = 221

diazonium salt
of 4-bromoaniline

Compound IV:
4-bromo-
benzonitrile

Compound V:
4-bromobenzoic acid
MW = 201, NE = 201

4. Compound I is in the solubility class of I, indicating that it is a saturated hydrocarbon, a haloalkane, an aryl halide, a diaryl ether, or a deactivated aromatic compound. Compound I is not an aldehyde or a ketone (negative 2,4-dinitrophenylhydrazine test), does not have an active hydrogen (negative acetyl chloride test), and does not have a multiple bond (negative potassium permanganate test). A positive silver nitrate test was given after heating, thus showing that the halogen is not very labile and probably attached to an aromatic ring. Compound I reacts with zinc and ammonium chloride and the filtrate from this reaction reduces Tollens reagent, thus indicating the presence of a nitro group.

Compound I was vigorously oxidized to produce compound II. Compound II still contains a bromine group and a nitro group. Compound II is also aromatic and a carboxylic acid. To come up with the neutralization equivalent of 145 ± 1, compound II must be a bromonitro-1,2-benzenedicarboxylic acid. Compound I is a bromonitronaphthalene, with both the bromo and nitro group on the same ring.

Nitro groups are reduced to amino groups with tin and hydrochloric acid, thus compound III contains an amino group. Compound III is an aniline or phenol (positive bromine water test). Additionally, compound III contains an active hydrogen (positive acetyl chloride test) and the amine is primary (reaction with benzenesulfonyl chloride, and soluble in base).

Oxidation of compound III produced compound IV, which does not contain bromine or nitrogen and has a neutralization equivalent of 82 ± 1. A compound that fulfills these criteria is a benzenedicarboxylic acid. Since compound IV loses water to produce an anhydride, the only possible structure for this compound is 1,2-benzenedicarboxylic acid (phthalic acid). Compound III is an aminobromonaphthalene, with both the amino and the bromo group on the same ring.

The ^{1}H NMR spectrum of compound III concurs with the structure of 1-amino-4-bromo-naphthalene, with a broad singlet with two hydrogens for the amino group at δ 3.96 and six hydrogens in the aromatic region in the range of δ 6.45–8.2. Thus, the other structures can be determined.

1-Bromo-4-nitronaphthalene (compound I) is vigorously oxidized to 3-bromo-6-nitro-1,2-benzenedicarboxylic acid (compound II).

Compound I:
1-bromo-4-nitronaphthalene

vigorous oxidation

Compound II:
3-bromo-6-nitro-1,2-benzenedicarboxylic acid
MW = 290, NE = 145

Treatment of 1-bromo-4-nitronaphthalene (compound I) with tin and hydrochloric acid reduces the nitro group to an amino group to produce 1-amino-4-bromonaphthalene (compound III), which is vigorously oxidized to 1,2-benzenedicarboxylic acid (phthalic acid) (compound IV).

Compound I:
1-bromo-4-nitronaphthalene

1. Sn, HCl
2. OH⁻

Compound III:
1-amino-4-bromo-napthalene

vigorous oxidation

Compound IV:
1,2-benzene-dicarboxylic acid, phthalic acid
MW = 166, NE = 83

5. Solid (compound I) does not contain a labile chlorine (negative silver nitrate test), an active hydrogen (negative acetyl chloride test), or a multiple bond (negative bromine test). However, it is an aldehyde or a ketone (positive 2,4-dinitrophenylhydrazine test).

Compound II contains chlorine and is a carboxylic acid.

Compound III is not a phenol (negative bromine water test), does not have a multiple bond (negative bromine in carbon tetrachloride test), and is not an aldehyde or a ketone (negative 2,4-dinitrophenylhydrazine test).

Compound IV, which is produced from the oxidation of compound I, is a carboxylic acid and contains chlorine. The appearance of two doublets with a total of four protons in the aromatic region of δ 7.37–7.81 indicates a p-disubstituted benzene ring. The signal at δ 7.45 is due to the –COOH. Compound IV contains chlorine and has a neutralization equivalent of 156 ± 1. The only possible structure for compound IV is 4-chlorobenzoic acid.

The neutralization equivalents for compounds II and III are considerably higher than the neutralization equivalent for compound IV, thus indicating that these two compounds may be twice the size of compound IV. Since compound II reacts with acetic anhydride to yield compound III, then compound II must be an alcohol. Compound II contains chlorine and is a carboxylic acid. From deductive reasoning, compound II must be 4,4′-dichlorobenzilic acid and compound I is 4,4′-dichlorobenzil.

In the presence of base, 4,4'-dichlorobenzil (compound I) undergoes a benzilic acid rearrangement to form 4,4'-dichlorobenzilic acid (compound II), which reacts with acetic anhydride to yield α-acetyl-4,4'-dichlorobenzilic acid (compound III).

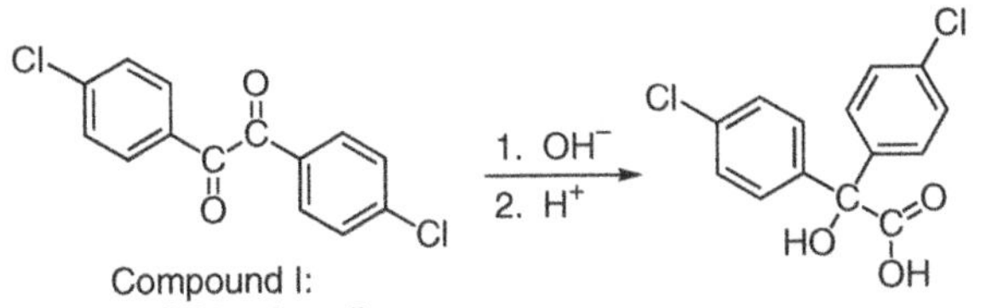

Compound I:
4,4'-dichlorobenzil

1. OH⁻
2. H⁺

Compound II:
4,4'-dichlorobenzilic acid
MW = 297, NE = 297

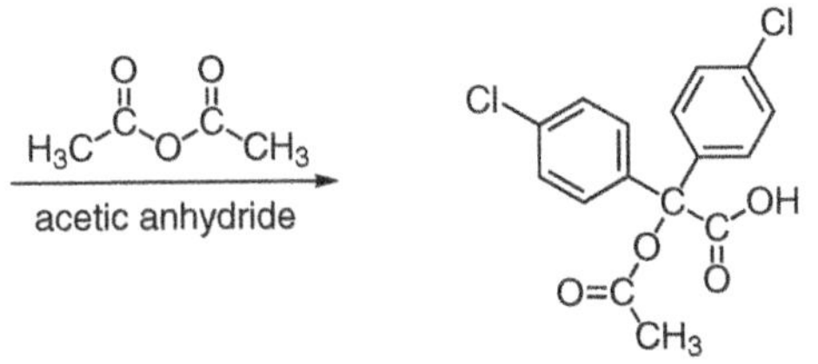

acetic anhydride

Compound III:
α-acetyl-4,4'-dichloro-
benzilic acid
MW = 339, NE = 339

4,4'-Dichlorobenzil (compound I) is oxidized to 4-chlorobenzoic acid (compound IV).

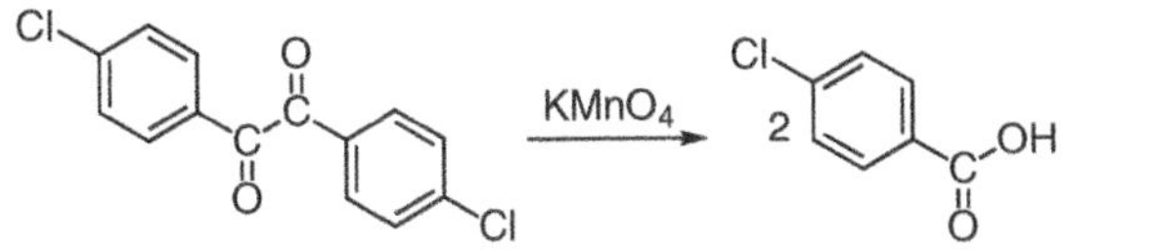

Compound I:
4,4'-dichlorobenzil

KMnO₄

Compound IV:
4-chlorobenzoic acid
MW = 156.5, NE = 156.5

PROBLEM SET 20

1. Compound I undergoes reaction with acetyl chloride to produce compound II, indicating that compound I has an active hydrogen. Compound I is oxidized to Compound III, which contains an aldehyde or ketone group (positive 2,4-dinitrophenylhydrazine test). Vigorous oxidation of I, II, or III yields an acid with a neutralization equivalent of 122 ± 1, which corresponds to benzoic acid. The difference in the neutralization equivalents of Compounds I and III is 28 (150–122), which is a carbon and an oxygen. Thus, compound I is 2-phenyl-2-hydroxyethanoic acid (mandelic acid) and compound III is 2-phenyl-2-oxoethanoic acid. Compound II is the acetyl derivative of compound I.

2-Phenyl-2-hydroxyethanoic acid (mandelic acid) (compound I) is reacted with acetyl chloride to yield 2-acetyl-2-phenyl-2-hydroxyethanoic acid (compound II).

Compound I:
2-phenyl-2-hydroxyethanoic acid,
mandelic acid
MW = 152, NE = 152

Compound II:
2-acetyl-2-phenyl-2-
hydroxyethanoic acid
MW = 194, NE = 194

2-Phenyl-2-hydroxyethanoic acid (mandelic acid) (compound I) is oxidized to 2-phenyl-2-oxoethanoic acid (compound III).

Compound I:
2-phenylhydroxyethanoic acid,
mandelic acid
MW = 152, NE = 152

Compound III:
2-phenyl-2-oxoethanoic acid
MW = 150, NE = 150

Vigorous oxidation of 2-phenyl-2-hydroxyethanoic acid (mandelic acid) (compound I), 2-acetyl-2-phenyl-2-hydroxyethanoic acid (compound II), or 2-phenyl-2-oxoethanoic acid (compound III) produces benzoic acid (compound IV).

Compound I:
2-phenylhydroxy-
ethanoic acid,
mandelic acid
MW = 152,
NE = 152

Compound II:
2-acetyl-2-phenyl-2-
hydroxyethanoic acid
MW = 194,
NE = 194

Compound III:
2-phenyl-2-
oxoethanoic acid
MW = 150,
NE = 150

Compound IV:
benzoic acid
MW = 122,
NE = 122

The ^{1}H NMR spectrum of 2-phenyl-2-hydroxyethanoic acid (mandelic acid) (compound I) shows the isolated methine proton at δ 5.22 and the aromatic and hydroxy protons at δ 6.93–7.88, with an integration of seven protons. The IR spectrum has a signal for the carbonyl stretching of the ketone at 1706 cm^{-1} and the –OH stretching of the carboxylic acid at 2400–3333 cm^{-1}.

2. Since compounds I, II, and III have neutralization equivalents, they are probably carboxylic acids. Zinc and hydrochloric acid reduce carbonyl groups in aldehydes and ketones to methylene groups and reduces nitro groups to amino groups.

    The ^{1}H NMR spectrum of compound II indicates a disubstituted benzene ring in the aromatic region of δ 7.72–δ 8.71 and the hydroxy peak of the carboxylic acid at δ 12.98. The groups are probably *meta*, due to the splitting pattern of the aromatic protons. Compound II is 3-nitrobenzoic acid, and compound I is then 4-(3-nitrophenyl)-2-oxobutanoic acid.

    4-(3-Nitrophenyl)-2-oxobutanoic acid (compound I) is vigorously oxidized to 3-nitrobenzoic acid (compound II).

Compound I:
4-(3-nitrophenyl)-2-
oxobutanoic acid
MW = 223, NE = 223

Compound II:
3-nitrobenzoic acid
MW = 167, NE = 167

    With zinc and hydrochloric acid, the nitro and the carbonyl groups are reduced in 4-(3-nitrophenyl)-2-oxobutanoic acid (compound I) to yield 4-(3-aminophenyl)butanoic acid (compound III).

Compound I: 4-(3-nitrophenyl)-2-
oxobutanoic acid
MW = 223, NE = 223

Compound III: 4-(3-aminophenyl)-
butanoic acid
MW = 179, NE = 179

3. The original sulfur-containing compound is in the solubility class of A$_2$, which includes phenols, enols, oximes, imides, sulfonamides, thiophenols, β-diketones, and nitro compounds. Oxidation of the original compound produces a sulfur-containing acid with a neutralization equivalent of 102 ± 1. The sulfur-containing acid was treated with superheated steam to yield a sulfur-free acid.

    The ^{1}H NMR spectrum of the sulfur-free acid shows five aromatic protons at δ 7.00–7.75, indicating a monosubstituted benzene. A singlet at δ 8.00 corresponds to the –OH of the carboxylic acid. Therefore, this spectrum corresponds to benzoic acid. The ^{1}H NMR spectrum of the original compound indicates an isolated methyl at δ 2.29, an isolated proton at δ 3.37, and a disubstituted benzene ring at δ 7.02. Since all of the aromatic protons are split, the compound is either *ortho* or *meta* disubstituted. The methyl group is probably attached to the benzene ring, and oxidation would convert the methyl to a carboxylic acid group to produce the sulfur-containing acid. By adding up the weights of C$_6$H$_4$ (a disubstituted benzene ring, –COOH, and –SO$_3$H), a value of 202 is obtained. Since it contains two acid groups, the neutralization equivalent is equal to 101.

In the second compound, the neutralization equivalent is much lower, indicating the presence of two acid groups. If the sulfur is in a sulfonic acid group, then the neutralization equivalent works out. Thus, the sulfur-containing acid is a carboxybenzenesulfonic acid. The original compound is a methylthiophenol and the final compound is benzoic acid. The answers are drawn for the *meta* disubstituted compounds.

Vigorous oxidation of 3-methylphenol (original compound) produced 3-carboxy-benzenesulfonic acid (sulfur-containing acid), which is treated with steam to yield benzoic acid (sulfur-free acid).

| 3-methylthiophenol | 3-carboxybenzenesulfonic acid (sulfur-containing acid) MW = 202, NE = 101 | benzoic acid (sulfur-free acid) |

4. Compound I contains a labile halogen (positive silver nitrate test), is a methyl ketone or has a methyl group next to a hydroxy on a CH (positive iodoform test), and has an active hydrogen (positive acetyl chloride test). The hydrolysis of compound I yielded a chlorine-free compound (compound II), which must be symmetrical since it was oxidized by periodic acid to a single compound (compound III). For oxidation by periodic acid to occur, the compound must have a hydroxy group next to a hydroxy group, a hydroxy group next to a carbonyl group, or a carbonyl next to a carbonyl. Since compound I was hydrolyzed to compound II and it is symmetrical, then the hydroxy groups must be on adjacent carbons. Compound III contains a methyl ketone or a methyl group next to a hydroxy on a CH (positive iodoform test). Several structures are possible.

The chlorine group is replaced by –OH in 4-chloro-5-hydroxy-2,7-octadione (Compound I) to yield 4,5-dihydroxy-2,7-octadione (compound II). Treatment of 4,5-dihydroxy-2,7-octadione (compound II) with periodic acid produces 3-oxobutanal (compound III).

Compound I:
4-chloro-5-hydroxy-2,7-octadiene

Compound II:
4,5-dihydroxy-2,7-octadiene

Compound III:
3-oxobutanal

Another set of answers could be 3-chloro-2-butanol (compound I), 2,3-butanediol (Compound II), and ethanal (compound III).

Compound I:
3-chloro-2-butanol

Compound II:
2,3-butanediol

Compound III:
ethanal

5. Compound A contains an active hydrogen (positive acetyl chloride test), but is not an aldehyde or a ketone (negative 2,4-dinitrophenylhydrazine test). Compound B is an aldehyde (positive Tollens reagent), but is not an aldose, an α-hydroxyaldehyde, an α-hydroxyketone, or an α-ketoaldehyde (negative Fehling's test). Compound C contains an active hydrogen (positive acetyl chloride test) and is an aldehyde or a ketone (positive phenylhydrazine test).

The ^{1}H NMR spectrum of compound B indicates an isolated methyl group at δ 2.42, a *p*-disubstituted benzene ring (4H) with doublets at δ 7.18 and δ 7.56, and an isolated aldehyde group at δ 9.81. From this data, compound B is 4-methylbenzaldehyde. Compound A must be 1,2-di(4-methylphenyl)-1,2-ethanediol.

Compound A:
1,2-di(4-methylphenyl)-
1,2-ethanediol

Compound B:
4-methylbenzaldehyde

Compound C:
4,4′-dimethylbenzoin

Compound D:
4,4′-dimethylbenzil

Compound E:
4-methylbenzoic acid
MW = 135, NE = 135

Treatment of 4,4′-dimethylbenzil (compound D) with *o*-phenylenediamine produced 2,3-di(4-tolyl)quinoxaline.

Compound D:
4,4′-dimethylbenzil

o-phenylenediamine

2,3-di(4-tolyl)quinoxaline

Catalytic hydrogenation of 4,4′-dimethylbenzoin (C) or 4,4′-dimethylbenzil (compound D) yielded 1,2-di(4-methylphenyl)-1,2-ethanediol (compound A).

Compounds C or D →(catalytic hydrogenation)→

Compound A:
1,2-di(4-methylphenyl)-
1,2-ethanediol

6. This problem is a very involved problem. The best way to figure out the structures is to list the characteristics of each compound and to list the types of reactions that are occurring.

Compound A does not have an active hydrogen (negative acetyl chloride test) or a multiple bond (negative bromine test). It is not a methyl ketone, nor does it have a methyl adjacent to an –CHOH (negative iodoform test). Compound A is not an aldehyde (negative Tollens test) and is probably a diaryl ketone since it reacts slowly with 2,4-dinitrophenylhydrazine. A nitro group is present (positive zinc and ammonium chloride test, with product giving a positive Tollens test).

Compound A undergoes reaction with hydroxylamine hydrochloride in pyridine, thus converting the carbonyl to an oxime in compound B. Phosphorus pentachloride rearranges the oxime group in compound B to an amide in compound C. Boiling with acid hydrolyzed the amide in compound C to an acid, compound D, and an amine, compound E.

Acid D undergoes reaction with thionyl chloride and ammonium hydroxide to form an amide. The amide was subjected to the conditions of a Hofmann rearrangement, bromine and base, to produce an amine, compound F. Compound F is an aniline derivative since diazotization of this compound followed by addition of 2-naphthol produces a red color. Treatment of compound F with tin and hydrochloric acid reduces the nitro group to an amine to form Compound G.

Compound E is an aniline derivative, since diazotization of this compound followed by addition of 2-naphthol produces a red color. When the diazotized solution is treated with base, a phenol derivative, Compound H, is formed. Oxidation of either Compounds E or H produced an acid with a neutralization equivalent of 83, which is easily converted to the anhydride. An acid that fulfills this criteria is 1,2-benzenedicarboxylic acid. Compound H does not contain an amino group since it does not undergo coupling with diazonium solutions.

From the ^{1}H NMR spectrum, compound F is a *p*-disubstituted benzene ring, with doublets (4H) at δ 6.58 and δ 7.88. From above, compound F contains both an amino group and a nitro group. The amino group appears at δ 5.68. The structure of compound F must be 4-nitroaniline.

In the ^{1}H NMR spectrum of compound H, six aromatic protons are present in the range of δ 7.1–8.2. An isolated methyl is present at δ 2.32 and an –OH appears at δ 5.12. Therefore, compound H is 1-hydroxy-2-methylnaphthalene and compound A is 2-methyl-1-(4-nitrobenzoylphenyl)naphthalene.

2-Methyl-1-(4-nitrobenzoylphenyl)naphthalene (compound A) is treated with hydroxylamine hydrochloride to form 2-methyl-1-(4-nitrobenzoylphenyl)naphthalene oxime (compound B). 2-Methyl-1-(4-nitrobenzoylphenyl)naphthalene oxime (compound B) undergoes a rearrangement in the presence of phosphorus pentachloride to yield 2-methyl-1-(4-nitrobenzoyl)aminonaphthalene (compound C), which is hydrolyzed to 4-nitrobenzoic acid (compound D) and the hydro-chloride salt of 1-amino-2-methylnaphthalene (compound E).

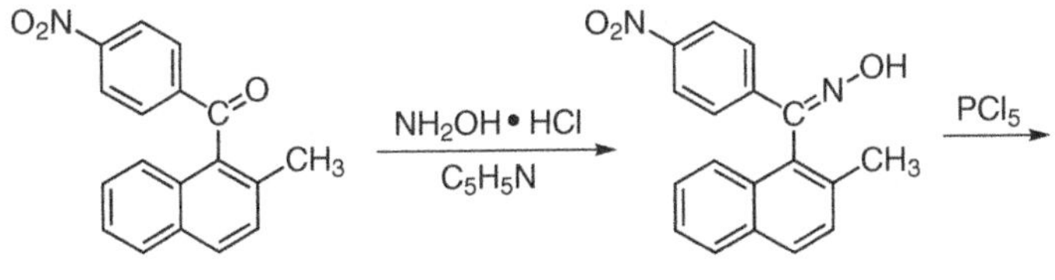

Compound A:
2-methyl-1-(4-nitrobenzoyl-
phenyl)naphthalene

Compound B:
2-methyl-1-(4-nitrobenzoyl-
phenyl)naphthalene oxime

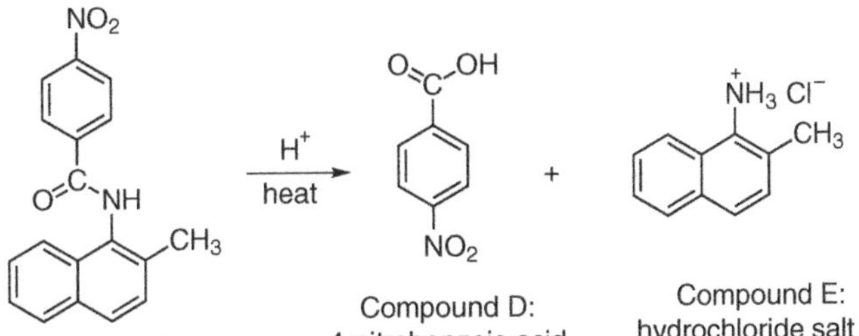

Compound C:
2-methyl-1-(4-nitrobenzoyl)-
aminonaphthalene

Compound D:
4-nitrobenzoic acid
MW = 167, NE = 167

Compound E:
hydrochloride salt of
1-amino-2-methyl-
naphthalene
MW = 193.5,
NE = 193.5

4-Nitrobenzoic acid (compound D) was treated with thionyl chloride to yield 4-nitrobenzoyl chloride, which was reacted with ammonia to form 4-nitrobenzamide. By reacting 4-nitrobenzamide with bromine and sodium hydroxide, a Hofmann rearrangement occurred, thus producing 4-nitroaniline (compound F). The nitro group in 4-nitroaniline (compound F) was reduced to an amino group to form 1,4-diaminobenzene (compound G).

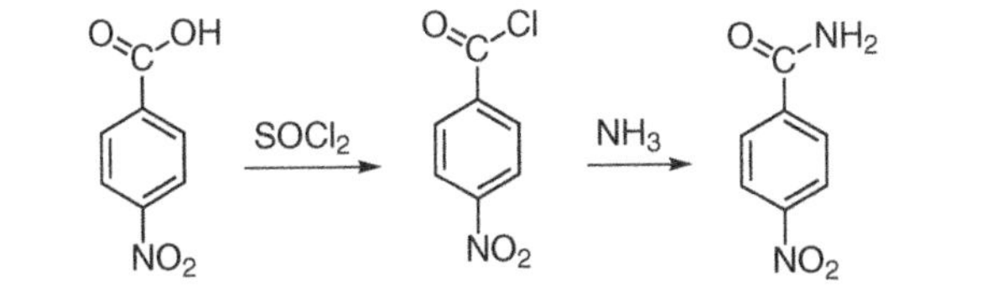

Compound D:
4-nitrobenzoic acid
MW = 167, NE = 167

4-nitrobenzoyl
chloride

4-nitrobenzamide

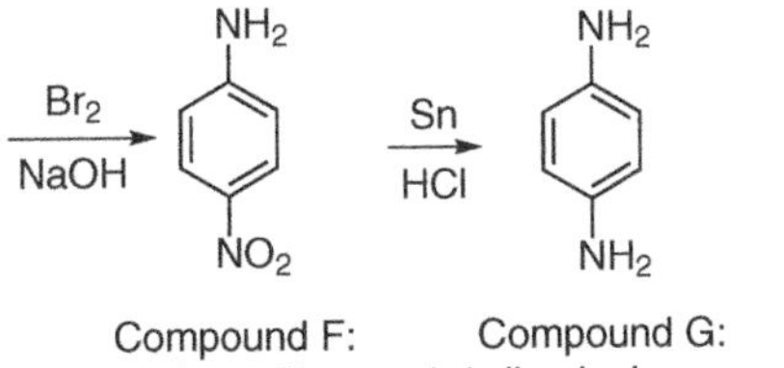

Compound F:
4-nitroaniline

Compound G:
1,4-diaminobenzene

The hydrochloride salt of 1-amino-2-methylnaphthalene (compound E) is treated with nitrous acid to yield 1-hydroxy-2-methylnaphthalene (compound H).

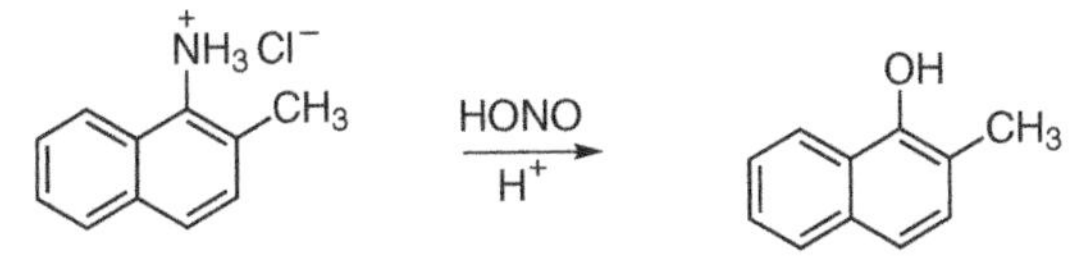

Compound E:
hydrochloride salt of
1-amino-2-methylnaphthalene
MW = 193.5, NE = 193.5

Compound H:
1-hydroxy-2-methylnaphthalene

The hydrochloride salt of 1-amino-2-methylnaphthalene (compound E) and 1-hydroxy-2-methylnaphthalene (compound H) are oxidized to 1,2-benzenedicarboxylic acid (phthalic acid), a compound that is readily converted to 1,2-benzenedicarboxylic anhydride (phthalic anhydride).

1,2-benzene-
dicarboxylic acid,
phthalic acid
MW = 166, NE = 83

1,2-benzenedicarboxylic
anhydride,
phthalic anhydride

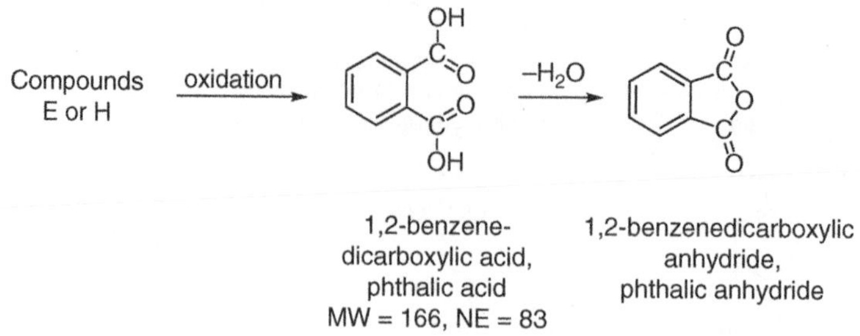